ADVANCES IN GENETICS

VOLUME 23

Contributors to This Volume

Michael T. Clegg
J. E. Endrizzi
Bryan K. Epperson
Wolfgang Hennig
R. J. Kohel
D. R. Miller
N. C. Mishra
A. Nasim
J. Phipps
E. L. Turcotte

ADVANCES IN GENETICS

Edited by

E. W. CASPARI

Department of Biology
University of Rochester
Rochester, New York

JOHN G. SCANDALIOS

Department of Genetics
North Carolina State University
Raleigh, North Carolina

VOLUME 23

1985

ACADEMIC PRESS, INC.
(Harcourt Brace Jovanovich, Publishers)
Orlando   San Diego   New York   London
Toronto   Montreal   Sydney   Tokyo

ACADEMIC PRESS, INC.
Orlando, Florida 32887

United Kingdom Edition published by
ACADEMIC PRESS INC. (LONDON) LTD.
24–28 Oval Road, London NW1 7DX

LIBRARY OF CONGRESS CATALOG CARD NUMBER: 47-30313

ISBN 0-12-017623-8

PRINTED IN THE UNITED STATES OF AMERICA

85 86 87 88   9 8 7 6 5 4 3 2 1

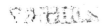

CONTENTS

v

Recent Developments in Population Genetics

MICHAEL T. CLEGG AND BRYAN K. EPPERSON

Genetics, Cytology, and Evolution of *Gossypium*

J. E. ENDRIZZI, E. L. TURCOTTE, AND R. J. KOHEL

CONTRIBUTORS TO VOLUME 23

Numbers in parentheses indicate the pages on which the authors' contributions begin.

MICHAEL T. CLEGG (235), *Department of Botany and Plant Sciences, University of California, Riverside, California 92521*

J. E. ENDRIZZI (271), *Department of Plant Sciences, University of Arizona, Tucson, Arizona 85721*

BRYAN K. EPPERSON (235), *Department of Botany and Plant Sciences, University of California, Riverside, California 92521*

WOLFGANG HENNIG (179), *Department of Genetics, Katholieke Universiteit, Toernooiveld, 6525 ED Nijmegen, The Netherlands*

R. J. KOHEL (271), *USDA, ARS, Cotton and Grain Crops Genetics Research, College Station, Texas 77841*

D. R. MILLER (1), *Division of Biological Sciences, National Research Council of Canada, Ottawa K1A OR6, Canada*

N. C. MISHRA (73), *Department of Biology, University of South Carolina, Columbia, South Carolina 29208*

A. NASIM (1), *Division of Biological Sciences, National Research Council of Canada, Ottawa, K1A OR6, Canada*

J. PHIPPS (1), *Division of Biological Sciences, National Research Council of Canada, Ottawa, K1A OR6, Canada*

E. L. TURCOTTE (271), *USDA, ARS, Cotton Research Center, Phoenix, Arizona 85040*

RECOVERY, REPAIR, AND MUTAGENESIS IN
Schizosaccharomyces pombe

J. Phipps, A. Nasim, and D. R. Miller

Division of Biological Sciences, National Research Council of Canada
Ottawa, Canada

I. Introduction

Living organisms are continually exposed to a wide variety of environmental agents, many of which are demonstrably toxic. Such ex-

1

posure results in DNA damage which, if not repaired by cellular systems, can lead to serious consequences. In addition, there is abundant evidence that suggests that repair, replication, and recombination are related to each other and may have several steps in common. These are just some of the reasons why biologists have taken a very keen interest in the study of microbial recovery and repair systems (for recent reviews see Kilbey, 1975; Resnick, 1979; Kunz and Haynes, 1981; Lemontt, 1980; Lawrence, 1982). Such studies, however, have largely been confined to *Escherichia coli* and *Saccharomyces cerevisiae*. For comparative purposes, studies using other microbes are of extreme importance and can be very revealing.

The present review deals with a description of such a biological system, i.e., the fission yeast *Schizosaccharomyces pombe*. This yeast has been studied genetically since 1950 and a great deal of information is now available. Aspects that have been well documented include its cell cycle, genetics, mutagenesis, and repair.

We owe to McCully and Robinow's improved techniques a major contribution to the knowledge of the structure and the ultrastructure of the dividing fission yeast (McCully and Robinow, 1971; Robinow, 1975, 1977), while the genetic fine structure and chromosome maps have been provided by Leupold and associates (Leupold and Gutz, 1965; Leupold, 1970; Ahmad and Leupold, 1973; Kohli *et al.*, 1977). The reviews by Gutz *et al.* (1974) and Egel *et al.* (1980) bring up to date the genetics of *S. pombe*. Extensive studies have also been devoted to its cell cycle. The physiological aspects investigated by Mitchison (1974, 1977) together with the genetic approach (Nurse, 1981; Beach *et al.*, 1982a), the timing of the cell cycle (Fantes, 1977, 1983), and the morphological and morphometric studies by Johnson *et al.* (1974, 1982) provide a wealth of useful information for future work in *S. pombe*.

Aside from these contributions, the development of *S. pombe* mutagenesis assays and their application to the evaluation of mutagenicity of various industrial compounds have been described by Loprieno *et al.* (1974), Loprieno (1981), and Loprieno and Abbondandolo (1980).

The interested reader should refer to detailed reviews for a more comprehensive coverage. Information about mating types and meiosis can be found in Gutz *et al.* (1974) and Egel *et al.* (1980). Suppressors are covered in Gutz *et al.* (1974), Hawthorne and Leupold (1974), Egel *et al.* (1980), and Kohli *et al.* (1980) and mitochondrial genetics by Wolf and collaborators (Lang *et al.*, 1975, 1976; del Giudice *et al.*, 1979; del Giudice and Wolf, 1980; Anziano *et al.*, 1983; Wolf *et al.*, 1982).

Studies of the repair and recovery mechanisms, however, lack a similarly detailed analysis, especially in the area of biochemical characterization. The purpose of this article is to present a detailed account of recovery, repair, and mutation processes in the fission yeast. Some of the aspects such as the cell cycle, macromolecular synthesis, and cell synchrony are briefly documented to give an idea of the potential and experimental usefulness of *S. pombe*. In addition, the differences from *S. cerevisiae* are discussed, particularly as they relate to evolutionary considerations.

II. Cell Cycle

The heterogeneity encountered in the group of unicellular fungi referred to as yeasts is well illustrated by the two genera *Saccharomyces* and *Schizosaccharomyces*. The shape, growth pattern, cell cycle, physiology, mitotic events, and chromosome number all differ.

A. Chromosome Number and DNA Content

It has been difficult to establish the karyotype of *S. pombe* with certainty since the size of the chromosomes approaches the limit of resolution of the light microscope. McCully and Robinow (1971) using light and electron microscopic techniques described an intranuclear spindle during karyokinesis and the existence of five to seven chromosomes in diploid cells. Later, Fischer *et al.* (1975) found eight chromosomes in haploids. Kohli *et al.* (1977) and Robinow (1977), however, have suggested a number of three for the haploids as a result of linkage group and microscopic analysis. During the nuclear division, which occurs inside the intact nuclear membrane, the three chromosomes are associated with three microtubular filaments. The chromosome number is considerably smaller than the figure of 17 or 18 reported for *S. cerevisiae*. In spite of this difference, the level of DNA expressed as a percentage of the dry weight is close for the two yeasts [0.41 $\pm$ 0.03% for *S. pombe* and 0.40 $\pm$ 0.09% for *S. cerevisiae* (Sokurova, 1974)] and so is the content per haploid spore (roughly 1.5 $\times$ 10^{-14} g) (Bostock, 1970) but their density differs: 13 $\times$ 10^6 for *S. pombe* against 50 $\times$ 10^6 to 12 $\times$ 10^9 for *S. cerevisiae*. This indicates a difference in the base composition of the DNA and in the DNA content per chromosome within the two organisms.

B. VEGETATIVE MITOTIC CYCLE

1. G_1 and G_2 Phases

Mitchison (1971) suggested the presence of two distinct cycles in *S. pombe:* the DNA division cycle involving mitosis and cell division as well as DNA replication and the growth cycle involving macromolecule and enzyme synthesis. During the vegetative phase, which is normally haploid, nuclear division begins at 0.75 of the cycle and septa formation at 0.85. DNA replication occurs early in the 150-min cell cycle and is completed within 10–15 min. The timing and the sequence of the cell division events are such that most of the cycle of *S. pombe* is occupied by the G_2 stage while most of the cycle of *S. cerevisiae* is occupied by the G_1 stage (Fig. 1). In asynchronous cultures, most of the cells are at the G_2 stage throughout the cycle. During log phase, 75% of the cells are in the G_2 phase. Cultures at the stationary phase contain variable proportions of G_1 cells (Bostock, 1970; Fabre, 1973).

This characteristic of a long G_2 phase is most important with respect to any repair-related event. During the G_2 phase, the presence of two DNA copies allows recombinational processes to proceed. In the G_1 cells, the predominance of other types of repair system may lead to a different response to DNA injuries.

Thus eukaryotic systems such as *S. pombe* and *Ustilago maydis,* in which the predominant phase during the cell cycle is G_2, provide useful tools for comparing modes of DNA repair, which may be different during G_1 and G_2, as discussed earlier by other workers (Unrau, 1975; Gentner, 1981).

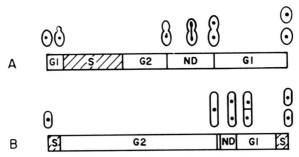

FIG. 1. Cell cycles of *S. cerevisiae* (A) and *S. pombe* (B) illustrating the difference in extent of the G_1 and G_2 phases of the two organisms. S, Period of DNA synthesis; G_1 and G_2, pre- and postreplicative gaps; ND, nuclear division. From Mitchison and Carter (1975).

2. Cell Cycle Regulation

The temporal pattern and the timing of all the events leading to vegetative cell division are strictly regulated in the fission yeast (for reviews see Streiblova, 1977; Nurse and Fantes, 1981; Nurse, 1981).

Any disturbance of the cell cycle that affects the recovery of cells with damaged DNA may also affect the mutational response. Not only is the presence of two DNA copies at the time of the lesion important, but also unscheduled DNA synthesis, possible blockage of DNA replication in G_2 (as a prerequisite for some repair steps), protein synthesis, or any delay in cell division.

As will be obvious in the next part of the article, the work that has been completed so far on repair in the fission yeast rarely made use of the various advantages offered by the numerous well-characterized cell cycle mutants. Approaching the discussion of cell cycle regulation with this in mind should lead to suggestions for further studies on the molecular mechanisms of repair using such mutants.

Two independent cycles, the growth cycle and the nuclear cycle, share responsibilities during mitosis (Mitchison, 1974). The independence of the two cycles is evident, for example, in cell cycle mutants, in which cell mass and macromolecular synthesis are not arrested. Each of the cycles progresses through a dependent sequence of events controlled by major rate-controlling steps, and are interrelated (Nurse, 1981; Fantes, 1982).

The difference between the mitotic control of the two yeasts, *Saccharomyces* and *Schizosaccharomyces*, lies mainly in the size requirement at mitosis peculiar to *S. pombe*. It therefore involves some kinds of controls that are not required in budding yeasts (Beach *et al.*, 1982a). The known lethal temperature-sensitive mutants of the cell cycle (Nurse *et al.*, 1976; Thuriaux *et al.*, 1978; Minet *et al.*, 1979) map over 26 genes. The mutants are distributed across the nuclear cycle. They allowed elucidation of the sequence of events progressing during nuclear and cell division and how this sequence is maintained. They also provided information on the absolute timing of the orderly sequences (Nurse, 1981; Fantes, 1983). A description of the various *cdc* (cell cycle temperature-sensitive) mutant alleles is to be found in Bonatti *et al.* (1972), Thuriaux *et al.* (1978), Nurse (1981), Nurse and Thuriaux (1980), and Nurse and Bisset (1981).

Cell cycle mutants could serve as very useful tools in studying the correlation between cell cycle events and DNA repair systems. Among them, let us mention the mutants of cell size (*wee*⁻ alleles) since they can provide good yields of cells at either G_1 or G_2 stages without any physiological change resulting from altered growth conditions. The wee^{-1} alleles have lost the capacity to regulate both the cell size at division and the control of the timing of division. They, however, express another completely independent regulation system that is usually cryptic in the wild type (Nurse, 1975; Fantes and Nurse, 1977; Thuriaux *et al.*, 1978; Nurse and Thuriaux, 1980; Fantes, 1981).

The *cdc 17* and *cdc 24* mutant strains also deserve to be mentioned as they might be relevant to repair studies. They have been found to be defective in the process of DNA synthesis and not in the chemical doubling (Nasmyth and Nurse, 1981). The *cdc 17* gene function was first described by Nasmyth (1977, 1979). It is the only gene of the cell cycle the product of which has been identified. This mutant is deficient in DNA ligase, a major enzyme for DNA synthesis, repair, and recombination processes. *S. pombe* is thought to possess only one DNA ligase, which therefore must be common to all the above-mentioned processes.

C. GROWTH CYCLE

DNA, RNA, and protein metabolism all influence processes of repair of damaged DNA. Some steps of these pathways require energy. Although this aspect has been rather neglected, it becomes important in cases in which the effects of toxic compounds are analyzed since many of these compounds are capable of altering both the DNA and the energy status of the cell (Ally *et al.*, 1984).

1. Macromolecular Cell Components

Because of the specific manner by which *S. pombe* divides, giving rise to two equal cells, all the cell components during the cell cycle have to double at one time or another. Earlier in this article, we briefly mentioned the control of DNA replication during the vegetative cycle. Meiotic DNA synthesis has been studied by Egel and Egel-Mitrani (1974). The ratio of RNA:DNA is very high in *S. pombe,* about 100:1 to 150:1 (Mitchison and Lark, 1962). The rate of ribosomal and polyadenylated messenger RNA synthesis doubles in steps just after DNA replication in the successive vegetative cycles (Wain and Staatz, 1973; Fraser and Moreno, 1976; Elliott, 1983a). It was suggested by Fraser and Moreno (1976) and Fraser and Nurse (1978) that the rates of RNA synthesis could be modulated through the process of gene concentration (which is basically the ratio between DNA and protein content). One way of altering the gene concentration is to make use of variable cell size mutants in the same organism. With such a system, Fraser and Nurse (1979) and Barnes *et al.* (1979) demonstrated that the point of the cycle coinciding with the doubling in rate of synthesis of RNAs is related to cell mass. The small cells exhibit a delay in the timing of the RNA doubling rate step. This is a compensation mechanism that allows the cells to contain similar mean concentrations of both total RNA and poly(A$^+$) mRNA irrespective of their sizes. The doubling of the rate of synthesis of RNAs occurs when cells of different sizes yield a very similar protein content, suggesting that the rate change occurs when the cell reaches a threshold size. Fraser and Nurse (1979) proposed that doubling is achieved through a mechanism of inhibitor dilution analogous to the model for DNA replication proposed for *E. coli* (Pritchard *et al.*, 1969). Such a mechanism will control and maintain a balanced exponential growth even in growth mutant cells of the *wee*$^-$ type. The question of the transcription maximum (or maximal rate of RNA synthesis) has been recently investigated by Elliott (1983b). According to the proposed model a component controlled by the nucleus may reach a threshold in concentration that blocks a further increase in RNA synthesis rate. The recent review by Egel *et al.* (1980) provides insight about informational suppression and transfer RNA.

Both linear and periodic synthesis has been reported for a number of enzymes (Bostock *et al.*, 1966; Mitchison and Creanor, 1969; Mitchison, 1973, 1977; Miyata and Miyata, 1978). However, the apparent stepwise pattern has been shown to reflect the stress suffered by the cells during selection and synchronization (Kramhoft *et al.*, 1976; Mitchison and Carter, 1975; Creanor *et al.*, 1983). It seems likely that the enzymes related to DNA synthesis mainly exhibit a periodic biosynthesis (Mitchison, 1981). Such a stepwise pattern may result from the control exerted by the mRNA content. It should be stressed that protein synthesis is not required for mitosis after the cell has passed what is known as the "start" but is required for normal septation and division.

2. Energy-Linked Processes

The interest in steps requiring energy during DNA repair is starting to increase (Verma *et al.*, 1982; Bianchi, 1982) since various pollutants can inhibit ATP synthesis, and this would presumably lead to disastrous consequences when the organisms are simultaneously exposed to mutagens or carcinogens. This topic, however, has not yet been extensively studied in yeast repair. The use of various well-characterized mutants of the phosphorylation pathway and the electron transport chain could prove very fruitful.

Two large and two small mitochondria (versus 50 in *S. cerevisiae*) are present in haploid cells of *S. pombe* (Davison and Garland, 1977). Although the whole cell oxygen uptake is small compared to that of bakers' yeast, the growth of the wild-type fission yeast is limited under anaerobic conditions. Low levels of oxido-reduction seem a requirement for growth in *S. pombe* (Labaille and Goffeau, 1975). Oxygen and carbon dioxide evolution during the growth cycle has been followed by Creanor (1978a,b). Oscillations have been observed that could not, however, be directly related to the periodic pool variations and ratios of b_T and $a + a_3$ cytochromes (Poole and Lloyd, 1974). The synthesis of the various cytochromes is under a common regulatory system that may require the mitochondrial protein synthesis apparatus (Poole and Lloyd, 1974). The use of chloramphenicol, which has been shown to inhibit specifically the mitochondrial protein synthesis in this yeast (Hamburger and Kramhøft, 1981), could prove to be a useful tool in such studies. Detailed investigations on the electron transport chain have been conducted (by Lang *et al.*, 1975; Burger *et al.*, 1975; Poole *et al.*, 1980). The amino acid sequence of cytochrome *c* has been established (Claisse and Simon-Becam, 1978; Simon-Becam *et al.*, 1978). Some work has been completed on energy-linked events contributing to cell division, for example, Ca^{2+} uptake (Boutry *et al.*, 1977) and mitochondrial transfer of nucleotides (Labaille and Goffeau, 1975).

There are some striking differences concerning the mitochondrial DNA of *S. pombe* and *S. cerevisiae*. As opposed to bakers' yeast, the mitochondrial DNA of fission yeast is not altered by the intercalative dye ethidium bromide (Bandlow and Kaudewitz, 1974). Its GC content is very high, 32.7 versus 17% in *S. cerevisiae* (Bostock, 1969), and its density is very close to that of the nuclear DNA: 1.26×10^7 as opposed to 5×10^7 for *S. cerevisiae*. Evidence of a common control of nuclear and mitochondrial DNA synthesis has been presented by del Giudice and Wolf (1980). Nonmitochondrial circular DNA has been described in the petite negative fission yeast as a 2-μ covalently closed DNA unit showing the same density as the nuclear DNA. These particles are closely associated with the mitochondrial DNA but their function is not known (del Giudice *et al.*, 1979).

D. MEIOTIC CYCLE

Whether under laboratory or natural environmental conditions, the meiotic cycle is initiated when the fission yeast is growing under unsuitable conditions (Egel and Egel-Mitani, 1974). Such conditions either favoring or, on the contrary, inhibiting the setting of the meiotic cycle have been discussed in a review by Crandall *et al.* (1977). Mainly, shortage or lack of a nitrogen source as well as aeration induce conjugation, while nitrogen compounds and anaerobiosis and also, consequently, respiratory poisons (Calleja, 1973) are inhibitory. Nutritional media favoring sexual reproduction have been described (Egel, 1971; Gutz *et al.*, 1974; Haber and Halvorson, 1975). Procedures for the inactivation of vegetative cells have also been developed for *S. pombe* (Munz and Leupold, 1979). Leupold (1955b) designed a procedure of spore visualization upon cell exposure to iodine vapors. The spores, which contain an amylose-like substance, stain dark blue. Reviews on mating systems of various fungi have already appeared (Leupold, 1970b; Crandall, 1977; Crandall *et al.*. 1977).

The following traits characterize the sexual cycle of the fission yeast.

1. Conjugation of G_1 stage cells occurs between opposite mating types of the haploid strains of *S. pombe*. After cytoplasmic and nuclear fusion, the zygote produces four spores through meiotic divisions. These germinate into haploid cells under appropriate conditions.

2. As an alternative, a zygote may produce a few diploid cells by vegetative divisions. These cells will undergo two meiotic divisions later when reaching the stationary phase, producing four haploid spores.

3. A few diploids are always found in the haploid cultures of *S. pombe*. These can be detected by using the dye phloxin B (Kohli *et al.*, 1977). Such cells probably originate from endomitosis (Leupold, 1955b).

4. Upon crossing with haploid heterothallic strains, diploid cells may generate rare triploid zygotes. Tetrad segregations characteristic of triploids are obtained from such zygotes (Leupold, 1956; Dietlevsen and Martelius, 1956).

1. Morphological and Physiological Aspects of the Sexual Cycle

a. Sex-Directed Agglutination. As in most unicellular fungi, the actual pattern of the mating process starts with sexual cell agglutination, or *flocculation*, which precedes copulation. The sexual agglutination has to be distinguished from the non-sex-directed flocculation occurring in wine or brewers' yeasts (Johnston and Reader, 1983). Although self-agglutination of homothallic strains is the rule under adverse nutritional conditions, the sex-directed flocculation of the heterothallic strains is induced. The initiation of the process requires a mixed culture of the opposite mating types for at least one or two generations. The various aspects of the sex-directed process have been investigated by Calleja *et al.* (1981a). It seems likely that diffusible substances comparable to pheromones excreted by *S. cerevisiae* are responsible for cell agglutination, according to the experiments performed by Egel and Egel-Mitani (1974). However, such substances have not been identified so far in *S. pombe*. Flocculation in this organism has been described in some detail by Calleja (1970) and Calleja and Johnson (1971). The same authors showed later that a protein component is likely to be involved in the process since irreversible deflocculation of competent cells in a culture can be obtained by a treatment with proteinases (Calleja, 1974). A quantitative analysis of flocculation has been developed for the fission yeast and compared to previously existing procedures (Calleja and Johnson, 1977; Yoshida and Yamagishima, 1978; Calleja *et al.*, 1981a).

There does not seem to be any morphological difference that could distinguish between incompetent and induced cells. However, sex hairs or filaments, first described by Poon and Day (1975), have been observed in the fission yeast (Calleja *et al.*, 1977), and physiological changes demonstrated by an alteration of the electrophoretic protein pattern were reported by Calleja *et al.* (1982).

b. Conjugation. Cytoplasmic fusion of mating cells is achieved through the formation of a conjugation tube. The process as a whole involves a series of morphogenetic events including the formation of the conjugation tube, cross-wall erosion, and cytoplasmic fusion. Detailed studies of these morphogenetic events have been presented by Calleja *et al.* (1977). It has been shown by Miyata and Miyata (1981) that in homothallic strains the four lined cells resulting from the last two residual divisions preceding conjugation only generate one zygote by either sister-cell or non-sister-cell copulation. The other remaining nonzygotic cells can conjugate, but only with cells from another set of four lined cells to form one zygote. Conjugation-specific enzymes have not been identified so far in *S. pombe*. Fleet and Phaff (1974) reported a fourfold increase in β-glucanase activity, an enzyme that may be involved in initial cell wall extension rather than in the lysis of cell wall. The localized specificity of the lytic enzyme(s) is intriguing and still not understood. However, it is clear that the lytic activity is compensated by an active biosynthesis of the cell wall material, which prevents complete cell lysis.

c. Nuclear Fusion.   Usually, nuclear fusion occurs in haploid strains after the nuclei have moved toward each other. In diploid cells, in which nuclear fusion is not always the rule, each individual nucleus may undergo meiosis, generating an ascus that contains eight spores. The term *twin meiosis* has been used for such an event by Gutz (1967).

d. Zygotes.   Zygote formation is immediately followed by meiotic division since the very same cultural conditions that favor zygote formation are also suitable to ascospore development. However, when inhibitory substances (such as yeast extract) are present in the sporulation media, some zygotes germinate into diploid cells, undergoing a number of mitotic divisions. Such cells, provided they contain compatible mating-type alleles and are maintained under suitable conditions, may enter the G_1 phase after mitotic divisions, initiate premeiotic DNA synthesis, and complete meiosis and sporulation (Egel *et al.*, 1980).

2. Mating Types

Both homothallism and various forms of heterothallism are known to occur in *S. pombe*. Leupold (1950) isolated the heterothallic strains representing the two opposite mating types (h^+ and h^- and an homothallic strain. The latter is self-fertile but also crosses with the h^+ and h^- strains. The wild-type strain h^{90} isolated by Leupold (1950) was characterized by the production of 90% ascospores. Leupold (1958) distinguished four mating types: one homothallic; two heterothallic($+$): $+N$ and $+R$; and one heterothallic($-$).

Gutz and Doe (1973) described a second heterothallic mating type: $h^- U$. Basically, in the heterothallic strains, only one type is expressed per clone while the homothallic strains exhibit a mixture of cells expressing one or the other type. Diploids can exhibit homo- and $+$ or $-$ heterozygote constitution at the mating-type locus. Homozygote diploids are further able to conjugate while the heterozygote types are only able to sporulate (Egel, 1978). Leupold (1958) had proposed a two-gene scheme in an effort to understand the genetic aspect of the mating type. New observations were made (Gutz and Doe, 1973; Meade and Gutz, 1976; Egel *et al.*, 1980) leading to the interpretation discussed in a recent review (Egel *et al.*, 1980). These models were based on the assumption that the different expressions of the mating type resulted from the various combinations of the two segments *mat 1* and mat 2 mapping on the linkage group II, each of the segments carrying the two sets of allele activities, minus (M) and plus (P). Each set of alleles could be arranged in various combinations and also be expressed in varying degrees.

Since that time, this interpretation has evolved (Egel, 1977, 1979, 1980, 1981). It has been recently shown (Egel and Gutz, 1981) that a third component, *mat 3,* maps in the mating-type region. Mutants of minus information have been isolated at this specific site (Bresch *et al.,* 1968; Egel, 1978, 1981). The recent interpretation is that *mat 2* and *mat 3* constitute two individual loci carrying the plus and minus information as a silent store. These may become expressed as a result of transposition of one or the other silent segment to *mat 1* locus, the *mat 1* locus thus being the site where either one of the h^+ or h^- mating-type functions can be expressed (Egel and Gutz, 1981). Two categories of events are of great genetic significance to mating-type expression: (1) the switching between + and − phenotypes that occurs in homothallic wild-type strains h^{90} and (2) the genetic instability of heterothallic strains.

a. *Gene Switching in Homothallic Strains.* The rate of switching from one mating type to the other is regularly very high in the homothallic strain h^{90}. This explains the high incidence of sporulation observed in this strain. The switching rate is controlled by genes located inside (Gutz and Fecke, 1979) or outside (Gutz *et al.,* 1975) the mating-type locus. Egel (1980) demonstrated that the target signal for switching is governed by genes located inside the mating-type locus, while the switching reaction itself is affected by genes located outside.

The mechanism of mating-type switching in *S. pombe* has now been characterized at the molecular level. Beach (1983) used a plasmid containing *mat P* (h^+) activity as a complementation probe as well as additional hybridization and heteroduplex analysis. This work elegantly confirms the genetic evidence presented in earlier studies that mating-type switching in *S. pombe* occurs by transposition of individual segments to the *mat 1* locus. Moreover, Beach also identified the cis *smt* locus affecting switching (Egel, 1980) as the site of a chromosomal double-stranded break at *mat 1*. This cut is apparently stabilized in quiescent cells and may also be inherited through mitotic division (Beach, 1983). These findings in *S. pombe* draw closer together the mating-type switching events occurring in the fission and the bakers' yeast. In both organisms the switching results from the copy transposition of a silent "cassette" into a locus where it can be expressed. In both cases the switching event is initiated by the occurrence of a relatively stable double-strand break at the site of the insertion. Genetic evidence for the presence of two tightly linked silent cassettes in the mating-type region has been presented by Egel (1984a), who also established the pedigree pattern of mating-type switching (Egel, 1984b).

b. *Genetic Instability of Heterothallic Strains.*   There is only one true heterothallic strain, the stable h^{-s} (Gutz et al., 1974), which does not switch at all. Egel *et al.* (1980) present a table of the various spontaneous mating-type interconversions observed. A model invoking temporal asymmetry as an explanation for unequal sex interconversion in a haploid clone has been proposed by Calleja *et al.* (1981b). It seems likely that both transposition and alteration of target signals are involved here (Egel, 1976a;b). The alterations affect the *smt* target signal ("switching of mating type"), now identified as a double-strand break site at the *mat 1* locus, a site that is probably recognized by recombination enzymes controlled by *swi* genes, according to Egel *et al.* (1980). Other modifications alter the "restrained" *r* switching signal (Egel, 1978).

3. Mutants of the Meiotic Cycle

Some 300 mutants of the meiotic cycle were isolated from the h^{90} strain by Bresch *et al.* (1968). These are controlled by at least 28 individual genes. Such mutants may affect any step from the sex-directed flocculation to sporulation. A table and a diagram summarizing the various genes known to be involved in the sexual cycle are to be found in Gutz *et al.* (1974) and Crandall *et al.* (1977), respectively. Information concerning mutants of meiosis and ascospore formation in the fission yeast has also been presented in Esposito and Esposito (1975).

Out of the mating-type mutants that have already been discussed, various classes of deficient clones can be distinguished. Briefly, *map 1*, *man 1* and *map 2, man 2*, respectively, prevent self-agglutination and a later stage during conjugation (Egel, 1973a). *mei 1, mei 2,* and *mei 3* (Egel, 1973b) block reproduction before premeiotic DNA synthesis and generate high rates of diploids while *mei 4* arrest occurs after the first meiotic division. Other mutants are blocked at the cell fusion (*fus 1*) (Bresch *et al.,* 1968). Some mutations affect sporulation: *spo 1–spo18* (Bresch *et al.,* 1968). All these mutations are recessive. In addition, 16 sterile mutants (*ste 2–6*), representing at least five sterility genes and genes resulting in phenotypic sterility, have been isolated (Girgsdies, 1982).

Special mention has to be made of mutations affecting the switching site. These can be expected to result in a lower rate of sporulation by decreasing the level of mating-type switching. Colonies originating from such strains exhibit sporulating and nonsporulating sectoring. They display a mixture of dark blue and white zones upon exposure to iodine vapors. The terms "speckled" and "mottled" are used to describe such mutants. Mottled strains have been isolated by Gutz and Doe (1975) and Calleja *et al.* (1979). They may result from the loss of the gene product of four unlinked genes (originally designated as *mmo*, "mating type modifier," by Gutz and Doe, 1973). Mutants obtained from the h^{90} strain map either inside (Gutz and Fecke, 1979) or outside (Gutz *et al.*, 1975) the mating-type locus. The mutants mapping inside the mating-type locus alter the target site at the mating-type locus whereas the ones mapping outside this locus exhibit a modification of *swi* 8–10 genes (Egel, 1980).

The mutant originally described as *mei 1-B 102* (Bresch *et al.*, 1968) is also worth mentioning. It has lost the function required to initiate premeiotic DNA synthesis although it is normal in conjugation (Egel and Egel-Mitani, 1974). An interesting feature of this mutation (defined as *mat 2-Pm* in Egel, 1976b) is its suppressibility by opal nonsense suppressors (Egel, 1978).

In addition, numerous lethal mutants seem to occur at the *mat 1* site (Crandall *et al.*, 1977). They may result from unrepaired breaks or sublethal lesions.

4. Meiotic Spontaneous and Induced Recombination

a. General Aspects of Recombinational Exchanges. It is well known that in yeasts meiotic inter- or intragenic exchanges largely outnumber the matching mitotic ones. Among fungi, double-strand breaks constitute the initiating step of gene conversion and mitotic crossing-over (Szostak *et al.*, 1983). Beach's data (1983) provide an excellent molecular example of such a recombination event occurring naturally as mating-type switching in *S. pombe*. It also indicates that spontaneous recombination can occur in the fission yeast as a constitutive process and that therefore some or all the enzymes required for recombinational repair are potentially present. Beach's work in our opinion, constitutes a significant step in our understanding of the mechanisms of spontaneous recombination in yeast.

Three extensive reviews or book chapters have dealt with recombination events in yeast: the various aspects of recombination have been described in Esposito and Wagstaff (1981) and Fogel *et al.* (1983), with little reference, however, to *S. pombe*. The review by Egel *et al.* (1980) is the only one that focuses on *S. pombe*. The reader interested in intragenic recombination and gene conversion can refer to Fogel *et al.* (1983) for a detailed account. The information presently available on recombination in fission yeast is somewhat limited as compared to data on *S. cerevisiae*. Some other fungi such as *Ascobolus* and *Sordaria* that exhibit unextended diplophase during their life cycle (as does *S. pombe*) display specific segregation patterns (Fogel *et al.*, 1983). It would be of interest to compare them to those found in *S. pombe*. Schematically, the process of recombination represents a strand exchange occurring between homologous regions of two chromatids, as deduced from the various models proposed to explain the formation of hybrid DNA. It is not our aim to discuss such structures, all of which are derived from the original Holliday model (Holliday, 1964). Such discussion is amply developed in Esposito and Wagstaff (1981) and Fogel *et al.* (1979).

It is worth mentioning that Fogel *et al.* (1983) suggests that the Radding revision of the Holliday model generally accounts for the data obtained from tetrad analysis of *S. cerevisiae* (mainly diploid) and *Ascobolus* (haploid). There is some evidence that this is also true for *S. pombe* (Munz and Leupold, 1979; Goldman and Smallets, 1979; Egel *et al.*, 1980). The exchange results in a heteroduplex DNA exhibiting mismatches. Mispaired bases are then repaired by mismatch repair, which restores the parental model or leads to gene conversion. Gene conversion is frequently associated with crossing-over. Sectored colonies, when originating from a single spore, represent the consequence of unrepaired mismatches leading to postmeiotic segregation (PMS) of the corresponding marker. There is evidence that mismatch repair is very potent in *S. pombe* since the levels of PMS are generally low (Egel *et al.*, 1980). It is not known whether correction arises before or after the resolution of the Holiday structure although the latter seems more likely (Fogel *et al.*, 1979). Whether any repair process influences the resolution of these structures is not known either.

In fact, the mechanisms for intragenic recombination and gene conversion are more complex than presented above and the recombinational exchange can produce a variety of DNA defects such as conformational changes, single-strand loops, and different types of mismatches (Fogel *et al.*, 1983). How *S. pombe* copes with intra- and intergenic recombination exchanges is of great interest, especially as these events relate to the various repair enzymes and pathways known to occur in this yeast.

b. Gene Conversion.   Gene conversion is the replacement of a given sequence of nucleotides by a novel one. The formation of hybrid DNA irrespective of what that entails in terms of segregation is considered a conversion event (Egel, 1980). The ratio between reciprocal and non-reciprocal exchange is roughly the same in mitotic and meiotic recombination events (Egel *et al.*, 1980). The nonreciprocal exchange is the most frequent type of exchange in meiotic cells of *S. pombe.* As in *S. cerevisiae*, it is detected by careful analysis of tetrad segregation patterns. Such studies have been performed by Gutz (1971), Goldman and Smallets (1979), Munz and Leupold (1979), and Thuriaux *et al.* (1980) with various aims. Spontaneous rates of intragenic recombination were analyzed in the wild type and various *rad* mutants by Grossenbacher-Grunder and Thuriaux (1981), and part of their results are presented in Table 2 (see Section III,C). Transfer of genetic information between tRNA genes coding for different serine isoacceptors—as a result of intergenic conversion—has been demonstrated by Munz *et al.*, 1982).

Most of the attention has been given to the following questions concerning meiotic recombination in *S. pombe:*

1. *Map expansion,* which is the greater occurrence of recombination between two widely separated sites than the sum of frequencies observed for the small intervening intervals. This effect had first been noticed in the structure maps of *S. pombe* (Leupold and Gutz, 1965). It has been reported for all genes mapped so far with the exception of the tRNA's coding *sup* genes (Hofer *et al.*, 1979). The properties of map expansion were analyzed by Egel *et al.* (1980).

2. *The marker effect,* which can be interpreted as an interference of a mutational alteration within a nucleotide sequence with either the formation of hybrid DNA at this site or the repair of the corresponding mismatch (Egel *et al.*, 1980). It has been studied in *S. pombe* by Gutz *et al.* (1974), Munz and Leupold (1979), and Thuriaux *et al.* (1980). In the latter work the unusually high frequency of PMS associated with the anticodon site *sup 3-e* is discussed. By contrast with the strong marker effect at the *sup 3-e* and *sup 9-e* anticodon site, the gene conversion frequency is not altered at the *sup 8-e* and *sup 10-e* primary anticodon site (with the exception of a secondary inactivating site mapping in *sup 8-e*: *sup 8-e, r139*) (Munz *et al.*, 1983). The authors suggest that in the former situation the poor recognition of the mismatches by the base excision repair system leads to the observed increase in PMS. The resulting market effect at *sup 3* and *sup 9* due to a decrease in repair efficiency would then be different from the one observed at *ade 6 M 26*, which probably results from a higher frequency of formation of hybrid DNA at the specific site.

3. *The site-specific induction of gene conversion.* This specificity was first reported by Gutz (1971) in a nonsense *ade 6 M 26* mutant of *S.*

pombe and further studied by Goldman (1974) and Goldman and Smallets (1979). It shows a strong marker effect for intragenic recombination and gene conversion. The frequency of conversions at this site upon crossing with nonidentical alleles yields roughly 15 times more prototrophs than obtained from *ade 6* intragenic crosses. The authors proposed a model to account for the regular occurrence of co-conversions depending upon bidirectional distances. The model involves the triggering of the intragenic recombination process at a preferential endonucleolytic cleavage site at or near the *ade 6* gene (Gutz *et al.*, 1974; Goldman, 1974; Goldman and Smallets, 1979). From their results it can be concluded that the observed marker effect results from the very lesion affecting the *M 26* site.

4. *Postmeiotic segregation,* which reflects a failure or misfunction of the mismatch repair otherwise very active in meiotic *S. pombe* cells (Egel *et al.*, 1979, 1980). UGA suppressor mutations *sup 3-e* and *sup 9-e* (Hofer *et al.*, 1979; Kohli *et al.*, 1980; Thuriaux *et al.*, 1980) increase the frequency of PMS up to 25% from the usual low 2% rate found in the wild type and therefore are thought to be defective in mismatch repair. The effect is base-pair dependent since *sup 3-i* exhibits a low frequency of PMS (Thuriaux *et al.*, 1980). Munz and Leupold (1979) found that the degree of heterogeneity of the genetic background does not influence the rate of PMS occurring among conversions. *Rad 2-44* and two mutator strains, *mut 1-4* and *mut 2-9*, displayed the same negative response.

c. *Meiotic Reciprocal Recombination.*   Reciprocal exchanges do not occur frequently at all in *S. pombe* (Aughern, 1964; Treichler, 1964; Egel *et al.*, 1980). The contribution of crossing-over to the total frequency of recombinational events has been studied in fission yeast by Ahmad and Leupold (1973), in an attempt to understand the map expansion effect. They showed that a positive correlation exists between distance from site and reciprocal crossing-over frequencies, an observation also made in the same organism by Treichler (1964).

Meade and Gutz (1978) showed that the mating type affects the frequency of meiotic crossing-over located between *his 7* and *mat 2* in a complex manner. There is a large increase in meiotic reciprocal recombination in *mat 1* homozygotes. A similar and even more dramatic effect is known for mitotic crossing-over frequencies arising at this site (Aughern and Gutz, 1968). Egel *et al.* (1980) suggest that such positive correlations between mating-type switching and *smt*-specific crossing-over at the mating-type locus may indicate the presence of a specific recombinational system. In the light of Beach's findings (Beach, 1983), it seems likely that it also reflects the presence at this locus of the relatively stable cut triggering recombinational events.

   d. Relationships between Mitotic and Meiotic Recombination. Due to the paucity of reciprocal recombination events, the brevity of the G_1 phase, and the fact that the haploid state is the rule in *S. pombe*, it is not surprising that mitotic crossing-over in this yeast is only poorly documented.

   We mentioned earlier that meiotic and mitotic crossing-over frequencies show similar trends at a specific site (Meade and Gutz, 1978). This is likely to result from the specific feature of the site lesion rather than from a similarity in the repair process since a mutant defective in mitotic but not meiotic recombination at the *ade 6* locus has been isolated (Goldman and Gutz, 1974). This illustrates that the two processes may be at least partly under separate control. It is interesting to note that the marker effect related to mismatch repair is almost absent from mitotic recombination events (Grossenbacher, 1976; Hofer *et al.*, 1979), which may suggest that base-pair mismatch repair is rather inefficient in mitotic cells. However, as discussed later in Section IV,D) the frequent occurrence of pure mutant clones or two-strand mutations (Nasim and Auerbach, 1967) clearly indicates the existence of a mechanism that can convert the initial one-strand lesion into a pure clone.

E. CELL SYNCHRONIZATION

   Techniques in cell synchronization involve either physical selection or age separation of cells from asynchronous cultures or induction of synchrony using physical or chemical agents (for detailed review, refer to Mitchison and Carter, 1975; Kramhøft and Zeuthen, 1975).

1. Selection Techniques

   Various procedures of harvest and fractionation of the cells according to their size and age using velocity sedimentation gradients have been described in Mitchison and Carter (1975). Mitchison and Vincent (1965) developed the now classical linear sucrose gradient separation. The best procedure for selection synchronization is with an elutriating rotor (Creanor and Mitchison, 1979). Some investigators have preferred using glucose or the carbohydrate supporting a given culture rather than sucrose in order to avoid physiological stress (Bostock *et al.*, 1966). Percoll has been tried (Dwek *et al.*, 1980). The difficulties encountered with age fractionation techniques have been recently discussed (Creanor *et al.*, 1983). The cells once selected are used as inoculum to start a new culture, the next four to five cycles of which will be synchronous.

2. Induction Methods

There are reports of various induction procedures, such as alternate starvation and reinoculation periods using limiting amounts of phosphate (Mitchison and Creanor, 1971), repetitive heat shifts, and the use of inhibitors of DNA synthesis (Mitchison, 1973), to further synchronize cell divisions as well as DNA and protein synthesis (Kramhøft et al., 1976, 1978). Various compounds have also been used that are believed to block the Mg^{2+} utilization by the cell thus inhibiting division since Mg^{2+} is essential to yeast division. Among these compounds, the antibiotic ionophore A23087 (Duffus and Patterson, 1974) and EDTA (Ahluwalia et al., 1978) both sequester magnesium; sodium pyrophosphate and citric acid alter the levels of Mg^{2+} in the cell (Walker and Duffus, 1979). The use of the wee^{-1} mutants after temperature shift and physical separation has been reported to provide good yields of nonstressed cells (Barale et al., 1982).

III. Recovery and Repair

A. RADIATION SENSITIVITY

The resistance of the wild-type fission yeast to UV and ionizing radiations is considerably higher than that of S. cerevisiae or E. coli (Table 1). This suggests that the repair systems of S. pombe may be more efficient than those of S. cerevisiae. The resistance depends on a number of variables that are more or less directly related to repair, for example, the growth stage of a cell or a population at the time of exposure, the postirradiation conditions, and the presence or absence of chemical inhibitors interfering with the recovery processes.

B. RADIATION RESISTANCE DURING THE CELL CYCLE

The inactivation curves resulting from UV and ionizing radiation lesions to the DNA of the wild-type fission yeast are shown in Fig. 3. The cells are far more resistant when irradiated at the G_2 stage than during the G_1 phase (Swann, 1962; Fabre, 1970, 1973; Gentner et al., 1978). Therefore, the inactivation curves obtained after synchronization at either stage will vary (Fig. 3). These variations explain the

TABLE 1
A Comparison of the Relative Radiation Sensitivity
of Different Organisms[a]

	Radiation dose giving approximately 37% survival (LD_{37})	
	---	---
Organism	Ultraviolet light (ergs mm^{-2})	Ionizing radiation (kR)
Escherichia coli K12	500	2[b]
Micrococcus radiodurans	6,000	150
Amoeba	—	120
Paramecium	—	350
Bodo marina	50,000	—
Saccharomyces cerevisiae	800	3
Schizosaccharomyces pombe	1,350	80
Ustilago maydis	1,300	—

[a] Adapted from Nasim and James (1978; see also references therein).

[b] Gunther and Kohn (1956).

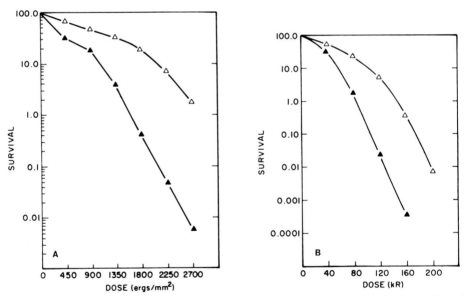

FIG. 2. Effects of caffeine on the UV (A) and γ-ray (B) inactivation of wild-type *S. pombe* 972 *h*⁻. (△) YEA medium; (▲) YEA plus 0.1% caffeine. From Nasim and Smith (1974).

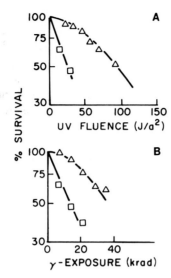

FIG. 3. The survival of G_1 (□) and G_2 (△) phase cells of the wild type *S. pombe* 972 h^- as a function of UV fluence (A) or ^{60}Co γ-ray exposure (B). The specific survival response toward the two DNA-damaging agents is thought to reflect the repair capabilities in relation to the presence of one or two copies of the genome at the time of the damage. From Gentner (1981).

bimodal shape of the inactivation curves of asynchronous cultures (Fabre 1970; Gentner and Werner, 1975). From these biphasic curves, which describe the heterogeneity of the population, a resistant part emerges representing the relative amount of G_2 cells in the analyzed culture (Fig. 2). Both the shoulder and the slope in general reflect the repair capabilities of the two classes of cells (Fabre, 1970). This appears to be true on a general basis. However, a study of the postirradiation inactivation kinetics emphasized the importance of the division delay in modifying the survival responses (Hannan *et al.*, 1976b). This point will be discussed in greater detail later in this section. Clearly, the relative variations in sensitivity and resistance during the cell cycle depend upon the repair capabilities of the cell at the time of the DNA injury and also on the type of lesion induced. This point has been clearly illustrated by Fabre (1970, 1971, 1973) after the first observation made by Swann (1962).

In the wild type, the fission yeast response to UV exposure parallels that obtained with γ- or X-irradiation: G_2 cells are more resistant than G_1 cells (Fig. 3). More precisely, the appearance of increased sensitivity coincides with nuclear division. These variations during the cell cycle differ among various groups of radiation-sensitive mutant strains. The double mutants show no cell cycle-related changes in sensitivity (Fabre, 1970, 1971, 1973). The radioresistance pattern was confirmed by Gentner and Werner (1975) and Gentner (1981), who showed that the radioresistance of the wild-type and recombinant-proficient G_2 cells is conferred by the double copy of the DNA present as well as by the existence of an additional DNA repair process specific to the G_2 prereplicative stage (Gentner, 1977). In the wild type, the shape of the inactivation curves resulting from the action of radiation reflects the percentage of G_1 cells versus G_2 cells in the population at the time of irradiation (Gentner *et al.*, 1978).

C. RADIOSENSITIVE MUTANTS

The identification of two major pathways for repair in *S. pombe* has been established by the isolation of radiosensitive mutants and by the observation that caffeine is a potent inhibitor of dark repair in the fission yeast (Fig. 2).

The need to search for new radiosensitive strains has already been stressed by Nasim and Auerbach (1967). The authors used such strains in an attempt to test the repair hypothesis supported by their data on mosaicism (Nasim and Auerbach, 1967). In 1966 Nasim isolated two moderately UV-sensitive mutants: *uvr-N₁* and *uvr-N₂* (Nasim, 1968). The isolation of another UV-sensitive mutant was reported by Haefner and Howrey (1967). Sixteen mutants mapping over 10 independent loci were selected and analyzed by Schüpbach (1971), who found that the sensitivity to UV is additive in strains mutated at two loci as compared to strains mutated at each of these loci. The high spontaneous lethality found in most of the radiosensitive mutants suggested that some of the gene functions involved in the repair of UV-induced lesions might also be responsible for the repair of spontaneously occurring lesions during DNA synthesis (Schüpbach, 1971). Pedigree analysis of individual cells performed by Nasim and Saunders (1968) confirmed this hypothesis for some of the strains described by Schüpbach.

The first yeast radiosensitive mutant blocked in UV mutability, *UVS1-1* (later designated *rad 1-1*), was described by Nasim (1968), who thus demonstrated the independence of lethal and mutational damages.

The isolation by Fabre (1971) of a double *rad 1–rad 13* mutant confirmed that more than one repair pathway for radioresistance was present in *Schizosaccharomyces pombe*. The presence of multiple repair pathways was confirmed by the study of the genetic control of radiation sensitivity carried out by Nasim and Smith (1974, 1975). The number of loci involved in radiation sensitivity was subsequently increased to 22 (as in *S. cerevisiae*) and several double mutants were constructed (Nasim and Smith, 1975). The mutants were shown to fall into three phenotypic classes: (1) mutants sensitive to UV but showing wild-type sensitivity to γ rays (*rad 2, rad 5, rad 12, rad 13*, for example); (2) mutants sensitive to γ rays but only slightly UV sensitive (like *rad 21, rad 22*); and (3) mutants sensitive to both kinds of radiation (like *rad 1, rad 3, rad 9*) (Fig. 4).

The pleiotropic nature of one of the radiosensitive mutants has been demonstrated by Duck *et al.* (1976a). In this mutant, *rad 4*, both temperature and radiation sensitivity resulted from a single gene mutation. The *rad 2-44* mutant has been found to induce mitotic intragenic recombination, probably by increasing the number of competent cells rather than the rate of recombination among competent cells. It does not show, however, any enhanced spontaneous mutation rate nor increase in intergenic recombination (Grossenbacher-Grunder, personal communication). Eighteen *rad* alleles representing unlinked genes

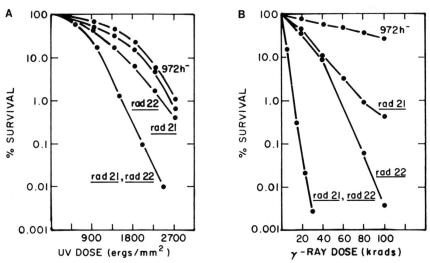

FIG. 4. Genetic control of radiation sensitivity in *Schizosaccharomyces pombe*. Ultraviolet (A) and γ-ray (B) inactivation curves for the wild type and various mutant strains. From Nasim and Smith (1975).

TABLE 2
Properties of the Radiation-Sensitive *Schizosaccharomyces pombe* Mutant Strains

S. pombe strain[a]	Sensitivity[b,c] to		Effect of caffeine on survival after treatment[b,c]		Response to liquid holding[b,d]	Dimers excision after 4–5 hr (%)[e]	Spontaneous intragenic recombination rate (10^{-6} events/division)[f]	Meiotic intragenic recombination (average prototrophs per 10^6 spores)[g]	UV-induced forward mutation (mutation frequency per 10^4 survivors)[h]
	UV	γ Ray	UV	γ Ray					
Wild type 972									
h⁻ RAD⁺	−	−	+++	+++	NLHE	95.9	10.0	873	1.02
rad 1-1	+++	++	−	+[i]	LHR	90.2	11.7	761	0.08
rad 2-44	++	−	−	++	NE			958	0.32
rad 3-136	+++	++++	−	−	NE	90.6	13.9	1013	0.024
rad 4-116	++	++	−	−	NE		6.2	657	3.58
rad 5-158	++	−	+++	+++	LHR			1129	2.51
rad 6-165	++	++	−	−	NLHE		11.7	1016	0.67
rad 7-185	+	−	−	++	NE				3.44
rad 8-190	++	++	−	−	NLHE		7.6	1001	0.01
rad 9-192	+++	++++	−	−	LHR		22.6	708	0.21
rad 10-198	+++	++	+++	−	NLHE	94.2	20.2	908	4.06

24

rad 11-404	+++	++	+++	+++	NE	86.2			1.35
rad 12-502	+++	-	+++	++	NE				0
rad 13	++	-	+++	+++	NLHE	75.1	6.5	745	76
rad 14 G	++	+	+++	++	NE				
rad 15 P	+++	+	+++	-	NLHE	61.3	120.4	640	6.35
rad 16 U	+++	+++	+++	+++	LHR	44.3	11.7		58.7
rad 17 W	++	++	+++	-	NLHE	74.7	13.5	1421	0.66
rad 18 X	+	++	-	-	NE		11.4	414	
rad 19 M 19	++	++	-	-				1195	
rad 20	-	+	-	-				824	1.05
rad 21	+++	++	+++	+++				491	
rad 22	-	+	+++	+++				642	

[a] Schüpbach (1971); Nasim and Smith (1975).

[b] Symbols: − No effect; + to ++++, increasing effect; LHR, liquid holding recovery; NLHE, negative liquid holding effect; NE, no effect.

[c] Nasim and Smith (1974).

[d] Shahin et al. (1973).

[e] Birnboim and Nasim (1974, 1975).

[f] Schüpbach (1971); Grossenbacher-Grunder and Thuriaux (1981).

[g] Grossenbacher-Grunder and Thuriaux (1981).

[h] Gentner et al. (1978).

[i] Gentner (1977).

TABLE 3
The Response of Double Mutants to Inactivation by UV Irradiation[a]

Parental strains used in the genetic cross[b]	Radiation sensitivity of the double mutant[c]
rad 1 × rad 10	+
rad 3 × rad 13	+
rad 2 × rad 13	−
rad 2 × rad 3	+
rad 3 × rad 9	−
rad 10 × rad 11	+
rad 10 × rad 13	−

[a] From Nasim and Smith (1974).

[b] The choice of mutants used in the cross is based on the caffeine response of individual mutants outlined in Table 2.

[c] The + indicates an enhanced sensitivity of the double mutant compared to any of the two single mutants used in the cross.

were tested for spontaneous mitotic and meiotic recombination as well as UV-induced mitotic recombination (Grossenbacher-Grunder and Thuriaux, 1981). The *rad 10* is a strong mutator strain (Loprieno, 1973a,b). The specific properties and responses to DNA injuries of the 22 radiosensitive mutants are summarized in Table 2, and Table 3 provides information about the double mutants.

D. PATHWAYS FOR DARK REPAIR

The principal DNA lesion caused by UV irradiation is the formation of pyrimidine dimers, essentially thymine dimers. Some way of dealing with these adducts is required for survival of irradiated organisms. During dark repair, they can be eliminated by the well-defined dimer excision process, which involves several steps, the final effect of which is to remove the altered DNA area and insert the correct sequence of bases using the matching duplex strand as a template. If DNA replication occurs before the correction has been completed, a gap results that can be repaired under appropriate conditions by one of the processes involving recombination. The breaks produced in either strand of the duplex by ionizing radiation fall into a similar category of lesions. The presence of a specific UV repair mechanism in the fission yeast was first suggested by Fabre (1970). Two independent pathways of recovery from UV-induced DNA injuries were suspected: one presumably controlling recombination (absent from the *UVS1*, or *rad 1-1*, deficient strain) and a second one controlled by the UVSA gene and thought to direct the removal of prerecombinational lesions by a process resembling excision (absent from *UVSA*, or *rad 13*, defective mutant) (Fabre 1970, 1973). The existence of at least two independent pathways was proved by the isolation of hypersensitive mutants deficient in both pathways (Haefner and Howrey, 1967; Fabre, 1971; Nasim and Smith, 1975).

Since biochemical evidence is largely lacking, it is difficult to be very specific about the precise nature of the different repair pathways. The situation is further complicated by the observation (Birnboim and Nasim, 1975) that all the radiation-sensitive strains tested were found to be able to excise pyrimidine dimers. There are, however, a number of indirect lines of evidence that suggest the presence of three repair pathways to cope with radiation damage.

It has now been demonstrated in *S. pombe* that there are at least two major dark repair pathways, one of which may be specifically inhibited by caffeine. In view of the fact that the major stage in the cell cycle is G_2, the second pathway has often been referred to as the "recombinational repair pathway." It is thought to involve some kind of an exchange mechanism.

1. Dimer Excision

The ability to excise dimers after UV exposure has been assayed biochemically (Birnboim and Nasim, 1974). Both wild-type and recombination-deficient strains are capable of dimer excision (Fabre and Moustacchi, 1973; Birnboim and Nasim, 1975). The following traits characterize the excision repair process in *S. pombe:*

1. The repair inhibitor caffeine does not alter the survival of the irradiated excision-proficient, recombination-deficient strains. Therefore, the "excision pathway" is not caffeine sensitive (Fabre, 1970; Nasim and Smith, 1974; Gentner, 1977), as opposed to the one in *E. coli,* for example (Setlow and Carrier, 1968).

2. Survival in the wild-type and "excision-deficient strains" is enhanced by a delay in cell division (Hannan *et al.,* 1976b).

3. The excision-deficient strains are only sensitive to UV and not to ionizing radiation (Fabre, 1972a; Gentner and Werner, 1975).

4. From the more extensive study conducted by Birnboim and Nasim (1975), it became clear that all the tested so-called excision-deficient strains were able to excise dimers after UV exposure. Seven UV-sen-

sitive recombination-deficient mutants as well as two excision-proficient strains and the wild type could equally excise 75–95% of the
induced dimers when maintained in complete medium. The rate of
excision was medium-dependent. It was inhibited in saline and phosphate buffer. These unexpected results prompted the authors to suggest that a multiple excision mechanism operates in S. pombe. In such
a case, a single gene mutation would likely give rise to only a limited
degree of UV sensitivity without completely abolishing the dimer excision capabilities (as in rad 13, for example). Independent pathways
for excision have been shown to exist in E. coli depending on patch size
from strains that are mutated in both patch size and excision rate
(Rothman, 1980). The pathway designated as excision repair (pathway
III in Gentner, 1977) is not involved in the repair of γ-ray damages
(Gentner and Werner, 1975) and is only operational for UV repair.

   5. Experiments conducted at higher UV doses show that extensive
excision still occurs at 600 J m^{-2} in the wild type of S. pombe as
opposed to haploid wild-type strains of S. cerevisiae. At low UV doses,
one strain of S. cerevisiae excised 54% of the 0.6–0.7% of dimers
formed; at 600 J m^{-2}, it excised 15.9% of the 2.3–2.5% of the pyrimidine dimers induced, while the corresponding figures for S. pombe are
90 and 75%, respectively. These data strongly suggest that the excision
capacity of S. cerevisiae is easily saturated, but that this is not the case
for S. pombe, thereby increasing the resistance of this organism
(Birnboim and Nasim, 1975).

2. "Recombinational Repair" as a Major Mechanism

   Any gap in a DNA strand can be considered a potential prerecombinational lesion. This type of lesion can be repaired through a postreplicative recombination process as well as prereplicative recombination. The latter form of repair takes place in the absence of DNA
synthesis in an already duplicated genome by sister chromatid exchange or a related mechanism. Both repair mechanisms require the
presence of a duplicated genome either at the time of irradiation or
during the first round of division following irradiation. The identification of the main pathway of repair for radiation injuries in S. pombe as
being related to a recombinational process has been essentially based
on the observation that it occurs only in the presence of a duplicated
genome. It has also been supported by the specific effect of the well-
known inhibitor caffeine, which unequivocally blocks the repair process that was independent of the excision process (Clarke, 1968; Schüpbach, 1971; Fabre, 1971).

a. *The Requirement for a Duplicated Genome, Pre- and Postreplication Repair.*   The observation by Fabre (1970,1973) that the wild type 972 h^- was more resistant to UV and ionizing radiation during the G_2 phase suggested the presence of a recombination pathway of repair in *S. pombe.* This hypothesis was further supported by the fact that the recombination-deficient, excision-proficient (*exc*$^+$) strains only showed variations in their resistance during the cell cycle in response to UV radiation. The *exc*$^-$ *rec*$^+$ strains showed the same variations as the wild type in response to the two kinds of radiation (Fabre, 1973; Nasim and Smith, 1974). It was concluded that, aside from excision, DNA lesions due to irradiation are repaired by a postreplicative mechanism acting on the lesions that are bypassed during DNA replication. The construction of double mutants (Fabre 1972a; Nasim and Smith, 1975) then strengthened the view that two independent repair processes were present in the wild-type *S. pombe.*

Since recombinational repair occurs only in cells possessing two DNA copies, it is efficient in G_2 cells possessing a duplicated genome and in G_1 cells after DNA synthesis has occurred (i.e., when they enter the G_2 phase). That prereplicative repair is active in G_2 cells before any DNA synthesis takes place was demonstrated by Gentner *et al.* (1978). It seems likely that this repair occurs as a recombinational exchange between sister chromatid strands of the two DNA copies (Gentner, 1977; Gentner and Werner, 1978) and is different from the recombination repair occurring in diploid cells in G_1 between homologous chromosomes. In G_1 cells DNA synthesis must first occur in order to obtain the necessary second DNA copy. In *S. pombe,* the pathway deficient in *rad 1-1* is responsible for both pre- and postrecombinational repair (Gentner *et al.,* 1978). In wild-type G_1 cells, the DNA synthesis occurs immediately after the transfer of the unirradiated cells into complete medium and the genome is duplicated after 4 hr, although cell division does not resume before 6 hr. In UV-irradiated cells, duplication of the genome resumes after 8 hr, as shown by the lag period that occurs before any caffeine sensitivity is detected (Gentner *et al.,* 1978).

b. *Caffeine as a Tool for Studying Recombinational Repair.*   It has been observed that only one of the two UV damage repair pathways in *S. pombe* is sensitive to caffeine (Clarke, 1968; Schüpbach, 1971; Fabre, 1971). The data presented by Nasim and Smith (1974), along with the finding that the excision- proficient UV-sensitive strain *rad 1-1* was only producing a limited amount of mitotic recombinants (Fabre, 1972a), reinforced the opinion that this pathway was of a recombinational nature. The various repair pathways identified in S. pombe were largely characterized by using caffeine. Screening for strains deficient in pathway III was done by selecting the radiation-sensitive mutants affected by the postirradiation plating on caffeine-supplemented complete agar media (Nasim and Smith, 1974).

The model of postirradiation kinetics of caffeine inactivation proposed by Gentner (1974) and Gentner and Werner (1975) provided more insight into the recombinational events. These researchers devised a procedure of sequential, delayed plating by incubating the irradiated cells in a liquid growing medium prior to plating on agar media with and without caffeine. The cells become progressively more resistant to the inhibitor, displaying an increasing survival rate with increasing plating delays (Fig. 5). The recovery is blocked by the presence of caffeine in the incubating medium, confirming its recombinational nature. This resistance to the caffeine effect, referred to by the authors as "recovery from the lethal enhancement by caffeine," is found both after UV and γ-irradiation to exhibit distinct properties (Gentner and Werner, 1975, 1976, 1977).

c. Properties of the Main Recombinational Repair Pathway. As summarized by Gentner (1981), the conditions for this recombinational pathway to function are as follows.

1. The cells must have genetic recombinational ability, i.e., they must not be mutated in one or the other gene products necessary for recombination.

2. A duplicated genome must be present at the time of irradiation, either in G_2 cells or in G_1 cells progressing toward nuclear division (Gentner et al., 1978).

3. The cells must be maintained in a medium capable of supporting growth (Shahin and Nasim, 1973; Gentner and Werner, 1975).

4. Protein synthesis is required for this type of recombinational repair to be active on both UV and γ-ray lesions (Gentner and Werner, 1976).

5. This pathway is thought to be responsible for UV mutagenesis (Gentner, 1977).

This repair is common both to UV- and ionizing radiation-induced damage and to strains blocked in RAD 1 (Fabre, 1972a, 1973). The synergistic interaction between combined UV- and γ-irradiation in the wild type has been explained by the difference in the repair kinetics of the two types of DNA damages (Gentner and Werner, 1978).

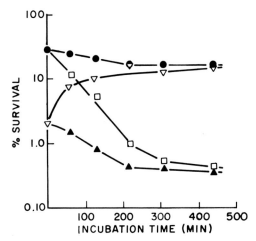

FIG. 5. Effect of caffeine on the survival of γ-irradiated (75 krad) *S. pombe* 972 h⁻. After irradiation, the cells are incubated in complete medium for increasing periods of time prior to plating on complete agar medium. The inhibitor caffeine is supplemented as follows: (●) incubated in YPG liquid medium, plated on YPG medium; (▽)incubated in YPG, plated on YPG plus caffeine; (□) incubated in YPG plus caffeine, plated in YPG; (▲) incubated in YPG plus caffeine, plated on YPG plus caffeine. Notes: The progressive enhancement of radiation killing by caffeine during repair is illustrated by the comparison between the curves (●) and (▽). The presence of caffeine after irradiation always results in an increase in the radiation-induced lethality, (□) versus (▽). A delay in the treatment with caffeine suppresses the killing potentiation. The data have been interpreted as an inhibition by caffeine of an early step in the repair of γ-ray-induced DNA injuries. From Gentner and Werner (1975).

3. A Second Prereplicative Repair Mechanism

The observation that the *rad 1-1* cells already blocked in the main recombinational pathway are still γ-ray sensitive and show lethal enhancement by caffeine (Nasim and Smith, 1974) leads one to the following observations concerning the properties of this second prereplicative repair pathway (Gentner, 1977).

1. It is specific for γ rays (and not UV).
2. It is constitutive rather than induced.
3. It does not require protein synthesis.

The characteristics of the three repair pathways identified so far in the fission yeast are summarized in Table 4. Table 2 presented the list of the 22 mutants first isolated by Schüpbach (1971) and Nasim and

TABLE 4

Characteristics of *Schizosaccharomyces pombe* Repair Pathways[a]

	Pathway		
Property	I	II	III
Requirement for duplicated genome	Yes	Yes	No
Inhibited by caffeine	Yes	Yes	No
Requirement for protein synthesis	Yes	Yes	(No)
Active on UV-induced damage	Yes	No	Yes
Active on γ-induced damage	Yes	Yes	No
Responsible for radiation-induced muta-genesis	Yes	Yes	No
Defective mutant	*rad 1*	Not known	*rad 13*

[a] From Gentner (1977).

Smith (1974), which fall into three phenotypic classes in terms of their response to UV and X and/or γ rays, and these can be further differentiated according to their sensitivity to caffeine and liquid holding response.

    Group 1: Mutants sensitive to UV and γ rays.

        No postirradiation caffeine sensitivity: *rad 3, rad 4, rad 6, rad 8, rad 9, rad 18, rad 19, rad 20.*

        Postirradiation caffeine sensitivity to both radiations: *rad 11, rad 16.*

        Post-UV-irradiation caffeine sensitivity only: *rad 10, rad 15, rad 17.*

        Post-γ-ray-irradiation caffeine sensitivity only: *rad 1-1.*

    Group 2: Mutants sensitive to UV only.

        Postirradiation caffeine sensitivity to both radiations: *rad 5, rad 13.*

        Postirradiation caffeine sensitivity to γ rays only: *rad 7, rad 12, rad 14.*

    Group 3: Mutants sensitive to γ rays only.

        Postirradiation caffeine sensitivity to both radiations: *rad 21, rad 22.*

Double mutants have been constructed (Nasim and Smith, 1974) and are listed in Table 3. In spite of the attempt by Nasim and Smith (1974) no triple mutant has been found so far that would confirm the existence of the third repair pathway as suggested by Gentner (1977). Such mutants should be sought in yeast strains already mutated in both pathways I and III, using increased sensitivity for caffeine following γ-irradiation. These triple mutations could be lethal, as has been suggested by Nasim and Smith (1974). However, the response to caffeine of the various radiation-sensitive strains strongly suggests the existence of more than two pathways (Nasim and Smith, 1974).

4. *Is the Main Pathway for Repair of Irradiation Injuries in Schizosaccharomyces pombe of a Recombinational Type?*

Suggestions had been made by Fabre (1972a) and further reinforced by Nasim and Smith (1974) that caffeine may block more than one repair pathway. Reservations were then raised by Grossenbacher-Grunder and Thuriaux (1981) about the recombinational nature of the caffeine-sensitive pathways in *S. pombe,* based on their results that, except for *rad 1-1,* mitotic and meiotic recombinational rates and frequencies of the caffeine-insensitive strains were not strikingly reduced compared to the wild-type and caffeine-sensitive ones (Table 2). These researchers proposed the following three possibilities to explain these findings: (1) the caffeine-sensitive pathways do not participate in mitotic and meiotic events; (2) multiple recombinational pathways are present in *S. pombe,* a situation similar to the one suggested for excision (Birnboim and Nasim, 1975), with the effect that the phenotype is not expressed when one function is lost since the others compensate; and (3) sister chromatid exchanges may not be as significant a repair pathway as suggested by some earlier workers (Fabre, 1971; Gentner, 1981).

In the absence of any studies conducted at the molecular level on the main pathway for repair of radiation damage it is difficult to conclude with certainty that a recombinational process is involved. However, to infer that it may not be recombinational in nature simply on the basis that the strains deficient in this pathway do not exhibit altered frequencies of the mitotic and meiotic recombinational events also carries the same degree of uncertainty. This rationale once again addresses the fundamental question of which component of mutational events can be attributed unequivocally to a given repair pathway.

In the fission yeast, it is particularly difficult to discuss the contribution of the repair processes toward mutations (although we make an attempt to do so in Section IV of this article), since one has to bear in mind that all the radiosensitive mutants tested displayed some excision capabilities. How much excision contributes to error-free repair or, on the contrary, to the induction of premutational and prerecombina-

tional lesions in each strain is not actually known. The number of
lesions could vary greatly from strain to strain and significantly alter
the frequency of mutations observed in strains showing otherwise an
equal recombinational ability.

Several points have to be taken into consideration when tentatively
relating repair processes to intra- and intergenic recombination rates
during mitosis and meiosis, as discussed by Resnick (1979) and Kunz
and Haynes (1981) for *S. cerevisiae*.

1. Spontaneous and induced genetic exchanges may be under the
control of independent repair processes, the former being of a constitutive and the latter of an inducible nature. Also, mitotic crossing-
over and gene conversion may not be induced by the same repair process, as indicated by the occurrence of *S. cerevisiae* mutant strains
influencing gene conversion but not crossing-over. Finally, the isolation of a mutant defective in mitotic but not meiotic recombination at
the *ade 6* locus (Goldman and Gutz, 1974) suggests at least a partial
independence of thse two mechanisms.

2. More than one repair pathway is likely to contribute to recombination events since these probably involve triggering DNA strand
breaks together with genetic exchange and further repair of mismatched regions of the DNA heteroduplex.

3. The stage at which recombination events are initiated is not necessarily limited to G_2.

4. The efficiency of recombination can be altered by DNA replication
and repair and the recombination competence of a given strain. Some
among the radiation-sensitive strains may display enhanced genetic
exchange frequencies related to greater or lesser ability to remove
prerecombinational lesions. The possible selection of the more resistant G_2 cells among survivors may also introduce some error in the
estimation of recombinational event frequencies.

In the light of observations (Grossenbacher-Grunder and Thuriaux,
1981) that are somewhat contradictory to the conclusions of Fabre,
Nasim, and Gentner, the possibility that the main repair pathway
present in *S. pombe* is not of a recombinational nature cannot be ruled
out. It could be of a type similar to the one monitored by the epistasis
group of genes under the control of *RAD 6* in *S. cerevisiae* (Lawrence
1982) and could be active on UV and γ-ray lesions. It would not be
difficult to assume that an epistatic group of genes would control two
separate functions: error-free repair processes and induced mutagenesis (similar to the *RAD 6* gene cluster in *S. cerevisiae*). Since the

rec⁻ strains have been selected on the basis of their insensitivity to potentiation of postirradiation killing by caffeine, then one might expect that, if the drug was a specific inhibitor of the error-free repair pathway, the selected strains would exhibit any level of spontaneous or induced recombination. However, the fact that the main repair pathway functions only in the presence of a duplicated genome together with the absence of caffeine effect on *rad 51* of *S. cerevisiae* (Hannan and Nasim, 1977), which is thought to lack recombinational functions (Resnick, 1979; Lawrence, 1982), strongly speaks in favor of the recombinational nature of the pathway.

In our present state of knowledge we propose that the main pathway of radiation repair in *S. pombe* should still be referred to as a caffeine-sensitive repair pathway or pathway I. The nomenclature proposed by Gentner (1977) (Table 4) could be adopted until the nature of each pathway is unequivocally established.

E. POSTIRRADIATION CONDITIONS AND RECOVERY

Whether analyzed from a genetic or a molecular point of view, the understanding of the repair mechanisms benefits from the observation of various physiological responses to postirradiation treatments.

1. Photorepair

A striking difference between *S. pombe* and other microorganisms including *S. cerevisiae* is the absence of photorepair in the wild type. The first report was made by Guglielminetti and Schüpbach (1969) and later confirmed by Fabre (1971). However, the UV-sensitive mutants *rad 1-1* exhibited higher survival rates after post-UV-irradiation exposure to visible light (Guglielminetti and Schüpbach, 1969; Fabre, 1971). The *rad 1-1* mutant survival is positively influenced to the same extent, whether illumination with visible light occurs after or before UV treatment. In addition, the increase in survival after postirradiation exposure to visible light was shown in the very same strains that exhibit positive liquid holding recovery.

Furthermore, in *rad 1-1* visible light exposure increases twofold the lag in cell division after irradiation. It is thought that this enhanced lag could account for the increase in survival exhibited by this strain under visible light illumination (Fabre, 1972b). This effect is clearly photoprotection and not photorepair. In the wild type, visible light alone enhances the survival of UV-irradiated cells when they are plated on caffeine-containing medium, implying a preference for the pathway that is not blocked by caffeine in this yeast, i.e., the pathway III (Fabre, 1972b). Visible light by itself does not lead to the splitting of pyrimidine dimers (Fabre, 1973). The photoreactivating enzyme is not present in *S. pombe*. Any light-triggered photoprotection is therefore an indirect effect.

2. *Liquid Holding Recovery*

Many recovery studies include an experimental procedure by which postirradiation plating is delayed. The cells are maintained in saline solution or buffer in the dark and plated after holding times varying from a few hours to several days. The percentage survival is affected in comparison to that obtained after immediate plating. In many micro-organisms liquid holding recovery is observed (Patrick and Haynes, 1964; Patrick *et al.*, 1964). However, in the fission yeast survival is increased, decreased, or unchanged depending on the nature of the DNA lesion, the repair capability of the strain, and the molecular mechanisms involved in the repair of the lesion. Whether or not the repair processes require protein synthesis for completion can be determined by the response of the cells to liquid holding (LH) conditions, since protein synthesis does not occur in nonnutrient media.

Again, *S. pombe* differs from *S. cerevisiae* in its response to liquid holding conditions. When incubated in nonnutrient media following irradiation, wild-type cells of *S. pombe* respond by an increase in lethality irrespective of the cell stage (Shahin and Nasim, 1973). As opposed to the radioresistance pattern observed in the wild type during G_1 and G_2, the liquid holding effect is not influenced by the genome size. It comes immediately to mind that this could reflect the different nature of major repair processes involved in UV-damage repair in the two genera. In *S. cerevisiae* the well-known liquid holding recovery (LHR) results in an increase in survival of UV-irradiated cells (Patrick *et al.*, 1964; Patrick and Haynes, 1964). By increasing the delay between UV-irradiation and the next DNA replication, incubation in nongrowth media favors excision repair. The contrasting negative liquid holding effect (NLHE) in *S. pombe* was first documented by Harm and Haefner (1969). A comprehensive study by Shahin and Nasim (1973) of the response displayed by the 22 *rad* mutants indicated that unirradiated G_1 cells of the wild type do not exhibit NLHE. This effect is only manifested by G_2 phase cells either unirradiated or after UV or γ-irradiation, when maintained under conditions that interfere with the normal progression of the so-called recombinational repair, that is, either in the nongrowth media or in the presence of caffeine. NLHE is not affected by the presence of caffeine when added in the nonnutritive solution (Shahin and Nasim, 1973). In addition, mutations blocking the caffeine-sensitive pathway(s) also abolish NLHE. This similarity in response to caffeine and LHE prompted the conclusion that they are both under the same genetic control. Gentner (1981) suggested that both caffeine-induced lethality and NLHE requires recombinational capability in order to maintain viability. The longer the cells are maintained under conditions blocking the caffeine-sensitive pathway, the less likely they are to overcome their DNA damage. Restoring their recombinational abilities later does not contribute to the survival.

Strains deficient in pathway I and/or II either largely exhibit LHR or are indifferent to LH. The spontaneous cell death is higher under LH among pathway III-deficient strains than in the wild type (Shahin, 1971). The pathway III-proficient strains exhibit the expected positive response (Shahin *et al.*, 1973). The behavior of such strains is comparable to that of the *E. coli K12* strain, which requires both recombination deficiency and the excision proficiency to express LHR (Shahin and Nasim, 1973).

3. Division Delay Following Irradiation

It seems unlikely that all observed effects of liquid holding can be attributed to the additional opportunity given to the cell to complete (or not to complete) its various repair pathways. It seems equally unlikely that liquid holding is the only form of delay that influences the survival of the cell. Hannan *et al.* (1976b) compared the wild type and nine radiation-sensitive mutants of *S. pombe* and found a correlation between naturally occuring postirradiation delay and enhanced survival. They pointed out that two quite different mechanisms might be active; in addition to allowing time for repair to proceed before normal DNA replication took place, an actual cessation of replication in progress would render the cell less vulnerable, since the replication center would then be less likely to encounter a partially excised region.

A comment on theoretical analyses may be in order at this point. Various attempts have been made to interpret damage and its repair in terms of molecular events and to predict the resulting survival curve shapes theoretically. The most popular is probably the $\alpha-\beta$ model of Chadwick and Leenhouts (1973), in which two types of potentially lethal damage are postulated, one requiring a single event (the α term) and the other an interaction of two events (the β term). These lead to an exponential or a shouldered region of the survival curve, respectively. If the second suggestion of Hannan *et al.* (1976b), that two mechanisms influence survival during liquid holding, is valid, an apparently exponential survival curve, as seen for example with certain sensitive strains irradiated in logarithmic growth conditions, would appear as a shouldered curve for resistant strains. There is indeed some evidence for this (Miller, 1970; Miller *et al.*, 1975; Hannan *et al.*, 1976b). In passing, we might note that according to this hypothesis recombination-deficient strains would exhibit nonshouldered survival curves following exposure to an agent producing double-strand breaks, as is indeed the case (Fabre, 1973).

Blattmann (1970) investigated the survival of wild-type *S. pombe* exposed to X rays, and ^{4}He, and ^{16}O ions. Two criteria were chosen: the probability of occurrence of the first three postirradiation mitotic divisions and the microcolony-forming ability. The survival curves of the probability for the first two divisions to occur were exponential but the ones recording microcolony-forming ability were more complex and exhibited shoulders.

Frankenberg and Frankenberg-Schwager (1981) observed in a diploid strain of *S. cerevisiae* that a shoulder in the dose–response curve (with a Poisson distribution) may be obtained as a result of the repair of lethal lesions during a restricted period following immediate plating. This amount of repair could be inversely related to the time taken by the replicon center to stop after irradiation. Also, in *S. cerevisiae*, a significant amount of DNA degradation has been shown to occur during division delay (Wilmore and Parry, 1976). Nevertheless, there seems to be more to division delay than the sole expression of excision repair capabilities. Some parameters, such as replicon center activity, are likely to greatly influence the survival as well as the mutability.

4. Dark Repair Inhibitors

Various dark repair inhibitors have been tested on *S. pombe* (Nasim and Smith, 1974). Among these inhibitors, caffeine is the one that has been most useful. Its systematic use as a tool for investigating the recombination processes is unique to *S. pombe*. It has been used not only to distinguish the caffeine-sensitive pathways of repair from the excision ones but also to isolate and select mutants strains deficient in these repair mechanisms (Nasim and Smith, 1974, 1975). The caffeine specificity in the fission yeast has also been used in an attempt to clarify the possible relationships between the caffeine-sensitive repair pathways and mutation induction (Loprieno and Schüpbach, 1971; Fabre, 1972a; Loprieno and Barale, 1974; Gentner *et al.*, 1978). This point will be discussed in Section IV. The specificity of caffeine toward the recombinational events in *S. pombe* deserves more attention, since the drug does not seem to affect the same pathways in all organisms. Its action on various organisms has been reviewed by Kihlman (1977) and Timson (1977). In bacteria, caffeine selectively prevents excision repair (Setlow and Carrier (1968), apparently by interfering with the activity of the excision enzymes (Lumb *et al.*, 1968). In irradiated *E. coli*, the study by Rothman (1980) emphasizes that rejoining as well as the incision rate is limited in the presence of caffeine.

In contrast to the above, excision is not inhibited by caffeine in yeast and mammalian tissues. The similarity in response between yeasts and mammalian tissue repair systems has been repeatedly pointed out (Hannan and Nasim, 1977; Reynolds and Friedberg, 1980). Potentiation of the lethal effect of UV-irradiation by caffeine has been observed in rodents (Rauth, 1967; Lehman, 1972). This sensitization has been attributed to the blockage of postreplication repair (Lehman, 1972; Trosko and Chu, 1973; Lehman and Kirk Bell, 1974; Nilsson and Lehman, 1975; van den Berg and Roberts, 1976). The various results obtained on UV-irradiated cell lines (Schroy and Todd, 1975), with the exception of HeLa cells (Roberts and Ward, 1973), and on normal human lymphocytes following X-ray and MMS treatment (Brogger, 1974) are all consistent with the hypothesis that caffeine may prevent the joining of newly formed short strands of DNA. This would therefore delay the filling of gaps during unscheduled DNA synthesis.

Few studies have been conducted at the molecular level. A mechanism that would reconcile otherwise apparently opposing sets of data dealing with caffeine effects on repair is proposed in the recent work of Gascoigne *et al.* (1981). UV-irradiation causes considerable reduction in the size of the newly replicated fragments and prevents their rejoining. Caffeine, on the other hand, increases the rate of DNA synthesis after UV irradiation or treatment with other DNA-damaging agents (probably by increasing the rate of replicon initiation). The increased rate of DNA synthesis probably reduces the time available for excision repair to restore the template DNA. DNA synthesis upon a damaged template during the next cell cycle would result in the introduction of new lesions, thereby increasing cell lethality. Thus, a unifying explanation for the potential killing effect by caffeine of cells with injured DNA would be that increased DNA synthesis upon a severely damaged DNA template causes lethality during the next cell cycle (Gascoigne *et al.*, 1981).

Caffeine then alters incision rate, rejoining, and DNA synthesis rates. Its potentiation of killing in cells and organisms in which DNA has been damaged by various agents is likely to reflect the operation of the major repair process(es) involved under the given circumstances. In *S. pombe,* in which excision is not the major route for repair of ionizing radiation-induced DNA injuries, caffeine selectively alters the repair mechanisms operating when two DNA copies are available, at either a pre- or post-replicative stage. *S. pombe* would prove to be an excellent tool for further studies of the molecular mechanisms for the action of this drug.

Recently, the organomercurial methylmercury chloride, which had been previously shown to exert a cytostatic effect on yeast (Phipps and Miller, 1982), has also been found to sensitize the wild-type cells of the fission yeast to UV irradiation. It competes with caffeine for the blockage of the main repair pathway since there is no additional lethal effect when the two drugs are supplemented in combination. However, the sensitization becomes synergistic when the cells are maintained under LH conditions, that is, when the main repair pathway of UV damage is not functional (Phipps *et al.,* 1984).

5. *Comparative Account*

S. cerevisiae and *S. pombe* recovery patterns differ in a number of important ways. The main difference is that *sensu stricto* there is no photorepair in the fission yeast. The major repair pathway for UV damage in *S. pombe* is not of an excision type but rather a caffeine-sensitive repair process that is active on both UV- and γ-ray-induced lesions and that may be recombinational. In the wild type, liquid holding after irradiation potentiates cell lethality as opposed to the well-known positive effect observed on the survival of *S. cerevisiae* under similar conditions.

The very specific effect of caffeine on the main repair pathway of the fission yeast has to be emphasized since it provides a powerful tool for experimental studies. When the molecular mechanism involved in the major repair pathway is elucidated, our knowledge of the exact role of the repair processes in mutagenesis will become much clearer. Also, more information needs to be gained concerning the cross-sensitivity of the 22 radiation-sensitive strains to a wide variety of chemical agents.

The observation that all the radiation-deficient strains tested display dimer excision capabilities should be stressed since, in addition to the suggestion that a multiple excision pathway operates in *S. pombe,* the hypothesis that a multiple recombinational pathway is also present has been proposed. Confirmation of these views may be of great interest in terms of the evolutionary aspects of cellular repair systems.

IV. Mutagenesis

A great deal of information has been gathered on mutagenesis of yeasts and the comment has been made that their eukaryotic organization renders the data obtained on yeasts more relevant to higher organisms than that obtained from prokaryotes (Kilbey, 1975). If one adds that yeast repair processes are believed to show enough similarities with those found in mammals to serve as a model in carcinogenic studies (Reynolds and Friedberg, 1980; Kunz *et al.*, 1980), one simultaneously stresses the importance of the relationships between repair and mutagenesis in yeasts. Among earlier reviews covering various aspects of mutagenesis in yeast, Moustacchi *et al.* (1977), Haynes *et al.* (1979), Lemontt (1980), Prakash and Prakash (1980), Esposito and Wagstaff (1981), Lawrence (1982), Hannan and Nasim (1984), and Fogel *et al.* (1983) can be cited. *Schizosaccharomyces pombe* has a number of well-characterized mutational systems (Table 5). The availability of these systems makes this yeast a very suitable organism for testing the mutagenicity of potential environmental mutagens and carcinogens. This feature combined with the availability of a large number of radiation-sensitive mutants provides the basis for investigating the correlations between repair systems and mutagenesis. We discuss this point in the next section. In addition, the mutational systems devised with the fission yeast is presented in Table 5. More detailed accounts of the different mutational tests can be found in the original references listed in the table and in various reviews (Loprieno *et al.*, 1977, 1983; Loprieno, 1981; Loprieno and Abbondandolo, 1980).

A. Mutant Enrichment and Mutational Systems

Several techniques have been developed for mutant enrichment. Nutrient limitation has been used to enhance mutation frequencies (McAthey and Kilbey, 1976, 1978). Inositol starvation inducing inositol-less death of the nonmutated cells upon transfer of the population onto complete media has been developed by Megnet (1964) and described in Gutz *et al.* (1974). Similarly, a 2-deoxyglucose procedure of enrichment has been developed (Megnet, 1965). The addition to the cultures of antibiotics (McAthey and Kilbey, 1976) or enzymatic methods taking advantage of the selective action of cell wall lytic enzymes combined with 2-deoxyglucose addition have also been proposed (Ferenczy *et al.*, 1975; Sipiczki and Ferenczy, 1978). 8'-Azaguanine, which induces anomalies in the selection of forward mutants in *S. pombe*, is

TABLE 5

Mutational Systems Using *Schizosaccharomyces pombe*[a]

Forward mutation tests. Forward mutation tests lead to loss of enzymatic function in a biosynthetic pathway or induction of resistance to antibiotics or metabolite analogs. Forward mutants can be further characterized by additional tests like phenotypic reversibility, chemically-induced revertibility and intragenic complementarity

1. *Forward mutations at the ade 6 and ade 7 loci.* Red pigmented colonies scored (Gutz, 1963) mutated at the *ade 6* and *ade 7 loci* (Leupold, 1955a, 1957; Munz and Leupold, 1970)

2. *Forward mutation at five loci.* White colonies resulting from any mutation at *ade 1, ade 3, ade 4, ade 5,* or *ade 9* locus are scored among the red *ade 6* (Heslot, 1960; Nasim and Clarke, 1965; Nasim, 1967) or *ade 7* colonies (Prendergast *et al.*, 1983). Spontaneous mutagenicity has been assessed (Friis *et al.*, 1971, Segal *et al.*, 1973). The addition of the *rad 10* mutation to the SP198 *ade 6-60* tester strain greatly enhances the sensitivity of the assay (Loprieno, 1981)

3. *Adenine 1 forward mutation.* Described by Heslot (1960). Complementary mutants map in two distinct regions, *ade 1A* and *ade 1B* of the *ade 1* gene. A noncomplementary third group maps in both regions (Ramirez *et al.*, 1974; Friis *et al.*, 1971; Flury and Flury, 1971). The *Ade* controls the two functions of an enzyme of the biosynthetic adenine pathway (Flury *et al.*, 1976)

4. *Methioine sensitivity.* Mutants isolated by Strauss (1979) are allelic to most of the loci controlling the adenine biosynthesis pathway

5. *Trichodermin resistance.* Suitable system for mutational studies in continuous culture (McAthey and Kilbey, 1976). More appropriate under these conditions than the *ade 6, ade 7* assays which may lead to erroneous data due to selectivity (Gutz *et al.*, 1974)

Microsomal and host-mediated systems

1. *Microsomal assay.* Associates the activating properties of microsomal fractions obtained from Swiss Albino mice with the *ade 6-60* forward mutation test at five loci (Loprieno *et al.*, 1976)

2. *Host-mediated assay.* Mutation screening of the *ade 6-60* strains (forward mutation test at five loci) is performed after intragastric or intraperitoneal injection of the yeast cells to animals exposed to mutagens (Barale *et al.*, 1981; Bauer *et al.*, 1980)

Mitotic inter- and intragenic conversion systems. The only mitotic system developed with *S. pombe,* since the fission yeast is mainly haploid. Mitotic gene conversion events are scored in the strains *ade 7-50/150, ade 7-50/151, ade 7-50/152* and *ade 7-50/275* (Loprieno *et al.*, 1974)

Reversion tests. The reversion assays are based on the restoration of a previously lost function. Strains have been developed that allow the simultaneous detection of frameshift or missense reversions and reversions at suppressible loci or occurring by suppressor mutations.

1. *Reversion to methionine dependence.* The revertants present a mutation at suppressor loci. The tester strain is *met 4, Dig h⁻* (Loprieno and Clarke, 1965; Loprieno, 1978).

TABLE 5 *(Continued)*

2. *Reversion at the arginine 1 locus.* The test has been used by Heslot (1960).
3. *Reversion of nonsense mutants.* The nonsense alleles of *S. pombe* can only be divided into two distinct groups (as opposed to *S. cerevisiae*) depending on their susceptibility to either of the suppressors *sup 3-e*, *sup 9-e*, *sup 8-e*, and *sup 10-e* or *sup 3-i* and *sup 8-i* (Hawthorne and Leupold, 1974; Kohli *et al.*, 1979; Janner *et al.*, 1979).
4. *Reversion to cycloheximide sensitivity.* Cycloheximide-resistant strains *cyh 1-C7* are reversed by a forward mutation of a second suppressor gene (Ibrahim and Coddington, 1976, 1978).

System based on the resistance to antibiotics and metabolite analogs. Mutants resistant to antimycin (Lang *et al.*, 1976); antimycin, funiculosine, and diuron (Burger *et al.*, 1977, Burger and Wolf, 1981); cyclohexamide, trichodermin, and anisomycin (Ibrahim and Coddingon, 1976; Berry *et al.*, 1979; Johnston and Coddington, 1982); and erythromycin and chloramphenicol (Michel and Schweyen, 1971; Seitz *et al.*, 1977a) have been isolated. A multiple drug resistance system could be developed taking advantage of the extra chromosomal mutator described by Seitz-Mayr and Wolf (1982) as in Seitz *et al.* (1977b).

System based on the survival as an indicator of potential mutagenicity. rad 3 and wild-type sensitivity are used in this test (Nestman *et al.*, 1982)

[a] The reader can refer to Loprieno (1981) and Loprieno *et al.* (1983) for information about the various chemicals tested using these systems.

therefore not suitable, at least for this yeast. However, it proved to be useful for the enrichment and simultaneous isolation of mutants at two loci (McAthey, 1976). In addition, the construction of strains whose genome contains beside a phenotypic marker an additional deficiency known to enhance mutations (like *rad 10*) (Loprieno, 1977, 1978) and *mut 1, mut 2,* or *mut 3* (Prendergast *et al.*, 1984), has proved useful.

A number of well-defined mutational systems using *S. pombe* strains are of current use in genetics and environmental mutagenesis testing. Some have been fully described and associated to the *S. cerevisiae*-based ones in Mortimer and Manney (1971), Gutz *et al.* (1974), and Zimmermann (1973).

B. MUTATORS

Mutators and antimutators represent a class of mutated strains that display altered spontaneous or/and induced mutation rates or both. Such strains are frequently found among radiosensitive mutants. Five genetically unlinked mutator alleles at the *ade 7* locus have been examined by a reversion to adenine independence (Munz, 1975). Earlier, Loprieno (1973a, 1973b) had studied the mutator properties of a radiosensitive mutant *rad 10-198*. The presence of *rad 2-44* and *rad 5-158* or *rad 10-198* was reported to enhance the intragenic recombination frequency of two *ade 6* alleles (Loprieno, 1970). *rad 10* has been largely used to enhance mutation enrichment and to increase the mutability of test systems. The mechanism responsible for the mutator activity in that strain is inhibited by caffeine (Loprieno, 1971). *rad 15-P* and *rad 21-45* seem to alter the mutation frequencies and *rad 8-190* the UV-induced ones. *rad 2-44* is active on both (Grossenbauer-Grunder and Thuriaux, 1981). Extrachromosomal mutator activity has been reported by Seitz-Mayr and Wolf (1982). It is thought to be of mitochondrial origin, as its effect is exclusively exerted on the mitochondrial DNA. It influences the rate of spontaneous mitochondrial drug resistance mutations (Seitz-Mayr and Wolf, 1982). This mutation has been reported to cause point mutations and also deletions, thus generating respiratory deficiency.

Several hypotheses have been advanced to explain the mechanism by which mutators arise and function, but no strong evidence has been yet obtained to support any one of these hypotheses.

C. RECOMBINATION EVENTS

Alterations of the genome at the chromosome level are often due to recombination events. Genetic exchange is an important feature in yeasts. As discussed in Resnick (1979), Fogel et al. (1979), Kunz and Haynes (1981), Esposito and Wagstaff (1981), and Fogel et al. (1983), repair processes most probably contribute to induced inter- and intragenic recombination in *S. cerevisiae*. No defined recombinogenic repair pathway has, however, ever been identified in this yeast. Although the pathway controlled by *RAD 52* is clearly involved in both the repair of DNA strand breaks and the induction of genetic exchanges (Resnick, 1979), the routes governed by *RAD 6* and *RAD 3* genes also alter the frequency of gene conversion and crossing-over (Kunz and Haynes, 1981; Lawrence, 1982).

Mitotic crossing-over, first discovered in yeasts by James (1954), occurs in diploid cells as reciprocal exchanges between homologous chromosomes. Such exchanges also occur during meiotic or premeiotic commitment at a high frequency. Since *S. pombe* is mainly in a haploid form, mitotic crossing-over is not as well documented in this yeast as it is in *S. cerevisiae*. Diploid strains of the type *mat-1⁻/mat 1⁻* present a high frequency of crossing-over localized between *his 7* and *mat 2* as reported by Leupold (1970a). Ahmed and Leupold (1973) presented a model of reciprocal recombination.

Genetic recombination in *S. pombe* has been mainly studied in relation to map expansion (Ahmad and Leuopold, 1973) and marker effect, as reviewed in Gutz *et al.* (1974) and Egel *et al.* (1980). Spontaneous mitotic intragenic recombination has been observed at the *ade 6* and *ade 8* (Augehrn, 1964), *ade 7* (Leupold, 1958), and *ade 9* (Heslot, 1960) loci. The spontaneous rates of mitotic intragenic recombinations have been estimated using a number of constructed parent diploid strains derived from the 22 known radiation-sensitive strains. The recombination rate was found to be close to 11×10^{-6} recombination events per division for all the strains examined except for the *hyperrec* (Grossenbacher-Grunder and Thuriaux, 1981). UV-induced prototrophic recombinants have also been studied. A plateau in mutation induction occurred around the 25% survival range and yielded 15 times more recombinants than the spontaneous rate (Grossenbacher and Thuriaux, 1981). Leupold (1957, 1958), conducting an extensive analysis of intragenic recombination at the *ade 7* locus induced by UV-irradiation, reported that the mutants are not randomly distributed on the chromosome map. "Hot spots" were shown to occur. Gutz (1961) observed that X-rays induce mutations at the same sites. Nitrous acid-induced mutants, however, have been shown to appear at a different "hot spot" (Gutz, 1961).

Minet *et al.* (1980) investigated the origin of a centromere effect on mitotic recombination. They established that intragenic recombination is distributed in a nonrandom manner on *S. pombe* chromosomes, as is intergenic crossing-over, and discussed the possible nature of this centromere effect.

Mitotic gene conversion of heteroallelic diploid strains has been studied by Bonatti *et al.* (1976). Site-specific induction of gene conversion during meiosis has been extensively studied using the suppressible *ade 6 M26* mutant (probably a nonsense mutant). This mutant exhibits a specific marker effect with regard to meiotic intragenic recombination and gene conversion. At the *M 26* site, wild-type convertants occur 12 times more frequently than in the opposite direction (Gutz, 1971). Goldman and Smallets (1979) presented evidence that this result is consistent with the occurrence of a site-specific single-strand break occurring at the *M 26* site and the concomitant presence of an intact template introducing new information as replication proceeds. Molecular evidence confirming this has been recently presented

(Beach, 1983) (see discussion in Section II,D). Munz and Leupold (1979) showed that the *rad 2-44* mutation does not alter the conversion pattern in the nonsense suppressor genes analyzed.

D. RELATIONSHIPS BETWEEN INDUCED MUTATIONS AND REPAIR

Generally speaking, the questions of the extent to which repair mechanisms contribute to mutagenesis and which of the known pathways are involved are still often discussed. Recovery processes also influence the mutation frequency. It has been suggested that the inducible repair processes responsible for liquid holding recovery in *S. cerevisiae* may involve error-free as well as error-prone processes (Eckardt *et al.*, 1978; Ferguson and Cox, 1980). It would be of great interest to compare relevant data from the fission yeast, which displays a negative liquid holding effect. If a component of the excision type of repair participates in mutagenesis, then *S. pombe* radiosensitive strains, which all exhibit the properties of excising dimers and have no photorepair, would be an experimental model of choice to study this problem. Several approaches have provided some insight into the role of repair in mutagenesis in the fission yeast.

1. Mosaicism and Lethal Sectoring

Chemically induced mutations occurring on one DNA strand (one-strand mutations) produce one mutated and one normal cell, resulting in the induction of sectored colonies. This phenomenon is referred to as mosaicism. However, besides the sectored colonies, some pure mutant clones are found. Among the four explanatory hypotheses proposed by Nasim and Auerbach (1967) and Nasim (1968), experimental data supported the repair hypothesis; i.e., a mutation occurs on one DNA strand altering the normal sequence of bases. The repair process would alter the mismatched bases of the opposite strand of the duplex resulting in a two-strand mutation and a pure mutant clone. Guglielminetti *et al.* (1967) and Abbondandolo and Bonatti (1970) were more in favor of the induction of pure clones through the mechanism of lethal sectorism; i.e., the mutated strand generates a mutant colony with the other strand suffering lethal damage and producing nonviable progeny. The independence of mutational and lethal damage had, however, been demonstrated earlier by Nasim and Saunders (1968). Abbondandolo and Simi (1971) suggested that the lethal sectoring hypothesis would be relevant at high levels of killing but a repair mechanism would be active at high survival rates.

The contribution of lethal sectoring was ruled out, for UV at least, by Hannan *et al.* (1976a) using a G_1 haploid population of *S. cerevisiae* and by James and Kilbey (1977) using pedigree analysis. In the same yeast, excision-deficient strains were shown to induce lower frequencies of two-strand mutants compared to one-strand mutants (James *et al.*, 1978). To the best of our knowledge, no exhaustive study has been conducted on *S. pombe* radiosensitive mutants. A *rad*-deficient strain (UV-sensitive) has been shown by Nasim (1968) to enhance mosaicism and reduce pure mutant clone incidence. However, although it can be assumed that a repair mechanism is responsible for the incidence of pure mutant clones among sectored colonies, no general conclusions can be drawn about the repair pathway that contributes to their induction. It would be interesting to be able to identify an heteroduplex mechanism of repair and its possible correlation with any one of the three repair pathways described in *S. pombe*.

When induced mosaic mutants are replated, most of the colonies are pure clones but sectored colonies can still be scored even after a number of cell generations (Nasim, 1967). This phenomenon (also found in *S. cerevisiae*) is known as replicating instability. It is a common response to a number of mutagens (Loprieno *et al.*, 1968). It has been proposed that replicating instability is expressed by the change from unstable to stable white and stable red. The mutation to the stable form in EMS-induced unstable mutants occurs at 10^4 times the spontaneous mutation rate (Nasim and James, 1971). The unstable colonies are independent of the mutant and nonmutant phenotype (James and Nasim, 1972). Mutations resulting from genetic instability were observed after 270 cell generations; they may result in forward mutations at high frequencies and in reverse or suppressor mutations at low frequencies (Nasim and Grand, 1973). The frequencies of secondary mosaicism were found to be reduced in strains exhibiting high UV sensitivity. Caffeine also strongly reduced the induction of mosaicism (Dubinin and Kurennaya, 1972), an observation which lends support to the repair hypothesis.

Moreover, no enhanced levels of replicating instability were observed in *three* different radiation-sensitive strains tested (Nasim, 1969). Also, a correlation has been found between the ability of mutagens to produce base-pair substitutions and replicating instability (Kuremaya *et al.*, 1982). Genetic instability of the *sup 3* gene in *S. pombe* has also been reported. The instability may result from heterologous exchanges (Amstutz *et al.*, 1982). Again, there is some evidence that some form of repair may be involved in replicating instability but further investigations are needed.

2. Cell Stage and Mutation Frequency

The comparison of mutation frequency during the G_1 and G_2 stages has been tentatively used to assess the role played by recombinational repair in mutagenesis. No ideal system exists to study the relationship between mutational events and cell cycle stages. The G_1 and G_2 phases in *S. pombe* offer some advantages in that they provide cells with different types of repair capabilities. Approaches such as blockage of recombination in individual strains with an appropriate inhibitor at the G_2 stage or comparison of the wild type with derived rec^- deficient strains are not completely satisfactory. There is some uncertainty associated with these approaches because of the lack of specificity of the inhibitors or the possibility that the absence of rec^+ function in a strain might be accompanied by other genotypic alterations. The generation of stationary G_1 and G_2 phase cells in minimal media containing limiting amounts of phosphates represents rather harsh conditions and therefore restricts the credibility accorded to the interpretation of the data. Recently, advantage has been taken of the cell cycle mutants of the wee^+ type to recover high yields of G_1 and G_2 cells synchronized by using the restrictive temperature (Barale et al., 1982).

To our knowledge, there is no report in the literature about the rate of spontaneous mutation during the different phases of the fission yeast cell cycle. A correlation between radiation-induced mutation rate and cell DNA content was observed by Nasim (1974). Investigating X-ray-induced reversions in an auxotrophic strain, Abbondandolo (1972) found that mutation frequency was dose related in the radiation-resistant G_2 cells while, on the contrary, it diminished with increasing dose in the G_1 cells. Most probably, the mutations induced in these cells were of the base exchange type, although *no recombinational events could have occurred* (Abbondandolo, 1975). This conclusion is reinforced by a study of 14 different auxotrophic mutants investigated at the G_1 and G_2 stages, which all showed the induction of reverse mutants at the G_1 phase by X rays (Rainaldi and Abbondandolo, 1975). More recently, Barale *et al.* (1982) analyzed the changes in X-ray and UV irradiation-induced frequencies during the nuclear cycle. The double mutant *ade 6-60 wee 1-50* was used. At low doses (70% survival) with both treatments (X-ray and UV irradiation) the mutation frequency began to increase from the early G_1 stage, reached a maximum at late G_1 or early S, and continuously decreased throughout the G_2 phase. At higher X-ray doses lethality due to unrepaired lesions altered the pattern. The authors suspected an error-free repair to be operational during the G_1 and G_2 phases while misreplication or an error-prone repair could be the cause of the increased mutation frequency during DNA replication.

Nitrous acid treatment resulted in mosaics forming 50% of the mutants induced in G_1 and over 80% of those formed in G_2. Mosaics were dose dependent in G_1 and unrelated to dose in G2 (Abbondandolo and Bonatti, 1970). Similar results were obtained with UV irradiation (Abbondandolo and Simi, 1971).

3. Mutation Rates in Radiosensitive Mutants

The mutation rates of the various radiosensitive mutants deficient in one of the three characterized repair pathways should provide some information on the contribution, if any, of these pathways to mutagenesis. Compared to the wild type, the *rad 10-198* strain is highly mutable (Loprieno *et al.*, 1975). The authors suggest that error-prone repair operates in the wild type while more DNA lesions associated with less repair leads to the high mutagenicity of the *rad* strain. On the other hand, *rad 1-1* and *rad 3* are deficient in pathway I and have been shown by Nasim (1968) to be refractory to UV mutability. From the work of Gentner *et al.* (1978) on the induction of forward mutations, it is apparent that X-ray-induced mutagenesis varies greatly among the various *rad* strains. Very high mutability in *rad 13, rad 16, rad 15,* and *rad 10* is accompanied by UV sensitization with caffeine. Extremely low mutability (*rad 2, rad 1, rad 3,* and *rad 8*) is associated with the absence of UV caffeine-sensitive repair. A group of mutants similar to the wild type in regard to UV mutability either show a response or do not respond at all to caffeine. However, the same strains assayed for spontaneous meiotic and mitotic recombination as well as UV-induced mitotic recombination do not display any clear reduction in mutation rates except for *rad 1-1*. Enhanced mitotic recombination was found for *rad 15-P* and *rad 21-45* (spontaneous) for *rad 8-190* (UV-induced) and *rad 2-44* (both induced and spontaneous) (Grossenbacher-Grunder and Thuriaux, 1981). It thus seems that only *rad 1-1* and *rad 3,* which are both deficient in repair pathway I, clearly show a reduction in UV mutability and mitotic crossing-over. The *rad 3* mutant, however, exhibits a normal mutation frequency in response to chemical mutagens (Nasim and Hannan, 1977), as does *rad 1-1* (Nasim, 1968). Since the dimer excision pathway is present in the two strains refractory to UV mutations, *rad 3* and *rad 1-1,* this pathway most probably functions error free during the repair of UV damage in *S. pombe.*

In the various repair-deficient strains, the number of premutational lesions repaired by an error-free system or evolving into lethal damage cannot be predicted, making it difficult to interpret the relations between mutation frequencies and repair deficiencies. Another unknown factor in the various mutants analyzed is the level of untargeted mutagenesis. Lawrence and Christensen (1978, 1979) and Lawrence (1982) suggest that a significant number of untargeted mutations are induced by UV irradiation. Therefore, due to possible variations in premutational lesions, untargeted mutations, and also division delay as well as other modifying experimental conditions, general conclusions cannot be drawn. Experiments of the kind carried out by Lawrence *et al.* (1982), which allow the estimation of the amount of untargeted mutation sites induced by UV irradiation could be very useful in *S. pombe*. A study of mutability induced by ionizing radiation in the 22 *rad*-deficient mutants paralleling the one produced on UV-induced recombination events by Grossenbacher-Grunder and Thuriaux (1981) would also be profitable.

4. Caffeine Effect on Mutations and Chromosomal Exchanges

Caffeine decreased the frequency of UV- and MNNG-induced forward mutations at the *ade 6* and *ade 7* loci and UV- and X-ray-induced reverse mutations of *his* mutants while spontaneous rates were not affected. Intergenic meiotic recombination was also reduced (Loprieno and Schüpbach, 1971). In constructed diploids, caffeine reduced the mitotic crossing-over frequency in the wild-type and pathway III-deficient strains but was inefficient in reducing this frequency in *rad 1-1* (Fabre, 1972a). Caffeine also reduced the frequency of a single crossover and decreased the probability for a second one in the genetic region close to the centromere (chromosome II). In the distal region, a decrease in the frequency of double crossovers was obtained (Lopieno and Barale, 1974).

The gene conversion rate between heterozygous sites located at different intervals in the *ade 7* locus was influenced by caffeine differently depending on the distance. Loprieno and Barale (1974) suggested a direct binding of caffeine to the DNA, introducing a higher stability. Although there is no clear-cut pattern for the effect of caffeine, it seems, however, that the following conclusions can be made:

1. The induction/repression of forward and reverse mutations is largely related to the presence/absence of the repair pathways sensitive to caffeine.

2. Meiotic and mitotic recombination events do not exhibit any close correlation with these pathways, indicated by the recombination frequencies displayed by the *rad* mutants.

3. Gene conversion is influenced by caffeine directly at the DNA level through initial lesions and not through repair.

Snow (1979) provided evidence that recombination events differ significantly in *S. pombe* and *S. cerevisiae.* The possibility exists that one recombinational pathway insensitive to caffeine operates at the chromosome level while another error-prone caffeine-sensitive pathway would be responsible for the induction of other mutations such as forward mutations.

5. *Conclusion*

A number of salient features emerge from the various data accumulated on both repair and mutagenesis using the radiosensitive mutants.

1. *rad 1-1* and *rad 3* are refractory to UV mutagenesis and are dimer excision-proficient (Nasim, 1968; Birnboim and Nasim, 1975; Nasim and Hannan, 1977). The one mutant that has been the most thoroughly investigated, i.e., *rad 1-1,* may be comparable to the *rev* strains of *S. cerevisiae* (Lemontt, 1980) in that respect.

2. In *rad 1-1* UV-induced mitotic crossing over is greatly reduced (Grossenbacher-Grunder and Thuriaux, 1981). Therefore, *rad 1-1* is excision proficient and truly deficient in a repair pathway contributing to UV-induced recombination. This strain exhibits a wild-type rate of spontaneous inter- and intragenic exchange frequency, which along with the small amount of UV-induced mitotic crossing-over is not altered by caffeine, nor does the drug cause any further sensitization to UV killing. This could be interpreted as meaning that caffeine inhibits a repair pathway controlling induced recombination in *S. pombe* (pathway I, absent from rad 1-1) and that a second independent non-caffeine-sensitive process participates in spontaneous genetic exchange (as suggested by Nasim and Hannan, 1977). The absence of UV mutability in the excision-proficient *rad 1-1* may indicate that the dimer excision pathway of *S. pombe* is essentially error-free. *rad 3* is very similar to *rad 1-1* although UV-induced mitotic crossing-over is not significantly reduced, suggesting that the mutant may be leaky.

3. Mosaic and complete mutants are induced in both strains at high rates (Nasim, 1968; Nasim and Hannan, 1977) in response to chemical treatment demonstrating that besides Pathway I a second, error-

prone, pathway exists in *S. pombe*. Which of the sensitive-strains are
deficient in this pathway as well as the type of repair involved remains
to be established. *Rad 13* and *rad 10,* which display high mutability
but no enhancement of either spontaneous or UV-induced recombination, may be proficient in this pathway, which is then also sensitive to
caffeine. However, many of these processes may be overlapping and
thus it is difficult to clearly demonstrate the number of independent
error-prone repair pathways possibly influencing mutation and recombination frequencies.

4. The hypothesis proposed by Grossenbacher-Grunder and Thuriaux (1981) that there are multiple recombinational pathways in *S.
pombe* similar to the earlier suggestions concerning the dimer excision
pathways (Birnboim and Nasim, 1975) deserves attention

V. Cloning and Transformation

The improvement of spheroplasting procedures (Stephen and Nasim,
1981) and the availability of a good vector (Beach and Nurse, 1981;
Beach *et al.,* 1982b) for cloning are very important features for future
work on the fission yeast.

A. SPHEROPLASTING PROCEDURES

In earlier studies it was observed that the production of protoplasts
from *S. pombe,* especially from stationary phase cells, is difficult. Neither glusulase nor zymolase seems to display full efficiency. Preliminary freezing and thawing associated with treatments with 2-mercap-
toethanol and detergent followed by snail enzyme digestion have also
been examinned (Luchkina and Bekker, 1975). High yields of protoplasts have been obtained by Poole and Lloyd (1976) by growing the
cells in the presence of glucose and small levels of 2-deoxyglucose, an
inhibitor of cell wall formation. However, further digestion of the remaining cell wall with glusulase was necessary. Partial or total cell
wall regeneration has been studied by Havelkova (1972), Kreger and
Kopecka (1978), and Havelkova and Kripalova (1980). Stephen and
Nasim (1981) have developed a new procedure in which the enzyme
mutanase (α-1,3-glucane glucanhydrolase) yielded spheroplasts after
30 min of exposure to the enzyme in both log and stationary phase
cultures.

B. Transformation Systems

Transformation of *S. pombe* was successfully achieved by Beach and Nurse (1981), who showed that the plasmid vector used in the transformation possesses special restriction sites suitable for the isolation of *S. pombe* genes and their subsequent transfer. With this technique, the cloning of the *LEU 2*[+] gene derived from *S. cerevisiae* into a *leu 1* strain of *S. pombe* has been achieved (Beach and Nurse, 1981). The vector used, pJDB248, incorporates pMB9, a bacterial plasmid, and the 2-µm plasmid and the *LEU 2*[+] gene of *S. cerevisiae*. The transformation frequency of the *leu 1-32 S. pombe* strains by this hybrid plasmid reaches 10^4 transformants per microgram of DNA, which is almost as high as the frequencies obtained with *S. cerevisiae* transformation of *leu 2* strain MC 16. The transformants are unstable. Other plasmids have been tested that provided only poor yields of transformants (pJDB207 and pJDB219). The authors further modified the hybrid plasmid by replacing the bacterial pMB9 sequence by pBR322 and removing from the 2-µm *S. cerevisiae* plasmid the *Eco*RI fragment in order to be able to use this vector for the construction of an *S. pombe* gene bank.

Already-known plasmids containing the *ars* sequences of *S. cerevisiae* together with newly constructed ones containing *PstI fragments from S. pombe* DNA into pDam 1 (containing pBR325 and the *LEU 2*[+] gene of *S. cerevisiae*) were tested for transforming capacities. Nine were found suitable with a level of transformation 10 times lower than that obtained with pJDB248, however. The pDB248 hybrid plasmid vector has been used for the construction of a *S. pombe* gene bank using the restriction enzyme *Sau*3A (Beach *et al.*, 1982b). It is able to propagate not only in *S. pombe* but also in *S. cerevisiae* and *E. coli*. Procedures for the recovery of the plasmids by reintroduction into the bacteria are described. Complementation using this transformation system has been achieved with *lys 1, his 2*, and *ade 6* genes (Beach *et al.*, 1982b). Two transformation systems have been developed, one using the *URA 3* gene of *S. cerevisiae* and the ribosomal 3-µm DNA molecule occurring in *S. pombe* (Fournier *et al.*, 1982) and the other using the 2-µm DNA from *S. cerevisiae* (Gaillardin *et al.*, 1983). It is of interest to note that in the latter case two alternatives in the transformation may be observed: either chromosomal integration in the *S. pombe* genome occurs, resulting in high instability, or a site-specific recombination event between two molecules of the foreign DNA produces a dimeric structure which is stable (Gaillardin *et al.*, 1983).

Inter- and intraspecific cytoplasmic transformation of yeasts was performed by Yoshida (1979). The restoration of respiratory function was obtained after treatment of the protoplasts from *S. cerevisiae rho*⁻ with mitochondria isolated from *S. cerevisiae rho*⁺*U* and *S. pombe*. In spite of the numerous differences in the mitochondrial DNA of the two yeasts, the incorporated organelles functioned normally in the *rho*⁻ transformed strain. A further cytological study (Yoshida and Takahashi, 1980) showed that the alien mitochondria were integrated through fusion rather than endocytosis in the recipient protoplasts.

C. Cloning

Cloning of a gene of *S. pombe* complementing a gene in *E. coli* was reported by Schmidt *et al.* (1979). It was used as a tool to gain understanding in the regulation of the galactose gene cluster in yeast. The galactose gene isolated from the fission yeast was implanted into the genome of an *E. coli* strain carrying a deletion in the galactokinase gene. A similar experiment with another gene of *S. pombe* complementing the *pyr B* mutation in *E. coli* (the gene for aspartate transcarbamylase) was performed by Yamamoto *et al.* (1981).

Use has been made of the gene banks constructed for *S. cerevisiae* and *S. pombe* to isolate and test by complementation the genes responsible for the "start" function of the mitotic cycle (Beach *et al.*, 1982a). It has thus been demonstrated that the *cdc 28* gene of *S. cerevisiae* functionally complements the *cdc 2* gene of *S. pombe*. Transformation of *S. cerevisiae cdc 28.4 leu 2.3* to *Leu*⁺ prototrophy using the plasmiol pcdc 2.3 (sp) in an attempt to test whether *cdc 2*⁺ would complement *cdc 28* failed, indicating that the *cdc 2* gene function of *S. pombe* cannot fully replace the *cdc 28* gene product of *S. cerevisiae*.

Expression of mitochondrial DNA from *S. pombe* cloned into *Eschirichia coli* mini cells has been successfully obtained by Wolf *et al.* (1982).

The development of the cloning and transformation techniques in *S. pombe* will probably facilitate better understanding of the repair systems in this organism. Specifically, the isolation and cloning of well-characterized repair genes from *S. cerevisiae* and the testing of their complementarity in the already defined radiation-sensitive mutants of *S. pombe* could prove very fruitful.

VI. Conclusion

It is quite remarkable that studies of repair and recovery using two yeasts, *S. cerevisiae* and *S. pombe,* reveal a number of important differences. The lack of photoreactivation in radiation-resistant microbes such as *S. pombe* makes sense, as there may always be very efficient dark repair systems that can cope with DNA damage. Similarly, as pointed out earlier, in haploid organisms where the life cycle is predominantly in G_2, the major repair system may be of the recombinational repair type.

We would like to emphasize that in order to understand the full impact and role of repair systems it is important to carry out comparative studies with related groups of microorganisms. In this respect *S. pombe* has already furnished a great deal of useful information and is certainly a very good model for compar tive studies.

Acknowledgments

We are very grateful to Teresa Brychcy (NSERC) for reviewing the manuscript. We also wish to thank Margaret Schade and staff (NRCC, Library), Celia Clyde (NRCC, Drafting), and Judi Meredith for typing the manuscript.

References

Abbondandolo, A. (1972). Induction of mutations by X-rays during defined nuclear stages in *Schizosaccharomyces pombe*. *Atti Assoc. Genet. Ital.* **17,** 50–52.

Abbondandolo, A. (1975). Mutation and nuclear stage in *Schizosaccharomyces pombe*. Part 1: An experimental approach to the role of recombination in mutation induction. *Mutat. Res.* **27,** 225–234.

Abbondandolo, A., and Bonatti, S. (1970). The production by nitrous acid of complete and mosaic mutations during defined nuclear stages in cells of *Schizosaccharomyces pombe*. *Mutat. Res.* **9,** 59–69.

Abbondandolo, A., and Simi, S. (1971). Mosaicism and lethal sectoring in G_1 cells of *Schizosaccharomyces pombe*. *Mutat. Res.* **12,** 143–150.

Ahluwalia, B., Dufus, J. H., Paterson, L. J., and Walker, G. M. (1978). Synchronization of cell division in the fission yeast *Schizosaccharomyces pombe* by ethylenediaminetetraacetic acid. *J. Gen. Microbiol.* **106,** 261–264.

Ahmad, A., and Leupold, U. (1973). On a possible correlation between fine structure map expansion and reciprocal recombination based on crossing over. *Mol. Gen. Genet.* **123,** 143–158.

Ally, A., Phipps, J., and Miller, D. R. (1984). Interaction of methylmercury chloride with cellular energetics and related processes. *Toxicol. Appl. Pharmacol.* **76,** 207–218.

Amstutz, H. P., Kohli, J., Leupold, U., Munz, P., and Thuriaux, P. (1982). Genetic instability of the *sup 3* gene in the fission yeast *Schizosaccharomyces pombe. Biol. Cell.* **43,** 18.

Anziano, P. Q., Perlman, P., Lang, B. F., and Wolf, K. (1983). Mitochondrial genome of the fission yeast *Schizosaccharomyces pombe. Curr. Genet.* **7,** 273–284.

Aughern, P. (1964). Untersuchungen über intragene Rekombinations-mechanismen und allele Komplementierung an Adeninmutanten von *Schizosaccharomyces pombe*. Ph.D. Thesis, University of Zürich.

Aughern, P., and Gutz, H. (1968). Influence of the mating type on mitotic crossing over in *Schizosaccharomyces pombe*. *Genetics* **60**, 158.

Bandlow, W., and Kaudewitz, F. (1974). Action of ethidium bromide on mitochondrial DNA in the petite negative yeast *Schizosaccharomyces pombe*. *Mol. Gen. Genet.* **131**, 333–338.

Barale, R., Zucconi, D., Romano, M., and Loprieno, N. (1981). Experimental model for *in vivo* detection of the formation of potentially oncogenic and mutagenic components. *Mutat. Res.* **85**, 264–265.

Barale, R., Rusciano, D., and Loprieno, N. (1982). Mutations induced by X-rays and UV radiation during the nuclear cell cycle in the yeast *Schizosaccharomyces pombe*. *Mutat. Res.* **92**, 39–48.

Barnes, A., Nurse, P., and Fraser, R. S. S. (1979). Analysis of the significance of a periodic cell cycle doubling in rates of macromolecular synthesis for the control of balanced exponential growth of fission yeast cells. *J. Cell Sci.* **35**, 41–52.

Bauer, C., Corsi, C., Leporini, C., Nieri, R., and Capetola, N. (1980). A mutagenic test *in vivo* combining the intrasanguinous and urinary assays. *Mutat. Res.* **74**, 291–302.

Beach, D. H. (1983). Cell type switching by DNA transposition in fission yeast. *Nature* **305**, 682–687.

Beach, D., and Nurse, P. (1981). High-frequency transformation of the fission yeast *Schizosaccharomyces pombe*. *Nature (London)* **290**, 140–142.

Beach, D., Durkacz, B., and Nurse, P. (1982a). Functionally homologous cell cycle control genes in budding and fission yeast. *Nature (London)* **300**, 706–709.

Beach, D., Pipe, M., and Nurse, P. (1982b). Construction of a *Schizosaccharomyces pombe* gene bank in a yeast bacterial shuttle vector and its use to isolate genes by complementation. *Mol. Gen. Genet.* **187**, 326–329.

Berry, C. H. J., Ibrahim, M. A. K., and Coddington, A. (1979). Characterization of ribosomes from drug resistant strains of *Schizosaccharomyces pombe* in a polyuridylic-acid directed cell free protein synthesizing system. *Mol. Gen. Genet.* **167**, 217–226.

Bianchi, V. (1982). Nucleotide pool unbalance induced in cultured cells by treatment with different chemicals. *Toxicology* **25**, 13–18.

Birnboim, H. C., and Nasim, A. (1974). Detection of pyrimidine dimers in hydrolyzates of yeast DNA by high voltage paper electrophoresis. *Mol. Gen. Genet.* **130**, 291–296.

Birnboim, H. C., and Nasim, A. (1975). Excision of pyrimidine dimers by several UV-sensitive mutants of *S. pombe*. *Mol. Gen. Genet.* **136**, 1–8.

Blattmann, H. (1973). Differential analysis of inactivation of yeast cells by X-rays helium-4 and oxygen-16 ions. *Experientia* **29**, 768.

Bonatti, S., Simili, M., and Abbondandolo, A. (1972). Isolation of temperature-sensitive mutants of *Schizosaccharomyces pombe*. *J. Bacteriol.* **109**, 484–491.

Bonatti, S., Meini, M., and Abbodandolo, A. (1976). Genetic effects of potassium dichromate in *Schizosaccharomyces pombe*. *Mutat. Res.* **38**, 147–150.

Bostock, C. J. (1969). Mitochondrial DNA in the fission yeast *Schizosaccharomyces pombe*. *Biochim. Biophys. Acta* **195**, 579–581.

Bostock, C. J. (1970). DNA synthesis in the fission yeast *Schizosaccharomyces pombe*. *Exp. Cell Res.* **60**, 16–26.

Bostock, C. J., Donachie, W. D., Masters, M., and Mitchison, J. M. (1966). Synthesis of enzymes and DNA in synchronous cultures of *Schizosaccharomyces pombe*. *Nature (London)* **210**, 808–810.

Boutry, M., Floury, F., and Goffeau, A. (1977). Energy dependent uptake of calcium by yeast. *Biochim. Biophys. Acta* **464**, 602–612.

Bresch, C., Müller, G., and Egel, R. (1968). Genes involved in meiosis and sporulation of a yeast. *Mol. Gen. Genet.* **102**, 301–306.

Brogger, A. (1974). Caffeine-induced enhancement of chromosome damage in human lymphocytes treated with methylmethanesulphonate, mitomycin C., and X-rays. *Mutat. Res.* **23**, 353–360.

Burger, G., and Wolf, K. (1981). Mitochondrially inherited resistance to antimycin and diuron in the petite negative yeast *Schizosaccharomyces pombe*. *Mol. Gen. Genet.* **181**, 134–139.

Burger, G., Lang, B., Bandlow, W., and Kaudewitz, F. (1975). Studies on the mechanism of electron transport in the BC-1 segment of the respiratory chain in yeast, part 2. Binding of antimycin to mitochondrial particles and the function of 2 different binding sites. *Biochim. Biophys. Acta* **396**, 187–201.

Burger, G., Lang, B., Backaus, B., Wolf, K., Bandlow, W., and Kaudewitz, F. (1977). Mutations to drug resistance in the CDB region of the mitochondrial genome. *In* "Mitochondria 1977: Genetics and Biogenesis of Mitochondria" (W. Bandlow *et al.*, eds.), pp. 205–212. de Gruyter, Berlin.

Calleja, G. B. (1970). Flocculation in *Schizosaccharomyces pombe*. *J. Gen. Microbiol.* **64**, 247–250.

Calleja, G. B. (1973). Role of mitochondria in the sex-directed flocculation of a fission yeast. *Arch. Biochem. Biophys.* **154**, 382–386.

Calleja, G. B. (1974). On the nature of the forces involved in the sex-directed flocculation of a fission yeast. *Can. J. Microbiol.* **20**, 797–803.

Calleja, G. B., and Johnson, B. (1971). Flocculation in a fission yeast an initial step in the conjugation process. *Can. J. Microbiol.* **17**, 1175–1177.

Calleja, G. B., and Johnson, B. F. (1977). A comparison of quantitative methods for measuring yeast flocculation. *Can. J. Microbiol.* **23**, 68–74.

Calleja, G. B., Yoo, B. Y., and Johnson, B. F. (1977). Fusion and erosion of cell walls during conjugation in the fission yeast (*Schizosaccharomyces pombe*). *J. Cell Sci.* **25**, 139–155.

Calleja, G. B., Johnson, B. F., Zucker, M., and James, A. P. (1979). The mating system of a homothallic fission yeast. *Mol. Gen. Genet.* **172**, 1–6.

Calleja, G. B., Johnson, B. F., and Yoo, B. Y. (1981a). The cell wall as sex organelle in fission yeast. *In* "Sexual Interactions in Eukaryotic Microbes," (D. H. O'Day and P. A. Horgen, eds.), pp. 225–259, Academic Press, New York.

Calleja, G. B., Zuker, M., and Johnson, B. F. (1981b). Temporal asymmetry of sex interconversion in a strain of the homothallic fission yeast *Schizosaccharomyces pombe*. *Curr. Microbiol.* **6**, 225–227.

Calleja, G. B., Johnson, B. F., and Walker, T. (1982). Sexual development in a homothallic fission yeast: Synthesis of readiness proteins resolved by gel electrophoresis. *Can. J. Biochem.* **60**, 693–704.

Chadwick, K. H., and Leenhouts, H. P. (1973). A molecular theory of cell survival. *Phys. Med. Biol.* **18**, 78–87.

Claisse, M. L., and Simon-Becam, A. W. (1978). Cytochrome c from *Schizosaccharomyces pombe:* Purification, spectral properties and amino acid composition. *Eur. J. Biochem.* **86**, 399–406.

Clarke, C. H. (1968). Differential effects of caffeine in mutagen-treated *Schizosaccharomyces pombe*. *Mutat. Res.* **5**, 33–40.

Crandall, M. (1977). Mating type interactions in microorganisms. *In* "Receptors and Recognition" (A. P. Cuatrecasas and M. F. Graeves, eds.), Series A, Vol. 3, pp. 45–100. Chapman and Hall, London, and Halsted Press/Wiley, New York.

Crandall, M , Egel, R., and MacKay, V. L. (1977). Physiology of mating in three yeasts. *In* "Advances in Microbial Physiology" (A. H. Rose and D. W. Tempest, eds.), pp. 307–398. Academic Press, New York.

Creanor, J. (1978a). Carbon dioxide evolution during the cell cycle of the fission yeast *Schizosaccharomyces pombe. J. Cell Sci.* **33**, 385–398.

Creanor, J. (1978b). Oxygen uptake during the cell cycle of the fission yeast *Schizosaccharomyces pombe. J. Cell Sci.* **33**, 399–411.

Creanor, J., and Mitchison, J. M. (1979). Reduction of perturbation in leucine incorporation in synchronous cultures of *Schizosaccharomyces pombe. J. Gen. Microbiol.* **112**, 385–388.

Creanor, J., Elliott, S. J., Bisset, Y. C., and Mitchison, J. M. (1983). Absence of step changes in activity of certain enzymes during cell cycle of budding and fission yeast in synchronous cultures. *J. Cell Sci.* **61**, 339–349.

Davison, M. T., and Garland, P. B. (1977). Structure of mitochondria and vacuoles of *Candida utilis* and *Schizosaccharomyces pombe* studied by electron microscopy of serial thin sectors and model building. *J. Gen. Microbiol.* **98**, 147–153.

del Giudice, L., and Wolf, K. (1980). Evidence for a joint control of nuclear and mitochondrial DNA synthesis in the petite negative yeast *Schizosaccharomyces pombe. In* "Endocytosymbiosis and Cell Biology," pp. 779–790. de Gruyter, Berlin.

del Giudice, L., Wolf, K., Sassone-Corsi, P., and Mazza, A. (1979). 2 micrometer covalently closed nonmitochondrial circular DNA in the petite negative yeast *Schizosaccharomyces pombe. Mol. Gen. Genet.* **172**, 165–170.

del Giudice, L., Wolf, K., Manna, F., and Massardo, D. R. (1983). Expression of cloned mitochondrial DNA from *Schizosaccharomyces pombe* in *Escherichia coli* mini cells. *Mol. Gen. Genet.* **191**, 91–98.

Dietlevsen, E., and Hartelius, V. (1956). A "gigas" mutant in *Schizosaccharomyces pombe* induced by X-ray. *C. R. Trav. Lab. Carlsberg, Ser. Physiol.* **26**, 41–49.

Dubinin, N. P., and Kurennaya, C. N. (1972). UV induced replication instability in fission yeast *Schizosaccharomyces pombe. Mutat. Res.* **16**, 249–264.

Duck, P., Nasim, A., and James, A. P. (1976). Temperature-sensitive mutant of *Schizosaccharomyces pombe* exhibiting enhanced radiation sensitivity. *J. Bacteriol.* **128**, 536–539.

Duffus, J. H., and Patterson, L. J. (1974). Control of cell division in yeast using the ionophore A-23187 with calcium and magnesium. *Nature (London)* **251**, 626–627.

Dwek, R. D., Kobrin, L. H., Grossman, N., and Ron, E. Z. (1980). Synchronization of cell division in microorganisms by percoll gradients. *J. Bacteriol.* **14**, 17–21.

Eckardt, F., Moustacchi, E., and Haynes, R. H. (1978). On the inducibility of error prone repair in yeast. *In* "DNA Repair Mechanisms" (P. C. Hanawalt, E. C. Friedberg, and C. F. Fox, eds.), pp. 421–423. Academic Press, New York.

Egel, R. (1971). Physiological aspects of conjugation in fission yeast. *Planta* **98**, 89–96.

Egel, R. (1973a). Genes involved in mating-type expression of fission yeast. *Mol. Gen. Genet.* **122**, 339–343.

Egel, R. (1973b). Commitment to meiosis in fission yeast. *Mol. Gen. Genet.* **121**, 277–284.

Egel, R. (1976a). The genetic instabilities of the mating type locus in fission yeast. *Mol. Gen. Genet.* **145**, 281–286.

Egel, R. (1976b). Rearrangements at the mating type locus in fission yeast. *Mol. Gen. Genet.* **148**, 149–158.

Egel, R. (1977). "Flip-flop" control and transposition of mating-type genes in fission yeast. *In* "DNA Insertion Elements, Plasmids and Episomes" (A. J. Burkhari *et al.*, eds.), pp. 447–455. Cold Spring Harbor Laboratory, Cold Spring Harbor, New York.

Egel, R. (1978). Orientation of "plus" genes at the mating-type locus in homothallic fission yeast. *Mol. Gen. Genet.* **161**, 305–309.

Egel, R. (1979). Site specific recombination and mating-type switching in homothallic fission yeasts. *Hereditas* **91**, 298–299.

Egel, R. (1980). Mating-type genetics in fission yeast: The advantage of mutants with lowered rates of switching. *In* "Molecular Biology of Yeast" (S. B. Hicks *et al.*, eds.). Cold Spring Harbor Laboratory, Cold Spring Harbor, New York.

Egel, R. (1981). Intergenic conversion and reiterated genes. *Nature (London)* **290**, 191–192.

Egel, R. (1984a). Two tightly linked silent cassettes in the mating-type region of *Schizosaccharomyces pombe*. *Current Genetics* **8**, 205–210.

Egel, R. (1984b). The pedigree pattern of mating-type switching in *Schizosaccharomyces pombe*. *Current Genetics* **8**, 205–210.

Egel, R., and Egel-Mitani, M. (1974). Pre-meiotic DNA synthesis in the fission yeast. *Exp. Cell Res.* **88**, 127–134.

Egel, R., and Gutz, H. (1981). Gene activation by copy transposition in mating-type switching of a homothallic fission yeast. *Curr. Genet.* **3**, 5–12.

Egel, R., Kohli, J., Thuriaux, P., and Wolf, K. (1980). Genetics of the fission yeast *Schizosaccharomyces pombe*. *Annu. Rev. Genet.* **14**, 77–108.

Elliott, S. G. (1983a). Coordination of growth with cell division regulation of synthesis of RNA during the cell cycle of the fisson yeast *Schizosaccharomyces pombe*. *Mol. Gen. Genet.* **192**, 204–211.

Elliott, S. G. (1983b). Regulation of the maximal rate of RNA synthesis in the fission yeast *Schizosaccharomyces pombe*. *Mol. Gen. Genet.* **192**, 212–217.

Esposito, M. S., and Esposito, R. E. (1975). Mutants of meiosis and ascospore formation. *In* "Methods in Cell Biology" (D. M. Prescott, ed), Vol. 11, Yeast Cells, pp. 303–326. Academic Press, New York.

Esposito, M. S., and Wagstaff, J. E. (1981). Mechanisms of mitotic recombination. *In* "The Molecular Biology of the Yeast *Saccharomyces*" (J. Stratheim *et al.*, eds.), pp. 341–370. Cold Spring Harbor Laboratory, Cold Spring Harbor, New York.

Fabre, F. (1970). UV sensitivity of the wild type and different UVS mutants of *Schizosaccharomyces pombe*. Influence of growth stage and DNA content of the cells. *Mutat. Res.* **10**, 415–426.

Fabre, F. (1971). A UV supersensitive mutant in the yeast *Schizosaccharomyces pombe*. Evidence for two repair pathways. *Mol. Gen. Genet.* **110**, 134–143.

Fabre, F. (1972a). Relation between repair mechanisms and induced mitotic recombination after UV-irradiation in the yeast *Schizosaccharomyces pombe*. *Mol. Gen. Genet.* **117**, 153–166.

Fabre, F. (1972b). Photoreactivation in the yeast *Schizosaccharomyces pombe*. *Photochem. Photobiol.* **15**, 367–373.

Fabre, F. (1973). The role of repair mechanisms in the variations of ultraviolet and γ-radiation sensitivity during the cell cycle of *Schizosaccharomyces pombe*. *Radiat. Res.* **56**, 528–539.

Fabre, F., and Moustacchi, E. (1973). Removal of pyrimidine dimers in cells of *Schizosaccharomyces pombe* mutated in different repair pathways. *Biochim. Biophys. Acta* **312**, 617–625.

Fantes, P. A. (1977). Control of cell size and cycle time in *Schizosaccharomyces pombe*. *J. Cell Sci.* **24**, 51–67.

Fantes, P. (1981). Isolation of cell size mutants of a fission yeast by a new selective method: Characterization of mutants and implications for division control mechanisms. *J. Bacteriol.* **146**, 746–754.

Fantes, P. A. (1982). Dependency relations between events in mitosis in *Schizosaccharomyces pombe*. *J. Cell Sci.* **55**, 383–402.

Fantes, P. A. (1983). Control of timing of cell cycle events in fission by the *wee* 1 + gene. *Nature (London)* **302**, 153–155.

Fantes, P., and Nurse, P. (1977). Control of cell size at division in fission yeast by a growth-modulated size control over nuclear division. *Exp. Cell Res.* **107**, 377–386.

Ferenczy, L., Sipiczki, M., and Szegedi, M. (1975). Enrichment of fungal mutants by selective cell wall lysis. *Nature (London)* **253**, 46–47.

Ferguson, L. R., and Cox, B. S. (1980). The role of dimer excision in liquid holding recovery of UV-irradiated yeast. *Mutat. Res.* **69**, 19–41.

Fischer, P., Binder, M., and Wintersberger, U. (1975). A study of the chromosomes of the yeast *Schizosaccharomyces pombe* by light and electron microscopy. *Exp. Cell Res.* **96**, 15–22.

Fleet, G. H., and Phaff, H. J. (1974). Glucanases in *Schizosaccharomyces*. Isolation and properties of the cell wall-associated β-(1,3)-glucanases. *J. Biol. Chem.* **249**, 1717–1728.

Flury, R., and Flury, U. (1971). Biochemical studies on mutants in the ADE-1 gene in *Schizosaccharomyces pombe*. *Heredity* **27**, 311–312.

Fluri, R., Coddington, A., and Flury, U. (1976). The product of the ADE-1 gene in *Schizosaccharomyces pombe*. A bifunctional enzyme catalyzing 2 distinct steps in purine biosynthesis. *Mol. Gen. Genet.* **147**, 271–282.

Fogel, S., Mortimer, R. K., Lusnak, K., and Taveres, F. (1979). Meiotic gene conversion. A signal of the basic recombination event in yeast. *Cold Spring Harbor Symp. Quant. Biol.* **43**, 1325–1341.

Fogel, S., Mortimer, R., and Lusnak, K. (1983). Meiotic gene conversion in yeast: Molecular and experimental perspectives. *In* "Yeast Genetics, Fundamental and Applied Aspects" (J. F. T. Spencer *et al.*, eds.), pp. 65–107. Springer Verlag, Berlin and New York.

Fournier, P., Gaillardin, C., de Lowencourt, L., Heslot, H., Lang, B. F., and Kaudewitz, F. (1982). r-DNA plasmid from *Schizosaccharomyces pombe:* Cloning and use in yeast transformation. *Curr. Gen.* **6**, 31–38.

Frankenberg, D., and Frankenberg-Schwager, M. (1981). Interpretation of the shoulder of dose–response curves with immediate plating in terms of repair of potentially lethal lesions during a restricted time period. *Int. J. Radiat. Biol.* **39**, 617–631.

Fraser, R. S. S., and Moreno, F. (1976). Synthesis of polyadenylated messenger RNA and ribosomal RNA during the cell cycle of *Schizosaccharomyces pombe* with an appendix: Calculation of the pattern of protein accumulation from observed changes in the rate of messenger RNA synthesis. *J. Cell Sci.* **21**, 497–521.

Fraser, R., and Nurse, P. (1978). Novel cell cycle control of RNA synthesis in yeast. *Nature (London)* **271**, 726–730.

Fraser, R., and Nurse, P. (1979). Altered patterns of RNA synthesis during the cell cycle a mechanism compensating for variation in gene concentration. *J. Cell Sci.* **35**, 25–40.

Friis, J., Flury, F., and Leupold, U. (1971). Characterization of spontaneous mutations of mitotic and meiotic origin in the *ade* 7 locus of *Schizosaccharomyces pombe*. *Mutat. Res.* **11**, 373–390.

Gaillardin, C., Fournier, P., Budar, F., Kudla, B., Gerbaud, C., and Heslot, H. (1983).

Replication and recombination of 2-μm DNA in *Schizosaccharomyces pombe. Curr. Genet.* **7**, 245–253.

Gascoigne, E. W., Robinson, A. C., and Harris, W. J. (1981). The effect of caffeine upon cell survival and postreplication repair of DNA after treatment of BHK 21 cells with either UV irradiation or *N*-methyl-*N*-nitrosoguanidine. *Chem.-Biol. Interact.* **36**, 107–116.

Gentner, N. E. (1974). Recovery from enhancement by caffeine of radiation induced inactivation. *Radiat. Res.* **59**, 16–17.

Gentner, N. E. (1977). Evidence for a second "prereplicative G₂" pathway mechanism specific for γ-induced damage in wild type *Schizosaccharomyces pombe. Mol. Gen. Genet.* **154**, 129–133.

Gentner, N. E. (1981). Both caffeine-induced lethality and the negative liquid holding effect, in UV- or γ-irradiated wild-type *Schizosaccharomyces pombe,* are consequences of interference with a recombinational repair process. *Mol. Gen. Genet.* **181**, 283–287.

Gentner, N. E., and Werner, M. M. (1975). Repair in *Schizosaccharomyces pombe* as measured by recovery from caffeine enhancement of radiation-induced lethality. *Mol. Gen. Genet.* **142**, 171–183.

Gentner, N. E., and Werner, M. M. (1976). Effect of protein synthesis inhibition on recovery of UV and γ-irradiated *Schizosaccharomyces pombe* from repair inhibition by caffeine. *Mol. Gen. Genet.* **145**, 1–5.

Gentner, N. E., and Werner, M. M. (1977). Slow UV recovery and fast γ recovery in wild type *Schizosaccharomyces pombe. Mol. Gen. Genet.* **154**, 123–128.

Gentner, N. E., and Werner, M. M. (1978). Synergistic interaction between UV and ionizing radiation in wild-type *Schizosaccharomyces pombe. Mol. Gen. Genet.* **164**, 31–37.

Gentner, N. E., Werner, M. M., Hannan, M. A., and Nasim, A. (1978). Contribution of a caffeine-sensitive recombinational repair pathway to survival and mutagenesis in UV-irradiated *Schizosaccharomyces pombe. Mol. Gen. Genet.* **167**, 43–49.

Girgsdies, O. (1982). Sterile mutants of *Schiz. pombe.* Analysis by somatic hybridization. *Curr. Genet.* **6**, 223–227.

Goldman, S. L. (1974). Studies on the mechanism of the induction of site specific recombination in the *ade 6* locus of *Schizosaccharomyces pombe. Mol. Gen. Genet.* **132**, 347–361.

Goldman, S., and Gutz, H. (1974). The isolation of mitotic rec⁻ mutants in *Schizosaccharomyces pombe. In* "Mechanisms in Recombination" (R. F. Grell, ed.), pp. 317–323. Plenum, New York.

Goldman, S. L., and Smallets, S. (1979). Site specific induction of gene conversion: The effects of homozygosity of the *ade 6* mutant M26 of *Schizosaccharomyces pombe* on meiotic gene conversion. *Mol. Gen. Genet.* **173**, 221–225.

Grossenbacher, A. M. (1976). Untersuchungen zur spontanen mitotische Rekombination in *ade 6* gen von *Schizosaccharomyces pombe.* Diplomarbeit Bern University.

Grossenbacher-Grunder, A. M., and Thuriaux, P. (1981). Spontaneous and UV-induced recombination in radiation-sensitive mutants of Schizosaccharomyces pombe. Mutat. Res. **81**, 37–48.

Guglielminetti, R., and Schüpbach, M. (1969). Meccanismi di "dark repair" e di fotoriattivazione in *Schizosaccharomyces pombe. Atti Assoc. Genet. Ital.* **14**, 182–184.

Guglielminetti, R., Bonatti, S., Loprieno, N., and Abbondandolo, A. (1967). Analysis of the mosaicism induced by hydoxylamine and nitrous acid in *Schizosaccharomyces pombe. Mutat. Res.* **4**, 441–447.

Gunther, S. E., and Kohn, H. I. (1956). The effect of x-rays on the survival of bacteria and yeasts. *J. Bacteriol.* **71**, 571–581.

Gutz, H. (1961). Distribution of X-ray- and nitrous acid-induced mutations in the genetic fine structure of the *ade 7* locus of *Schizosaccharomyces pombe. Nature (London)* **191**, 1124–1125.

Gutz, H. (1963). Untersuchungen zur Feinstruktur der Gene *ade 7* und *ade 6* von *Schizosaccharomyces pombe.* Lind. Habilitationschrift, Technische Universität, Berlin.

Gutz, H. (1967). "Twin" meiosis and other ambivalence in the life cycle of *Schizosaccharomyces pombe. J. Bacteriol.* **92**, 1567–1568.

Gutz, H. (1971). Site-specific induction of gene conversion in *Schizosaccharomyces pombe. Genetics* **69**, 317–337.

Gutz, H., and Doe, F. I. (1973). Two different h^- mating types in *Schizosaccharomyces pombe. Genetics* **74**, 563–565.

Gutz, H., and Doe, F. J. (1975). On homo- and heterothallism in *Schizosaccharomyces pombe. Mycologia* **67**, 748–759.

Gutz, H., and Fecke, H. C. (1979) Probable promoter mutations in the mating-type region of *Schizosaccharomyces pombe. Hoppe–Seyler's Z. Physiol. Chem.* **360**, 274.

Gutz, H., Heslot, H., Leupold, U., and Loprieno, N. (1974). *Schizosaccharomyces pombe. In* "Handbook of Genetics" (R. C. King, ed.), Vol. 1, pp. 395–446. Plenum, New York.

Gutz, H., Meade, J. H., and Walker, S. (1975). Mating-type modifier genes in *Schizosaccharomyces pombe. Genetics* **80**, 38.

Haber, J. E., and Halvorson, H. O. (1975). Methods in sporulation and germination of yeasts. *Methods Cell Biol.* **12**, 45–69.

Haefner, K., and Howrey, L. (1967). Gene-controlled UV-sensitivity in *Schizosaccharomyces pombe. Mutat. Res.* **4**, 219–221.

Hamburger, K., and Kramøhft, B. (1981). The effect of choramphenicol on respiration, fermentation and growth in *Schizosaccharomyces pombe. J. Gen. Microbiol.* **120**, 279–282.

Hannan, M. A., and Nasim, A. (1977). Caffeine enhancement of radiation killing in different strains of *Saccharomyces cerevisiae. Mol. Gen. Genet.* **158**, 111–116.

Hannan, M. A., and Nasim, A. (1984). Repair mechanisms and mutagenesis in microorganisms. In "Repairable Lesions in Microorganisms" (A. Hurst and A. Nasim, eds.). Academic Press, New York.

Hannan, M. A., Duck, P. D., and Nasim, A. (1976a). UV induced lethal sectoring and pure mutant clones in yeast. *Mutat. Res.* **36**, 171–176.

Hannan, M. A., Miller, D. R., and Nasim, A. (1976b). Changes in UV-inactivation kinetics and division delay in *Schizosaccharomyces pombe.* strains during different growth phases. *Radiat. Res.* **68**, 469–479.

Harm, W., and Haefner, K. (1969). Decreased survival resulting from liquid-holding of UV-irradiated *Escherichia coli* and *Schizosaccharomyces pombe. Photochem. Photobiol.* **8**, 179–192.

Havelkova, M. (1972). Experimental inhibition of cell wall formation and of reversion in *Nadsonia elongata* and *Schizosaccharomyces pombe. Protoplasma* **75**, 405–419.

Havelkova, M., and Kripalova, J. (1980). Karyokinesis in growing and reverting protoplasts of *Schizosaccharomyces pombe. Folia Microbiol. (Prague)* **25**, 219–227.

Hawthorne, D. C., and Leupold, U. (1974). Suppressor mutations in yeast. *Curr. Top. Microbiol. Immunol.* **64**, 1–47.

Haynes, R. H., Barclay, B. J., Eckardt, F., Cardman, O., Kunz, B., and Little, J. G.

(1979). Genetic control of DNA repair in yeast. *Proc. Int. Congr. Genet., 14th, 1978,* pp. 172–189.

Heslot, H. (1960). *Schizosaccharomyces pombe:* Un nouvel organisme pour l'étude de la mutagénèse chimique. *Abh. Dtsch. Akad. Wiss. Berlin, Kl. Med.,* No. 1, pp. 98–105.

Hofer, F., Hollenstein, H., Janner, F., Minet, M., Thuriaux, P., and Leupold, U. (1979). The genetic structure of nonsense suppressors in *Schizosaccharomyces pombe.* I *sup 3* and *sup 9. Curr. Genet.* **1,** 45–61.

Holliday, R. (1964). A mechanism for gene conversion in fungi. *Genet. Res.* **5,** 282–304.

Ibrahim, M. A. K., and Coddington, A. (1976). Genetic studies on cycloheximide resistant strains of *Schizosaccharomyces pombe. Heredity* **37,** 179–191.

Ibrahim, M. A. K., and Coddington, A. (1978). Genetic studies on revertants to sensitivity from a cycloheximide resistant strain of *Schizosaccharomyces pombe. Mol. Gen. Genet.* **162,** 213–220.

James, A. P. (1954). Evidence of radiation induced somatic crossing over in diploid yeast. *Genetics* **39,** 974.

James A. P., and Kilbey, B. J. (1977). The timing of UV mutagenesis. A pedigree analysis of induced recessive mutation. *Genetics* **87,** 237–248.

James, A. P., and Nasim, A. (1972). The nature of replicating instability in yeast. *Mutat. Res.* **15,** 125–133.

James, A. P., Kilbey, B. J., and Prefontaine, G. J. (1978). The timing of UV-mutagenesis in yeast: Continuing mutation in an excision-defective *(rad 1-1)* strain. *Molec. Gen. Genet.* **165,** 267.

Janner, F., Flury, F., and Leupold, U. (1979). Reversion of nonsense mutants induced by 4-nitroquinoline 1-oxide in *Schizosaccharomyces pombe. Mutat. Res.* **63,** 11–20.

Johnson, B. F., Yoo, B. Y., and Calleja, G. B. (1974). Cell division in yeasts. II. Template control of cell plate biogenesis in *Schizosaccharomyces pombe. In* "Cell Cycle Controls" (M. Padilla, I. L. Cameron, and A. Zimmerman, eds.), pp. 153–166. Academic Press, New York.

Johnson, B. F., Calleja, G. B., Yoo, B. Y., Zucker, M., and McDonald, I. J. (1982). Cell division key to cellular morphogenesis in the fission yeast, *Schizosaccharomyces. Int. Rev. Cytol.* **75,** 167–208.

Johnston, J. R., and Reader, H. P. (1983). Genetic control of flocculation. *In* "Yeast Genetics, Fundamental and Applied Aspects" (J. F. T. Spencer *et al.,* eds.), pp. 205–224. Springer Verlag, Berlin and New York.

Johnston, P. A., and Coddington, A. (1982). Multiple drug resistance in the fission yeast *Schizosaccharomyces pombe.* Evidence for the existence of pleiotropic mutations affecting energy-dependent transport systems. *Mol. Gen. Genet.* **185,** 311–314.

Kihlman, B. A. (1977). "Caffeine and Chromosomes." Elsevier, Amsterdam.

Kilbey, B. J. (1975). Mutagenesis in yeast. *Methods Cell Biol.* **12,** 209–231.

Kohli, J., Hottinger, H., Munz, P., Strauss A., and Thuriaux, P. (1977). Genetic mapping in *Schizosaccharomyces pombe* by mitotic and meiotic analysis and induced haploidization. *Genetics* **87,** 471–489.

Kohli, J., Kwong, T., Altruda, F., Soll, D., and Wahl, G. (1979). Characterization of a UGA suppressing serine transfer RNA from *Schizosaccharomyces pombe* with the help of a new *in vitro* assay system for eukaryotic-suppressor transfer RNA. *J. Biol. Chem.* **254,** 1546–1551.

Kohli, J., Altruda, F., Kwong, T., Rafalski, A., Wetzel, R., Soell, D., Wahl, G., and Leupold, U. (1980). Nonsense suppressor transfer RNA in *Schizosaccharomyces pombe. Cold Spring Harbor Monogr. Ser.* **98,** 407–419.

64   J. PHIPPS *et al.*

Kramhøft, B., and Zeuthen, E. (1975). Synchronization of the fission yeast *Schizosaccharomyces pombe* using heat shocks. *Methods Cell Biol.* **12**, 373–387.

Kramhøft, B., Nissen, S. E., and Zeuthen, E. (1976). The cell cycle in heat and selection synchronized *Schizosaccharomyces pombe*. *Carlsberg Res. Commun.* **41**, 15–25.

Kramhøft, B., Hamburger, K., Nissen, S. B., and Zeuthen, E. (1978). The cell cycle and glycolytic activity of *Schizosaccharomyces pombe* synchronized in defined medium. *Carlsberg Res. Commun.* **43**, 227–239.

Kreger, D. R., and Kopecka, M. (1978). Nature of the nets produced by protoplasts of *Schizosaccharomyces pombe* during the first stage of wall regeneration in liquid media. *J. Gen. Microbiol.* **108**, 269–274.

Kunz, B. A., and Haynes, R. H. (1981). Phenomenology and genetic control of mitotic recombination in yeast. *Annu. Rev. Genet.* **15**, 57–89.

Kunz, B. A., Hannan, M. A., and Haynes, R. H. (1980). Effect of tumor promoters on ultraviolet light induced mutation and mitotic recombination in *Saccharomyces cerevisiae*. *Cancer Res.* **40**, 2323–2329.

Kurennaya, O. N., Chernova, O. Y., and Tarasov, V. A. (1982). Induction of replicating instability by different types of mutagens in fission yeast *Schizosaccharomyces pombe*. *Genetika* **18**, 409–412.

Labaille, M. F., and Goffeau, A. (1975). Combined effects of an inhibitor of mitochondrial transfer of nucleotides and of an inhibitor of respiration on the growth of the yeast *Schizosaccharomyces pombe*. *Arch. Int. Physiol. Biochim.* **83**, 379–380.

Lang, B., Burger, G., Wolf, K., Bandlow, W., and Kaudewitz, F. (1975). Studies on the mechanism of electron transport in the BC-1 segment of the respiratory chain in yeast, part 3 isolation and characterization of an antimycin resistant mutant *ant 8* in *Schizosaccharomyces pombe*. *Mol. Gen. Genet.* **137**, 353–364.

Lang, B., Burger, G., Wolf, K., Bandlow, W., and Kaudewitz, F. (1976). Characterization of respiratory deficient and antimycin resistant mutants of *Schizosaccharomyces pombe* with extra chromosomal inheritance. *In* "Genetics, Biogenesis and Bioenergetics of Mitochondria" (W. Bandlow *et al.*, eds.), pp. 379–387. de Gruyter, Berlin.

Lawrence, C. W. (1982). Mutagenesis in *Saccharomyces cerevisiae*. *Adv. Genet.* **21**, 173–254.

Lawrence, C. W., and Christensen, R. B. (1978). Ultraviolet light induced mutagenesis in *Saccharomyces cerevisiae*. In "DNA Repair Mechanisms" (P. C. Hanawalt, E. C. Friedberg, and C. F. Fox, eds.), pp. 437–440. Academic Press, New York.

Lawrence, C. W., and Christensen, R. B. (1979). Absence of relationship between UV-induced reversion frequency and nucleotide sequence at the CYC_1 locus of yeast. *Mol. Gen. Genet.* **177**, 31–38.

Lawrence, C. W., Christensen, R. B., and Schwartz, A. (1982). Mechanisms of UV mutagenesis in yeast. *In* "Molecular and Cellular Mechanisms of Mutagenesis" (J. F. Lemontt and W. W. Generoso, eds.), pp. 109–120. Plenum, New York.

Lehman, A. R. (1972). Effects of caffeine on DNA synthesis in mammalian cells. *Biophys. J.* **12**, 1316–1325.

Lehman, A. R., and Kirk-Bell, S. (1974). Effects of caffeine and theophylline on DNA synthesis in unirradiated and UV-irradiated mammalian cells. *Mutat. Res.,* **26**, 73–82.

Lemontt, J. F. (1980). Genetic and physiological factors affecting repair and mutagenesis in yeast. *Basic Life Sci.* **15**, 85–120.

Leupold, U. (1950). Die Vererbung von Homothallie und Heterothallic bei *Schizosaccharomyces pombe*. *C.R. Trav. Lab. Carlsberg, Ser. Physiol.* **24**, 381–480.

Leupold, U. (1955a). Versuche zur genetischen Klassifizierung adeninabhängiger Mutanten von Schizosaccharomyces pombe. *Arch. Julius Klaus-Stift Vererbungsforsch. Sozialanthropol. Rassenhyg.* **30**, 506–516.

Leupold, U. (1955b). Metodisches zur Genetik von Schizosaccharomyes pombe. *Schweiz. Z. Allg. Pathol. Bakteriol.* **18**, 1141–1146.

Leupold, U. (1956). Some data on polyploid inheritance in Schizosaccharomyces pombe. *C.R. Trav. Lab. Carlsberg, Ser. Physiol.* **26**, 221–251.

Leupold, U. (1957). Physiologisch-genetische Studienan adeninabhängigen Mutanten von Schizosaccharomyces pombe. Ein Beitrag zum Problem der Pseudoallelie. *Schweiz. Z. Allg. Pathol. Bakteriol.* **20**, 535–544.

Leupold, U. (1958). Studies on recombination in Schizosaccharomyces pombe. *Cold Spring Harbor Symp. Quant. Biol.* **23**, 161–170.

Leupold, U. (1970a). Genetic studies on nonsense suppressors in Schizosaccharomyces pombe. *Heredity* **25**, 493 (abstr.).

Leupold, U. (1970b). Genetical methods for Schizosaccharomyces pombe. *Methods Cell Physiol.* **4**, 169–177.

Leupold, U., and Gutz, H. (1965). Genetic fine structure in Schizosaccharomyces pombe. *Genetics Today, Proc. Int. Congr., 11th, 1963*, pp. 31–34.

Loprieno, N. (1971). UV radiation-sensitivity and mutator activity in yeast Schizosaccharomyces pombe. *Proc. Int. Congr. Radiat. Res. 4th, 1970*, Vol. 1, pp. 2–10.

Loprieno, N. (1973a). UV radiation sensitivity and mutator activity in yeast Schizosaccharomyces pombe. *In* "Advances in Radiation Research, Biology and Medicine" (J. F. Duplan and A. Chapiro, eds.), Vol. 1, pp. 391–395. Gordon & Breach, New York.

Loprieno, N. (1973b). A mutator gene in the yeast Schizosaccharomyces pombe. *Genetics* **73**, 161–164.

Loprieno, N. (1977). The use of yeast cells in the mutagenic analysis of chemical carcinogens. *Colloq. Int. C.N.R.S.* **256**, 315–331.

Loprieno, N. (1978). Use of yeast as an assay system for industrial mutagens. *Chem. Mutagens* **15**, 25–53.

Loprieno, N. (1981). Mutagenicity of selected chemicals in yeast-mutation induction at specific loci. *Environ. Sci. Res.* **24**, 139–150.

Loprieno, N., and Abbondandolo, A. (1980). Comparative mutagenic evaluation of some industrial compounds. *In* "Short-Term Test Systems for Detecting Carcinogens" (K. H. Norpoth and R. C. Garner, eds.), pp. 333–356. Springer-Verlag, Berlin and New York.

Loprieno, N., and Barale, R. (1974). Genetic effect of caffeine. *Mutat. Res.* **26**, 83–87.

Loprieno, N., and Clarke, C. H. (1965). Investigations on reversions to methionine independence induced by mutagens in Schizosaccharomyces pombe. *Mutat. Res.* **2**, 312–319.

Loprieno, N., and Schüpbach, M. (1971). On the effect of caffeine on mutation and recombination in Schizosaccharomyces pombe. *Mol. Gen. Genet.* **110**, 348–354.

Loprieno, N., Abbondandolo, A., Bonatti, S., and Guglielminetti, R. (1968). Analysis of genetic instability induced by nitrous acid in Schizosaccharomyces pombe. *Genet. Res.* **12**, 45–54.

Loprieno, N., Barale, R., Bauer, C., Baroncelli, G., Bronzetti, G., Cammellini, A., Cinci, A., Corsi, G., Leporini, C., Nieri, R., Nozzolini, M., and Serra, C. (1974). The use of different test systems with yeasts for the evaluation of chemically induced gene conversions and gene mutations. *Mutat. Res.* **25**, 197–217.

Loprieno, N., Barale, R., Baroncelli, S., Cammellini, A., Melani, M., Nieri, R., Nozzolini,

M., and Rossi, A. (1975). Mutations induced by X-radiation in the yeast *Schizosaccharomyces pombe. Mutat. Res.* **28**, 163–173.

Loprieno, N., Barale, R., Baroncelli, S., Bauer, C., Bronzetti, G., Cammellini, A., Cercignani, G., Corsi, C., and Gervasi, G. (1976). Evaluation of the genetic effects induced by vinyl chloride monomer under mammalian metabolic activation studies *in vitro* and *in vivo. Mutat. Res.* **40**, 85–96.

Loprieno, N., Barale, R., Baroncelli, S., Bartsch, H., Bronzetti, G., Cammellini, A., Corsi, C., Frezza, D., Nieri, R., Leporini, C., Rosellini, D., and Rossi, A. M. (1977). Induction of gene mutations and gene conversions by vinyl chloride metabolites in yeasts. *Cancer Res.* **36**, 253–257.

Loprieno, N., Von Borstel, R. C., Herrera, L., and de Serres, F. J. (1981). Mutagenesis assays with yeasts and molds. *IARC Sci. Publ.*, 135–156.

Loprieno, N., Barale, R., Von Halle, E. S., and von Borstel, R. C. (1983). Testing of chemicals for mutagenic activity with *Schizosaccharomyces pombe. Mutat. Res.* **115**, 215–223.

Luchkina, L. A. and Bekker, M. L. (1975). Methods of obtaining sphoeroplasts from yeast cells. *Prikl. Biokhim. Mikrobiol.* **11**, 264–268.

Lumb, J. R., Sideropoulos, A. S., and Shankel, D. M. (1968). Inhibition of dark repair of ultraviolet damage in DNA by caffeine and 8-chlorocaffeine. Kinetics of inhibition. *Mol. Gen. Genet.* **102**, 108–111.

McAthey, P. (1976). Anomalies in the selection of mutants of *Schizosaccharomyces pombe* resistant to 8-azaguanine. *Mol. Gen. Genetic.* **149**, 239–242.

McAthey, P., and Kilbey, B. J. (1976). Trichodermin: A selective agent for screening forward mutations in continuous cultures of *Schizosaccharomyces pombe. Biol. Zentralbl.* **95**, 415–421.

McAthey, P., and Kilbey, B. J. (1978). Mutation in continuous cultures of *Schizosaccharomyces pombe.* II. Effect of amino acid starvation on mutational response and DNA concentration. *Mutat. Res.* **50**, 175–180.

McCully, E. K., and Robinow, C. F. (1971). Mitosis in the fission yeast *Schizosaccharomyces pombe:* A comparative study with light and electron microscopy. *J. Cell Sci.* **9**, 475–507.

Meade, J. H., and Gutz, H. (1976). Mating-type mutations in *Schizosaccharomyces pombe.* Isolation of mutants and analysis of strains with an h− or + phenotype. *Genetics* **83**, 259–273.

Meade, J. H., and Gutz, H. (1978). Influence of the *mat-1* M allele on meiotic recombination in the mating type region of *Schizosaccharomyces pombe. Genetics* **88**, 235–238.

Megnet, R. (1964). A method for the selection of auxotrophic mutants of the yeast *Schizosaccharomyces pombe. Experientia* **20**, 320–321.

Megnet, R. (1965). Screening of auxotrophic mutants of *Schizosaccharomyces pombe* with 2-deoxyglucose. *Mutat. Res.* **2**, 328–331.

Michel, R., and Schweyen, R. J. (1971). Inhibition of mitochondrial protein synthesis invivo by erythromycin in *Schizosaccharomyces pombe* and *Saccharomyces cerevisiae. Mol. Gen. Genet.* **111**, 235–241.

Miller, D. R. (1970). Theoretical survival curves for radiation damage in bacteria. *J. Theor. Biol.* **26**, 383–398.

Miller, D. R., Glazebook, A. D., Mullen, B. M., and Bramall, L. (1975). Prediction of bacterial survival after ultraviolet irradiation. *L. H. Gray Conf., 6th, 1974* pp. 179–189.

Minet, M., Nurse, P., Thuriaux, P., and Mitchison, J. M. (1979). Uncontrolled septation in a cell division cycle mutant of the fission yeast *Schizosaccharomyces pombe. J. Bacteriol.* **137**, 440–446.

Minet, M., Grossenbacher-Grunder, A., and Thuriaux, P. (1980). The origin of a centromere effect on mitotic recombination. *Curr. Genet.* **2**, 53–60.

Mitchison, J. M. (1971). "The Biology of the Cell Cycle."

Cambridge Univ. Press, London and New York.

Mitchison, J. M. (1973). The cell cycle of a eukaryote. *Soc. Gen. Microbiol.* **23**, 189–208.

Mitchison, J. M. (1974). Sequences pathways and timers in the cell cycle. *In* "Cell Cycle Controls" (G. M. Padilla, I. L. Cameron, and A. Zimmerman, eds.), pp. 125–142. Academic Press, New York.

Mitchison, J. M. (1977). Enzyme synthesis during the cell cycle. *In* "Cell Differentiation in Microorganisms, Plants and Animals, International Symposium" (L. Nover and K. Mothes, eds.), pp. 377–401.

Mitchison, J. M. (1981). Changing perspectives in the cell cycle. *In* "The Cell Cycle" (P. C. L. John, ed.), pp. 1–10. Cambridge Univ. Press, London and New York.

Mitchison, J. M., and Carter, L. A. (1975). Cell cycle analysis. *Methods Cell Biol.* **11**, 201–220.

Mitchison, J. M., and Creanor, J. (1969). Linear synthesis of sucrase and phosphatase during the cell cycle of *Schizosaccharomyces pombe. J. Cell Sci.* **5**, 373–391.

Mitchison, J. M., and Creanor, J. (1971). Induction of synchrony in the fission yeast *Schizosaccharomyces pombe. Exp. Cell Res.* **67**, 368–374.

Mitchison, J. M., and Lark, K. G. (1962). Incorporation of ^{3}H-adenine into RNA during the cell cycle of *Schizosaccharomyces pombe. Exp. Cell Res.* **28**, 452–455.

Mitchison, J. M., and Vincent, W. S. (1965). Preparation of synchronous cell cultures by sedimentation. *Nature (London)* **205**, 987–989.

Miyata, M., and Miyata, H. (1978). Relationship between extracellular enzymes and cell growth during the cell cycle of the fission yeast. *J. Bacteriol.* **136**, 558–564.

Miyata, H., and Miyata, M. (1981). Mode of conjugation in homothallic cells of *Schizosaccharomyces pombe. J. Gen. Appl. Microbiol.* **27**, 365–371.

Mortimer, R. K. and Manney, T. R. (1971). Mutation induction in yeast. In "Chemical Mutagens: Principles and Methods for their Detection" (A. Hollaender, ed.), Vol. 1, pp. 289–310. Plenum, New York.

Moustacchi, E., Chanet, R., and Heude, M. (1977). Ionizing and ultraviolet radiation: Genetic effects and repair in yeast. *In* "Research in Photobiology" (A. Castellani, ed.), pp. 197–206. Plenum, New York.

Munz, P. (1975). On some properties of 5 mutator alleles in *Schizosaccharomyces pombe. Mutat. Res.* **29**, 155–158.

Munz, P., and Leupold, U. (1970). Characterization of ICR-170 induced mutations in *Schizosaccharomyces pombe. Mutat. Res.* **9**, 199–212.

Munz, P., and Leupold, U. (1979). Gene conversion in nonsense suppressors of *Schizosaccharomyces pombe. Mol. Gen. Genet.* **170**, 145–148.

Munz, P., Amstutz, H., Kohli, J., and Leupold, U. (1982). Recombination between dispersed serine transfer RNA genes in *Schizosaccharomyces pombe. Nature* **300**, 225–231.

Munz, P., Dorsch-Hasler, K., and Leupold, U. (1983). The genetic fine structure of nonsense suppressors in *Schizosaccharomyces pombe.* II. *sup. 8* and *sup. 10. Curr. Genet.* **7**, 101–108.

Nasim, A. (1967). The induction of replicating instability by mutagens in *Schizosaccharomyces pombe. Mutat. Res.* **4**, 753–763.

Nasim, A. (1968). Repair mechanisms and radiation induced mutations in fission yeast. *Genetics* **59**, 327–333.

Nasim, A. (1969). X-Ray induced replicating instabilities in yeast. *Genetics* **60**, Suppl. 2, 2–43.

Nasim, A. (1974). Radiation induced mutation rate and DNA content in *Schizosaccharomyces pombe*. *Mutat. Res.* **24**, 211–212.

Nasim, A., and Auerbach, C. (1967). The origin of complete and mosaic mutants from mutagenic treatment of single cells. *Mutat. Res.* **4**, 1–14.

Nasim, A., and Clarke, C. H. (1965). Nitrous acid induced mosaicism in *Schizosaccharomyces pombe*. *Mutat. Res.* **2**, 395–402.

Nasim, A., and Grand, C. (1973). Genetic analysis of replicating instabilities in yeast. *Mutat. Res.* **17**, 185–190.

Nasim, A., and Hannan, M. A. (1977). Induction of mutations by chemicals and gamma rays in mutants of yeast refractory to UV-mutagenesis. *Can. J. Genet. Cytol.* **19**, 323–330.

Nasim, A., and James, A. P. (1971). Replicating instabilities in yeast, evidence from single cell isolation. *Genetics* **69**, 503–506.

Nasim, A., and James, A. P. (1978). Life under conditions of high irradiation. *In* "Microbial Life in Extreme Environments" (D. J. Kushner, ed.), pp. 410–439. Academic Press, New York.

Nasim, A., and Saunders, A. S. (1968). Spontaneous frequencies of lethal sectoring and mutation in radiation sensitive strains of *Schizosaccharomyces pombe*. *Mutat. Res.* **6**, 475–478.

Nasim, A., and Smith, B. P. (1974). Dark repair inhibitors and pathways for repair of radiation damage in *Schizosaccharomyces pombe*. *Mol. Gen. Genet.* **132**, 13–22.

Nasim, A., and Smith, B. P. (1975). Genetic control of radiation sensitivity in *Schizosaccharomyces pombe*. *Genetics* **79**, 573–582.

Nasmyth, K. (1977). Temperature sensitive lethal mutants in the structural gene for DNA ligase in the yeast *Schizosaccharomyces pombe*. *Cell* **12**, 1109–1120.

Nasmyth, K. A. (1979). Genetic and enzymatic characterization of conditional lethal mutants of the yeast. *Schizosaccharomyces pombe* with a temperature mutant DNA ligase. *J. Mol. Biol.* **130**, 273–284.

Nasmyth, K., and Nurse, P. (1981). Cell division cycle mutants altered in DNA replication and mitosis in the fission yeast *Schizosaccharomyces pombe*. *Mol. Gen. Genet.* **182**, 119–124.

Nestmann, E. R., Stephen, E. R., Kowbel, D. J., and Nasim, A. (1982).Differential survival as an indicator of potential mutagenicity using repair deficient strains of *Saccharomyces cerevisiae* and *Schizosaccharomyces pombe*. *Can. J. Genet. Cytol.* **24**, 771–775.

Nilsson, K., and Lehman, A. R. (1975). The effect of methylated oxypurines on the size of newly synthesized DNA and the production of chromosome aberations after UV-irradiation in chinese hamster cells. *Mutat. Res.* **30**, 255–266.

Nurse, P. (1975). Genetic control of cell size at cell division in yeast. *Nature (London)* **256**, 547–551.

Nurse, P. (1981). Genetic analysis of the cell cycle. *In* "Genetics as a Tool in Microbiology" (S. W. Gower and D. A. Hopwood, eds.), pp. 291–315. Cambridge Univ. Press, London and New York.

Nurse, P., and Bissett, Y. (1981). Gene required in G_1 for commitment to cell cycle and in

G_2 for control of mitosis in fission yeast. *Nature (London)* **292**, 558–560.

Nurse, P., and Fantes, P. (1981). Cell cycle controls in fission yeast—A genetic analysis. *In* "The Cell Cycle" (P. C. L. John, ed.), pp. 85–98. Cambridge Univ. Press, London and New York.

Nurse, P., and Thuriaux, P. (1980). Regulatory genes controlling mitosis in the fission yeast *Schizosaccharomyces pombe*. *Genetics* **96**, 627–637.

Nurse, P. Thuriaux, P., and Nasmyth, K. (1976). Genetic control of the cell division cycle in the fission yeast *Schizosaccharomyces pombe*. *Mol. Gen. Genet.* **146**, 167–178.

Patrick, M. H. and Haynes, R. H. (1964). Dark recovery phenomena in yeast. II. Conditions that modify the recovery process. *Radiat. Res.* **23**, 564–579.

Patrick, M. H., Haynes, R. H., and Uretz, R. B. (1964). Dark recovery phenomena in yeast. I. Comparative effect with various inactivating agents. *Radiat. Res.* **21**, 144–168.

Phipps, J., and Miller, D. R. (1982). Quelques aspects de la toxicité génétique du methylmercure chez les levures. *C. R. Hebd. Seances Acad. Sci.* **295** (III), 683–686.

Poole, R. K., and Lloyd, D. (1974). The development of cytochromes during the cell cycle of a glucose repressed fission yeast *Schizosaccharomyces pombe*. *Biochem. J.* **138**, 201–210.

Poole, R. K., and Lloyd, D. (1976). Fractionation by differential and zonal centrifugation of sphoeroplasts prepared from a glucose-repressed fission yeast *Schizosaccharomyces pombe* 972 h⁻. *J. Gen. Microbiol.* **93**, 245–250.

Poole, R. K., Lloyd, D., and Chance, B. (1980). The reaction of cytochrome oxidase with oxygen in the fission yeast *Schizosaccharomyces pombe* 972 h⁻ *studies at subzero temperatures and measurement of apparent oxygen affinity*. *Biochem. J.* **184**, 555–564.

Poon, N. H., and Day, A. W. (1975). Fungal fimbriae I. Structure, origin and synthesis. *Can. J. Microbiol.* **21**, 537–546.

Prakash, L., and Prakash, S. (1980). Genetic analysis of error-prone repair systems in *Saccharomyces cerevisiae*. *In* "DNA Repair and Mutagenesis in Eukaryotes" (F. J. de Serres, W. M. Generoso, and M. D. Shelby, eds.), pp. 141–158. Plenum, New York.

Prendergast, J. A., Kamra, O. P., and Nasim, A. (1984). The effect of spermine on spontaneous and UV-induced mutations in *Schizosaccharomyces pombe*. *Mutat. Res.* **125**, 205–211.

Pritchard, R. H., Barth, P. T., and Collins, J. (1969). Gene dosage effects in polyploid strains of *Saccharomyces cerevisiae* containing *gua-1* wild type and mutant alleles. *J. Bacteriol.* **124**, 1041–1045.

Rainaldi, G., and Abbondandolo, A. (1975). Mutation and nuclear stage in *Schizosaccharomyces pombe*. Part 2. Reverse mutations induced by X-rays in the absence of recombination. *Mutat. Res.* **27**, 235–240.

Ramirez, C., Friis, J., and Leupold, U. (1974). The *ade 1* gene of *Schizosaccharomyces pombe*. A complex locus controlling steps in purine biosynthesis. In Gutz *et al.* (1974) (unpublished).

Rauth, A. M. (1967). Evidence for dark-reactivation of ultraviolet light damage in mouse L cells. *Radiat. Res.* **31**, 121–138.

Resnick, M. A. (1979). The induction of molecular and genetic recombination in eukaryotic cells. *Adv. Radiat. Biol.* **8**, 175–217.

Reynolds, R. J., and Friedberg, E. C. (1980). The molecular mechanism of pyrimidine dimer excision in *Saccharomyces cerevisiae*. I. Studies with intact cells and cell-free systems. *In* "DNA Repair and Mutagenesis in Eukaryotes" (F. J. de Serres, W. M. Generoso, and M. D. Shelby, eds.), pp. 121–139. Plenum, New York

Roberts, J. J., and Ward, K. N. (1973). Inhibition of post replication repair of alkylated DNA by caffeine in chinese hamster but not in HeLa cells. *Chem.-Biol. Interact.* **7**, 241–264.

Robinow, C. F. (1975). The preparation of yeasts for electron microscopy. *Methods Cell Biol.* **11**, 2–22.

Robinow, C. F. (1977). The number of chromosomes in *Schizosaccharomyces pombe*. Microscopy of stained preparations. *Genetics* **87**, 491–498.

Rothman, R. H. (1980). Dimer excision in *Escherichia coli* in the presence of caffeine. *J. Bacteriol.* **143**, 520–524.

Schmidt, O., Hovemann, B., Mao, J., Silvermann, S., and Soll, D. (1979). Speciic transcription of cloned eukaryotic transport RNA genes in *Xenopus* germinal vesicle extracts. *Exp. Biol.* **38**, 297.

Schroy, C. B., and Todd, P. (1975). Potentiation by caffeine of ultraviolet light damage in cultured human cells. *Mutat. Res.* **33**, 347–356.

Schüpbach, M. (1971). The isolation and genetic classification of UV-sensitive mutants of *Schizosaccharomyces pombe*. *Mutat. Res.* **11**, 361–371.

Segal, E., Munz, P., and Leupold, U. (1973). Characterization of chemically induced mutations in the *ade 1* locus of *Schizosaccharomyces pombe*. *Mutat. Res.* **18**, 15–24.

Seitz, G., Lueckmann, G., Wolf, K., Kaudewitz, F., Boutry, M., and Goffeau, A. (1977a). Extrachromosomal inheritance in *Schizosaccharomyces pombe*. Part 6. Preliminary genetic and biochemical characterization of mitochondrially inherited respiratory deficient mutants. *In* "Mitochondria 1977: Genetics and Biogenesis of Mitochondria" (W. Bandlow *et al.*, eds.), pp. 149–160. de Gruyter, Berlin.

Seitz, G., Wolf, K., and Kaudewitz, F. (1977b). Extrachromosomal inheritance in *Schizosaccharomyces pombe*. Part 4. Isolation and characterization of mutants resistant to chloramphenicol and erythromycin using the mutator properties of *Ana-R-8*. *Mol. Gen. Genet.* **155**, 339–346.

Seitz-Mayr, G., and Wolf, K. (1982). Extrachromosomal mutator inducing point mutations and deletions in mitochondrial genome of fission yeast. *Proc. Natl. Acad. Sci. U.S.A.* **79**, 2618–2622.

Setlow, R. B., and Carrier, W. L. (1968). The excision of pyrimidine dimers *in vivo* and *in vitro*. *In* "Replication and Recombination of Genetic Material" (W. J. Peacock and R. B. Brock, eds.), pp. 134–141. Aust. Acad. Sci., Canberra.

Shahin, M. M. (1971). Dark repair and liquid holding recovery in *Schizosaccharomyces pombe*. *Can. J. Genet. Cytol.* **13**, 645.

Shahin, M. M., and Nasim, A. (1973). The effect of radiation sensitivity and cell stage on liquid holding response in *Schizosaccharomyes pombe*. *Mol. Gen. Genet.* **122**, 331–338.

Shahin, M. M., Gentner, N. E., and Nasim, A. (1973). The effect of liquid holding in *Schizosaccharomyces pombe*. strains after gamma and ultraviolet irradiation. *Radiat. Res.* **53**, 216–225.

Simon-Becam, A. W., Claisse, M., and Lederer, F. (1978). Cytochrome *c* from *Schizosaccharomyces pombe* part 2 amino acid sequence. *Eur. J. Biochem.* **86**, 407–416.

Sipiczki, M., and Ferenczy, L. (1978). Enzymic methods for enrichment of fungal mutants I. Enrichment of *Schizosaccharomyces pombe* mutants. *Mutat. Res.* **50**, 163–173.

Snow, R. (1979). Maximum likelihood estimation of linkage and interference from tetrad data. *Genetics* **92**, 231–246.

Sokurova, E. N. (1974). Nucleic acid content in yeast cells belonging to different systematic groups. *Microbiology (Engl. Transl.)* **42**, 908–911.

Stephen, E. R., and Nasim, A. (1981). Production of protoplasts in different yeasts by mutanase. *Can. J. Microbiol.* **27**, 550–553.

Strauss, A. (1979). Allelism of methionine sensitive mutants of *Schizosaccharomyces pombe* to loci involved in adenine biosynthesis. *Genet. Res.* **33**, 261–268.

Streiblova, E. (1977). Yeasts as a model for cell cycle studies. *Mol. Biol. Genet. Mikrobiol.* **42**, 173–182.

Swann, M. M. (1962). Gene replication, ultraviolet sensitivity and the cell cycle. *Nature (London)* **193**, 1222–1227.

Szostak, J. W., Orr-Weaver, T. L., Rothstein, R. J., and Stahl, F. W. (1983). *Cell* **33**, 25–35.

Thuriaux, P., Nurse, P., and Carter, B. (1978). Mutants altered in the control coordinating cell division with cell growth in the fission yeast *Schizosaccharomyces pombe*. *Mol. Gen. Genet.* **161**, 215–220.

Thuriaux, P., Minet, M., Munz, P., Zbaren, P., and Leupold, U. (1980). Gene conversion in nonsense suppressors of *Schizosaccharomyces pombe*. II. Specific marker effects in *sup 3*. *Curr. Genet.* **1**, 89–95.

Timson, J. (1977). Caffeine. *Mutat. Res.* **47**, 1–52.

Treichler, H. J. (1964). Genetische Feinstruktur und intragene Rekombinations mechanismen in *ade 2* locus von *Schizosaccharomyces pombe*. Ph.D. Thesis, Zurich University.

Trosko, J. E., and Chu, E. H. Y. (1973). Inhibition of repair of UV-damaged DNA by caffeine and mutation induction in chinese hamster cells. *Chem.-Biol. Interact.* **6**, 317–332.

Unrau, P. (1975). The exclusion of pyrimidine dimers from the DNA of mutant and wild-type strains of *Ustilago*. *Mutat. Res.* **29**, 53–65.

van den Berg, H. W., and Roberts, J. J. (1976). Inhibition by caffeine of post-replication repair in chinese hamster cells treated with cis-platinum (II) diaminedichloride. The extent of platinum binding to template DNA in relation to the size of low molecular weight nascent DNA. *Chem.-Biol. Interact.* **12**, 375–390.

Verma, A., Sharma, R., and Jain, V. K. (1982). Energetics of DNA repair in UV-irradiated peripheral blood leukocytes from chronic myeloid leukemia patients. *Photochem. Photobiol.* **36**, 627–632.

Wain, W. H., and Staatz, W. D. (1973). Rates of synthesis of ribosomal protein and total nucleic acid through the cell cycle of the fission yeast *Schizosaccharomyces pombe*. *Exp. Cell Res.* **81**, 269–278.

Walker, G. M., and Duffus, J. H. (1979). An investigation into the potential use of chelating agents and antibiotics as synchrony inducers in the fission yeast *Schizosaccharmyces pombe*. *J. Gen. Microbiol.* **114**, 391–400.

Wilmore, P. J., and Parry, J. M. (1976). Division delay and DNA degradation after mutagen treatment of the yeast, *Saccharomyces cerevisiae*. *Mol. Gen. Genet.* **145**, 287–291.

Wolf, K., Lang, B. F., del Giudice, L., Anziano, P. Q., and Perlman, P. S. (1982). *Schizosaccharomyces pombe*. A short review of a short mitochondrial genome. In "Mitochondrial genes" (P. P. Slonimski, P. Borst, and G. Attardi, eds.), pp. 335–360. Cold Spring Harbour Lab., New York.

Yamamoto, M., Asakura, Y., and Yanangiola, M. (1981). Cloning of a gene from the fission yeast *Schizosaccharomyces pombe* which complements *Escherichia coli* pyrb. the gene for aspartate trans carbamylase EC-2.1.3.2. *Mol. Gen. Genet.* **182,** 426–429.

Yoshida, K. (1979). Interspecific and intraspecific mitochondria-induced cytoplasmic transformation in yeasts. *Plant Cell Physiol.* **20,** 851–856.

Yoshida, K., and Takahashi, I. K. (1980). Cytological studies on mitochondrial-induced cytoplasmic transformation in yeasts. *Plant Cell Physiol.* **21,** 497–509.

Yoshida, K., and Yanagishima, N. (1978). Intra- and intergeneric mating behaviour of ascoporogenous yeasts. I. Quantitative analysis of sexual agglutination. *Plant Cell Physiol.* **19,** 1519–1533.

Zimmermann, F. K. (1973). Detection of genetically active chemicals using various yeast systems. *Chem. Mutagens* **3,** 209–239.

GENE TRANSFER IN FUNGI

N. C. Mishra

Department of Biology, University of South Carolina
Columbia, South Carolina

ADVANCES IN GENETICS, Vol. 23

I. Introduction

Transfer of a heritable character by a naked DNA molecule is called *transformation*. Since Avery and co-workers (1944) first elucidated its mechanism, this genetic process had captured the imagination of biologists because of the immense possibilities it promises in directing the changes in the genetic makeup of organisms (including humans) and in alleviating human diseases by possible gene therapy. More realistically, it provides a means for probing into the genetic chemistry of eukaryotes for the first time by permitting isolation of eukaryotic genes for their extensive biochemical, genetic, and molecular characterizations; that alone can provide a basis for the understanding of the relevance of their genomic organization and of control mechanisms involved in their expression and future manipulation. Transformation as a genetic process for transfer of genetic information has been well established in bacteria (Hotchkiss, 1955, 1977); however, the first conclusive evidence for the occurrence of gene transfer in higher organisms was provided by McBride and Ozer (1973). In the beginning of the 1970s Mishra, in collaboration with Szabo and Tatum, undertook such studies in *Neurospora crassa,* a haploid fungus, because of the advantages that fungi offer in genetic studies. These authors provided evidence for the occurrence of gene transfer in *Neurospora* (Mishra *et al.*, 1973). Soon these studies were followed by transformation experiments in other fungi, and unambiguous evidence for gene transfer has been provided in *Neurospora* (Mishra, 1979a,b; Case *et al.*, 1979), yeast (Hinnen *et al.*, 1978), and *Podospora* (Stahl *et al.*, 1981). In this article, I wish to review these and other recent data regarding the nature of evidence for gene transfer in fungi and the characterization of the transformants, and discuss the use of gene transfer as a genetic system to seek answers to several questions concerning the regulatory control and other basic mechanisms of eukaryotes.

II. Fungi as an Ideal Eukaryotic Genetic System

Fungi offer an unique system for the study of genetics and biochemistry of eukaryotes. It is of interest to mention here that in the past the study of fungi has been rather rewarding in providing answers to the problems in basic genetics and biochemistry regarding the nature of the gene action and the control of metabolic pathways (Beadle and Tatum, 1941; Srb and Horowitz, 1944). Studies of fungi have been particularly important in elucidating the process of genetic recombination (see Stahl, 1980), and in the emergence of yeast as the major system for the cloning of a eukaryotic gene (Fink *et al.*, 1978).

Fungi can be cultivated and manipulated using a combination of techniques used in the study of bacteria. Fungi possess a spectrum of nuclear, organellar (mitochondrial), and extranuclear as well as extraorganellar (i.e., plasmid) genetic systems. Both the nuclear and the extranuclear genetics are very well established and thus provide an unique opportunity for the characterization of the plasmid genetic elements and their interactions with the nuclear and the organellar genetic systems. A variety of growth habits (unicellular to multicellular) provides an opportunity for the study of the problems in development and differentiation of higher organisms using both genetic and biochemical approaches. The occurrence of haploid and diploid stages as well as the ability to form heterokaryons facilitates analysis of dominance and recessiveness and of allelelism by complementation. These studies are further aided by the availability of a large number of mutants, whose linkage relationship and biochemical basis have already been elucidated (Esser and Kuenen, 1967; Mortimer and Hawthorne, 1966; Perkins and Barry, 1977; Fincham *et al.*, 1978; Petes, 1980; Perkins *et al.*, 1982). The fact that tetrad analysis of sexual spores can easily clarify the events taking place during the process of meiosis makes this class of organisms most suitable for genetic and biochemical studies. Prospects for the use of fungi in such studies have been further increased by the development of techniques for (1) the construction of biologically active artificial chromosomes (Dani and Zakian, 1983; Murray and Szostak, 1983) and (2) the isolation of individual chromosomes (Schwartz and Cantor, 1984).

An individual fungal chromosome is almost of the same size (or somewhat smaller) as a prokaryotic chromosome (such as *Escherichia coli*); the fungal DNA sequences are not reiterated and thus have another feature closer to the prokaryotic genome. However, fungi are similar to higher eukaryotes in their chromosome structure (the nu-

TABLE 1
Genome Complexity[a]

Organism	No. of base pairs ($\times 10^6$)	DNA (pg)	No. of chromosomes	Unique sequence (%)	Repeat sequence (%)	Reference
Escherichia coli	3.7	0.004	1	99.7	0.3	Cairns (1963); Britten and Kohne (1968)
Saccharomyces cerevisiae (yeast)	13.5	0.015	17	93	7	Bicknell and Douglas (1970); Lauer et al. (1977)
Neurospora crassa	27	0.03	7	93	7	Perkins and Barry (1977); Krumlauf and Marzluf (1979)
Aspergillus nidulans	26	0.028	8	97	3	Timberlake (1980a,b)
Homo sapiens	2900	3.2	23	70	30	McCarthy (1969); Britten and Kohne (1968); Britten and Davidson (1971)

[a] Data presented for a haploid complement.

cleosomal organization and the presence of histones), transcriptional machinery, and processing of mRNA (capping and polyadenylation), and perhaps in the basic mechanism responsible for differential gene expression. Apart from these similarities, fungi are different from the higher eukaryotes in the nature of transcripts due to the absence of intervening sequences in the fungal transcript (except for ribosomal and transfer RNA and the yeast actin gene). Also, fungi are characterized by the lack of abundance of repeated DNA sequences (Hudspeth *et al.*, 1977; Krumlauf and Marzluf, 1979; Free *et al.*, 1979; Timberlake, 1980a; Cox and Penden, 1980). Some of the features of the fungal genome are summarized in Table 1.

III. Gene Transfer in Higher Organisms

A. Problems of Gene Transfer in Higher Organisms and Nature of Evidence for Gene Transfer

Transformation is a well-established method in bacteria. However, in eukaryotes the occurrence of such a genetic process could not be substantiated on firm ground until recently. The earlier evidence for gene transfer in eukaryotes was mostly based on the frequency of transformation, which was very low and could not be distinguished from the frequency of reversion. Use of a deletion mutant as a recipient in such studies would have enabled one to distinguish between transformation and reversion; however, in the absence of a recipient carrying deletions, it was not possible to decide with confidence whether or not the appearance of a new character in the recipient was due to transformation by the donor DNA or due to reversion of the mutant gene in the recipient. In the early 1960s, DNA-mediated gene transfer was reported in mammalian cells (Szybalska and Szybalski, 1962; Szybalski, 1963). However, poor understanding of the mechanics of DNA precipitation by calcium phosphate (required for transformation) hindered the reproducibility of these experiments (see Scangos and Ruddle, 1981; Loyter *et al.*, 1982).

McBride and Ozer (1973) designed experiments that, for the first time in the history of eukaryotic gene transfer, enabled them to identify the gene product of the transformants. They used a heterologous gene transfer system in which they treated a recipient mouse cell line defective in hypoxanthine guanosine phosphoribosyl transferase ($HGPRT^-$) with the metaphase chromosomes from the wild type (i.e., $HGPRT^+$)

Chinese hamster cells and they were able to show that the HGPRT enzyme produced in the transformed mouse cells was indeed characteristic of the donor Chinese hamster cells. The mouse HGPRT and the chinese hamster enzymes are distinguishable by their electrophoretic mobility on a zymogram and by their elution profile on ion-exchange chromatography. Thus, due to a clever design of the experiment, these authors could conclusively establish the occurrence of gene transfer in mammalian cells in tissue culture even though the frequency of transformation was very low ($1/10^6$ to $1/10^7$).

Since the genetic analysis of a higher eukaryote is complicated by its diploidy, Mishra initiated gene transfer experiments in the lower eukaryote *Neurospora crassa,* a haploid fungus (Mishra *et al.,* 1973). The evidence for gene transfer in *Neurospora* was based again on the design of the experiments. These authors showed that only the wild-type donor DNA (allo-DNA) was effective in transformation of a mutant recipient to the wild-type phenotype. The DNA of the mutant recipient (iso-DNA), when used as a donor, was not effective in transformation. In all cases the transforming effect of the donor DNA was abolished by treatment with DNase. These studies using a total DNA preparation have been now confirmed in several laboratories (Szabo and Schablik, 1982; Schablik *et al.,* 1977; Case *et al.,* 1979; Wootton *et al.,* 1980; Radford *et al.,* 1981). Mishra has provided conclusive evidence for genetic transformation by transfer of a temperature-sensitive character from a donor to recipient strain of *Neurospora.* Mishra (1979a,b) clearly demonstrated the dependence of the phenotypic expression of a character on the nature of the DNA used as donor; in such experiments (as discussed later in Section XII) the wild-type donor DNA produced the wild-type transformants whereas the mutant DNA produced transformants with the mutant phenotype. Thus, these experiments (Mishra, 1979a,b) provided unambiguous evidence for the occurrence of gene transfer in *Neurospora.* Hinnen *et al.* (1978) used a recombinant DNA plasmid in the transformation of yeast and, by the use of the method of Southern (1975) hybridization, they were able to identify the presence of donor DNA in the transformants. By following the methods of Hinnen *et al.* (1978) such demonstration of the presence of the donor DNA in the transformants has now been achieved in *Neurospora* (Case *et al.,* 1979), in *Aspergillus* (Yelton *et al.,* 1984) and in *Podospora* (Stahl *et al.,* 1981). In summary, the unambiguous evidence for gene transfer in a eukaryote is based on the demonstration of (1) a donor DNA-encoded gene product (McBride and Ozer, 1973; McBride and Peterson, 1978), (2) a donor DNA-specific expression of a phenotype among transformants (Mishra, 1979a,b), (3) physical presence of the donor DNA in the transformed cells, and (4) the fate of donor DNA after its entry into recipient cell (Hinnen *et al.*, 1978; Case *et al.*, 1979; Stahl *et al.*, 1981).

B. METHODS OF GENE TRANSFER

Three distinct methods of gene transfer are currently prevalent in fungi; these are described below.

1. *Gene Transfer by Treatment with a Total DNA Preparation (i.e., without a Vector)*

Fungal cells are treated with total DNA, either in the form of naked molecules or trapped in liposomes (Radford *et al.*, 1981), prepared from the donor strain and then examined for the appearance of the transformants by complementation of a defective function in the recipient cells. This method has been extensively used in the genetic transformation of *Neurospora* (Mishra *et al.*, 1973; Case *et al.*, 1979).

2. *Gene Transfer by Treatment with an Enriched DNA Preparation (i.e., Cloned DNA)*

Fungal cells are treated with a DNA preparation that has been previously cloned and amplified via a plasmid or other vectors in *E. coli*. This method offers several distinct advantages, such as the fact that the frequency of transformation is increased and that the transformants can be easily examined for the presence of donor DNA, unambiguously identifying the products of transformation. This method can also be used to amplify the copy number of a particular gene, which can then be examined for its structure and for identification of the different control signals and other domains within a gene by the methods of DNA sequencing. Thus, a fungal or other eukaryotic gene (or its transcript) can be analyzed for the presence of exons, introns, and flanking regions including nontranscribable or nontranslatable sequences as well as promoters, ribosome binding sites, terminator sequences, and sites for capping and polyadenylation.

3. *Gene Transfer via Cell Fusion*

In this method, spheroplasts of fungal cells are fused with polyethylene glycol (PEG), which allows an opportunity for gene transfer via genetic recombination (including diploidy and/or aneuploidy). This method is very useful when the gene transfer via total DNA treatment has failed and the method for molecular cloning of fungal genes via a selectable vector is not yet available. This method facilitates genetic recombination between unrelated fungal species, which cannot naturally undergo cell fusion. It has been successfully used to construct strains of fungi with increased yield of the antibiotic cephalosporin C. This mode of gene transfer has been recently reviewed by Elander (1980) and will not be discussed any further in this article. It is sufficient to say that the cell fusion method of gene transfer itself represents an advancement in the genetic engineering of the industrial fungi and that this method can also achieve a direct transfer of cloned genes from bacteria to fungal or other eukaryotic cells (Kingsman *et al.*, 1979; Schaffner, 1980).

IV. Methodology

A. MOLECULAR CLONING

Recombinant DNA technology has emerged as an integral part of transformation experiments in eukaryotes, particularly as a means to isolate and enrich donor DNA (for reviews on this subject, see Vosberg, 1977; Sinsheimer, 1977; Morrow, 1979; Gorbstein, 1980; Watson *et al.*, 1983). For the purpose of molecular cloning, the eukaryotic DNA is inserted into a vector (a plasmid, a bacteriophage, or a cosmid) by a method dubbed as "shotgun" (Sinsheimer, 1977). Eukaryotic DNA is broken down into small fragments (usually 10–40 kb in length) by hydrodynamic shearing or by a partial digestion with a restriction endonuclease and then joined into a vector. The ligation is facilitated by the generation of cohesive ends following digestion with a particular restriction endonuclease (Cohen *et al.*, 1972). The recombinant vector is then used to transfect *Escherichia coli* (Mandel and Higa, 1970). The presence of a recombinant plasmid is usually detected by the acquisition of a particular drug resistance by the transformed bacteria. In such experiments the plasmid (pBR322) containing the genes for resistance to tetracycline (*tet*) and ampicillin (*amp*) is usually used (Bolivar *et al.*, 1977). The eukaryotic DNA fragment is inserted either at the *Eco*RI (or *Hin*dIII, *Bam*HI, or *Sal*I) sites in the *tet* gene or at the *Pst*I site in the *amp* gene. The insertion of the eukaryotic DNA at either of the *Hin*dIII, *Bam*HI, or *Sal*I (but not at the *Eco*RI) sites inactivates the *tet* gene whereas the insertion at the *Pst*I site inactivates the *amp* gene. Such insertional inactivation of the antibiotic resistance marker is used to detect the presence of a recombinant plasmid in the transformed bacteria, which possess resistance to only one of the antibiotics. Since most of the plasmids used in gene transfer studies possess a relaxed mode of replication and do not require continuous protein synthesis for replication, the number of the plasmid molecules in the bacterium is amplified by adding chloramphenicol to the growth medium. The addition of chloramphenicol inhibits the growth of the bacterium but not that of the plasmid. The presence of a specific eukaryotic gene in the recombinant plasmid can be identified by a variety of methods that includes complementation of a defective function and immunological screening or hybridization with a specific probe (tRNA, rRNA, mRNA, cDNA).

1. Vectors

Choice of vectors depends much on the nature of the gene to be cloned. A vector must be easily isolatable, must possess both selectable characters such as drug resistance and unique cloning sites, and must be capable of amplification and maintenance in the presence of a large insert of foreign DNA. Different kinds of vectors have been used for the cloning of a fungal gene; their main features are described below.

a. Plasmids. These vectors were initially developed because of their small size, easy replication, and important identifiable characters (Bolivar *et al.*, 1977; Bolivar, 1978; Bolivar and Backman, 1979; Sinsheimer, 1977). The most commonly used plasmids (such as pBR322) have been obtained after modification of naturally occurring plasmids. These modifications include (1) a reduction in size, (2) the removal of factors for conjugal transer, (3) the possession of a minimal number of cleavage sites (such as *Eco*RI, *Hind*III, *Bam*HI, and *Pst*I), for a particular restriction endonuclease, (4) the selectivity of many characters for easy identification, as reported by Cohen and Chang (1974), and (5) a relaxed mode of replication, so that the plasmid number per cell can be amplified from 20–25 copies/cell to 3000 copies/cell after a 12-hr treatment with chloramphenicol. The origin and the genealogy of the multipurpose plasmid vector pBR322, discussed elsewhere (Bolivar *et al.*, 1977; Glover, 1980), is presented in Fig. 1.

Self-ligation is a problem during construction of a recombinant plasmid, and if no measures are taken to prevent it, only 15–20% of the transformants contain recombinant plasmids. Several methods are available that select for transformants with recombinant plasmids either by enriching such transformants or by preventing self-ligation of plasmids. The mixture of transformed and nontransformed bacteria

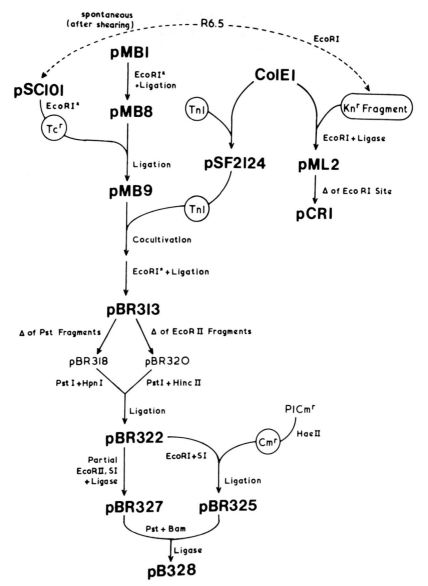

FIG. 1. Genealogy of the plasmid vectors. (Reproduced from Glover, 1980, with permission of authors and publisher.)

are first grown in the presence of ampicillin, which kills nontrans-formed cells. After the removal of ampicillin, tetracycline is added, which stops the growth of bacterial cells carrying hybrid plasmids; however, the bacteria carrying nonrecombinant plasmids (with a func-tional *tet* gene) continue to grow in the presence of tetracycline. At this point, cycloserine is added to induce the lysis of the growing bacterial cells carrying nonrecombinant plasmids (Schleif and Wensink, 1981). Finally, all drugs are removed and the surviving bacterial cells are the ones harboring the recombinant plasmid. This is an effective way to ensure that the bacterial transformants carry only the recombinant plasmids.

There are at least two other means available that can prevent self-ligation of the linearized plasmid molecules and ensure that the plas-mids capable of transformation of the host are the only ones that are recombinant plasmids. One method requires removal of a 5'-phosphate by a bacterial alkaline phosphatase (Ullrich *et al.*, 1977) at the co-hesive ends of a plasmid generated after cleavage with a restriction endonuclease. In our experience it is the most effective method (> 99%) to obtain the recombinant plasmid (Schablik *et al.*, 1983). The other method involves ligation using oligonucleotidyl tails. In this method a plasmid is cleaved with a restriction enzyme and then tailed with either poly(T) or poly(G) whereas the eukaryotic DNA segments are tailed with either poly(A) or poly(C); the presence of noncomplemen-tary tails prevents self-ligation. This method is quite effective (75% or more). When cloned in this manner, the recombinant plasmid contains a *Pst*I site on either side of the inserted eukaryotic DNA segment. This provides a means for the isolation of the inserted DNA segment by cleavage of the recombinant plasmid with *Pst*I enzyme.

The *E. coli* strain used as recipient in a transformation experiment carries the following genes: (1) *recA*$^-$ to avoid recombination of the ColE1 DNA plasmid at the homologous site in the *E. coli* chromosome, (2) *hsdR*$^-$ to prevent restriction of foreign DNA, and (3) *hsdM*$^+$ to allow proper methylation of the cloned hybrid plasmid DNA, so that plasmid DNA can be transferred to a wide variety of auxotrophic *hsdR*$^+$ strains to permit complementation analysis. The fungal DNA is usually cleaved into appropriate size fragments for insertion into the plasmid by digestion with *Eco*RI or *Bam*HI restriction endonuclease since these sites usually occur once per 6–10 kb of eukaryotic DNA. Restriction enzymes such as *Hpa* or *Xba* should not be used for the generation of eukaryotic DNA segments since these enzymes generate DNA segments of very small sizes. These enzymes should, instead, be used for the characterization of the eukaryotic DNA insert after molecular cloning. Construction of hybrid plasmids can also be facilitated by the use of restriction enzymes that generate identical cohesive termini but with different recognition sequences, such as *Sal*I and *Xho*I, which produce 5' TCGA termini, and *Bam*HI, *Bgl*II and *Mbo*I, all of which produce a 5' GATC cohesive end. This procedure permits a greater flexibilty in the subcloning of the restriction enzyme fragments of the eukaryotic DNA inserts (Szostak and Wu, 1979).

   b. Bacteriophage.   Development of bacteriophage as cloning vectors posed several problems (Murray and Murray, 1974) but these were finally solved and several defective λ phage lacking the nonessential region have been found to be good vectors. The interesting features of λ phage is that the recombinant phage can be detected based on its ability to package *in vitro.* Then only those λ DNA with an eukaryotic DNA insert can be packed into the virion. Although λ phage vector has several advantages this has not been extensively used for cloning fungal genes. Likewise, another bacterial virus (M13), which has several distinct advantages in molecular cloning and in DNA sequencing (see Zinder and Boeke, 1982), has not been utilized as a vector in the cloning of fungal genes.

   c. Cosmid.   Cosmids were developed as cloning vectors in order to accomodate larger DNA inserts. It is essentially a plasmid (pBR322) that contains the cohesive *(cos)* ends of the λ bacteriophage so that it can be packed into a virion. Thus, it has the packaging requirement of a λ phage but replication properties characteristic of a plasmid (Collins and Hohn, 1978). Since the cosmid genome is comparatively smaller in size than the bacteriophage, it can accommodate larger (30–45 kb) eukaryotic DNA fragments as inserts (Ish-Horowicz and Burke, 1981).

   d. Eukaryotic Plasmids.   Recently, several plasmids have been found to occur in fungi. These have been used in combination with the bacterial plasmids to yield efficient vectors. In yeast a naturally occurring plasmid has been described (Cameron *et al.,* 1977). This is approximately 2 μm (6 kb) in circumference and occurs in 50–100 copies/cell. This has been used for cloning purposes directly in yeast (Beggs, 1978). Recently, a portion of the 2-μm circles has been joined to the plasmid pBR322. Such a hybrid plasmid can multiply both in *E. coli* and in yeast (Fink *et al.,* 1978; Gerbaud *et al.,* 1979). The use of the 2-μm plasmid as a cloning vector has been recently reviewed (Hollenberg, 1981; Hinnen and Mayhack, 1981; Gunge, 1983).

   In senescent strains of *Podospora anserina,* a naturally occurring plasmid (circles of 0.75 μm) was found that originates from mitochondrial DNA (Esser *et al.,* 1980, 1983; Kuck *et al.,* 1981). In combination with a bacterial plasmid it can be used as hybrid vector for cloning in both prokaryotes and eukaryotes (Stahl *et al.*, 1981). A class of plasmids has been recently found to occur in *Neurospora* that can be used to develop an entirely new vector, at least for this filamentous fungus (Collins *et al.*, 1981). These authors have described a novel class of plasmids from mitochondria of certain wild type *Neurospora* strains that show no detectable sequence homology to mitochondria DNA. The plasmid exists as an oligomeric series of circular molecules with a monomeric length of 3.6 kb. Each monomeric circle contains 6 *Eco*RI sites clustered within a region of 1 kb and five or more *Pst*I sites in a 0.4-kb region. The plasmid also produces a major transcript 3.3–3.4 kb in length and it has been implied that it contains specific sites for the initiation and termination of transcription. These authors have identified two additional plasmids in mitochondria of *Neurospora*; these are 4.2 and 5 kb long; none of these three plasmids show any detectable sequence homology; also, the clustering of restriction sites seen in the first plasmid is not seen in the last two plasmids. These authors have suggested the evaluation of these plasmids as vectors in a *Neurospora* transformation system. As discussed later, Stohl and Lambowitz (1983) have indeed utilized one of these plasmids to construct a shuttle vector. Recently in my laboratory, Niagro (1981) has described the occurrence of a 1.7-μm circular plasmid-like mitochondrial DNA from a *Neurospora* strain that can be utilized in combination with an *E. coli* plasmid (pBR322) to act as a shuttle vector.

2. DNA Preparation

The hybrid plasmid DNA is purified by standard procedures, which consist of deproteinization and banding in a CsCl ethidium bromide gradient. Kado and Liu (1981) have described a rapid method for plasmid DNA preparation that in our experience is the most convenient method. Total chromosomal DNA is usually prepared by a modification of the procedure described by Marmur (1961), Cryer *et al.* (1975), Hautala *et al.* (1977), and Specht *et al.* (1982), which consists of deproteinization by a mixture of phenol, chloroform, and isoamyl alcohol (25:24:1); the DNA fiber is spooled after alcohol precipitation. The removal of RNA and protein is accomplished by treatment with RNase and proteinase K, respectively. The DNA is finally purified by banding in a CsCl ethidium bromide gradient. A new method (Metzenberg and Baisch, 1981) for DNA extraction from *Neurospora* has been developed that facilitates the removal of protein by precipitation of DNA in one step following treatment with alcoholic perchlorate. A high molcecular weight (25-kb) DNA is obtained by this method without the involvement of any laborious steps such as CsCl gradients or preparation of nuclei (Hautala *et al.*, 1977; Marmur, 1961; Cryer *et al.*, 1975). We have found the *Neurospora* DNA prepared by the above method (Metzenberg and Baisch, 1981) contained a large amount of RNA, which could be removed by spooling of DNA fibers immediately after alcoholic precipitation (Schablik *et al.*, 1983).

3. Construction of a Genomic Library

A genomic library of a fungus is usually constructed in *E. coli* as a collection of individual bacterial colonies harboring recombinant plasmids containing fungal DNA segments as inserts. Fungal DNA is obtained as pieces of ~10 kb size by random hydrodynamic shearing or by partial digestion with restriction endonuclease, ligated into a vector, and then introduced into *E. coli* by transformation. The presence of a recombinant plasmid (containing fungal genes as inserts) is usually detected by the loss of resistance to an antibiotic due to the insertional inactivation of the drug resistance (*tet* or *amp*) marker.

Genomic libraries of several fungi are now available. The number of the *E. coli* colonies needed to contain the whole genome of a fungus is determined by the Poisson equation and therefore is a function of the size of the fungal genome and of the DNA segments into which the genome is fragmented. Clarke and Carbon (1978) determined the number of *E. coli* colonies that represent the total yeast genome to be 5000, provided the size of yeast DNA segments is 10 kb long. The number of *E. coli* colonies representing a particular fungal genomic library can be calculated by the Poisson equation $P = [1-(T/G)]^N$ where P is the probability of a particular fungal DNA sequence, T is the average size of the inserted DNA (10 kb), G is the fungal genome size, and N is the total number of colonies containing a recombinant plasmid.

B. Methods of DNA Treatment during Fungal Transformation

In a typical transformation experiment an auxotrophic strain is used as recipient and is treated with either a total or enriched DNA prepared from a prototrophic donor strain; the treated culture is then examined for the acquisition of prototrophy. The recipient culture is usually digested with an enzyme to remove the fungal cell wall since spheroplast production facilitates DNA uptake. The complex fungal cell wall seems to hinder the uptake of DNA. Therefore the cells are usually treated with glusulase or zymolyase to remove the complex cell wall. The spheroplasts thus obtained are treated with the DNA preparation. DNA uptake is facilitated by addition of $CaCl_2$ and PEG to the transformation mixture.

In *Neurospora*, certain cell wall-deficient mutants have been used to facilitate the uptake of DNA without spheroplast formation (Mishra *et al.*, 1973). Alteratively, *Neurospora* conidia grown in sorbitol have been used to augment the uptake of DNA (Wootton *et al.*, 1980). In the case of *Neurospora*, a cell-wall-less strain called *slime* can be used without glusulase treatment. However, the fact that the genetic analysis of the *slime* strain of *Neurospora* is difficult has made the use of this strain less attractive in transformation experiments (N. C. Mishra, unpublished; G. Szabo, personal communication). An osmotic strain of *Neurospora* can be grown as spheroplasts in media of high osmotic pressure and used as a recipient in transformation experiments. In transformation of *Neurospora*, usually the germinating conidia are treated with glusulase and then with $CaCl_2$ and PEG. This treatment is accompanied by a brief exposure to a high (42°C) and then to a low (0°C) temperature in order to facilitate DNA uptake (Case *et al.*, 1979).

In the case of yeast, the recipient cells are also treated with an enzyme (zymolyase) to remove the cell wall before treatment with DNA. The spheroplast production can be easily monitored by microscopic examination of the treated cells. Usually, 1–3 hr of treatment is enough to produce spheroplasts suitable for use in a transformation experiment. Exact protocols for DNA treatment have been extensively described for yeast (Hinnen *et al.*, 1978; Clarke and Carbon, 1978; Kingsman *et al.*, 1979), *Neurospora* (Case *et al.*, 1979; Case, 1981), and *Podospora* (Tudzynski *et al.*, 1980). Kingsman and collaborators have devised an easy method that yields transformation of yeast at very high frequency. Their method consists of treating the yeast spheroplasts with partially lysed bacterial protoplasts harboring hybrid plasmids. This mixture is then treated with PEG (40%) for 30 min and subsequently analyzed for the presence of the transformants by plating an aliquot on appropriate medium (Kingsman *et al.*, 1979).

C. DETECTION OF TRANSFORMANTS

The transformants are detected by one of the following methods.

1. Complementation

The simplest way to detect a genetic transformation is to examine for complementation of a defective biochemical function by the donor DNA in mutant *E. coli* cells as recipients (Clarke and Carbon, 1975;

Ratzkin and Carbon, 1977; Carbon *et al.*, 1977; Kushner *et al.*, 1977; Vapnek *et al.*, 1977). A number of mutants that showed transformation by this method is described in Table 2 (also see Petes, 1980). From these experiments the important fact has emerged that only 25% of the cloned fungal genes can be expressed in *E. coli.* Genes that fail to function in *E. coli* and do not make enough transcript can be expressed when introduced back into fungal cells by transformation. A number of fungal genes have been cloned in this manner. Yeast *his-4* and *arginine permease* (*can-1*) genes have been cloned directly in yeast (Hinnen *et al.*, 1979; Broach *et al.*, 1979); likewise, the *qa-3* gene of *Neurospora*, which cannot be expressed in *E. coli,* was found to be expressed upon reintroduction into *Neurospora* (Schweizer *et al.*, 1981a,b). Such a limited expression of fungal genes in *E. coli* seems to be related to differences in the nature of the control signals for the initiation of transcription and translation in the eukaryotic and prokaryotic systems; this problem of gene expression is further discussed in Section IX of this review.

2. Assay for Enzyme Activity

The molecular cloning of a fungal gene in *E. coli* can easily be detected if the fungal gene can complement a defective function in the host cell. Whenever such a complementation does occur it is important to establish that the complementation is due to expression of the fungal gene and not due to some kind of suppression of the host defective function (Ratzkin and Carbon, 1977; Harashima *et al.*, 1981; Cihlar and Sypherd, 1982). This can be determined by showing that the enzyme produced is characteristic of the fungal gene. A number of yeast and *Neurospora* genes have been shown to produce enzymes in *E. coli* that are characteristic of their fungal origin in terms of physicochemical and immunological properties (i.e., their heat lability, elution profile on chromatography, and antigenicity) (Carbon *et al.*, 1977; Kushner *et al.*, 1977).

3. Immunological Method for the Screening of Cloned DNA in Transformants

It is expected that a majority of the eukaryotic genes cloned in *E. coli* cannot produce a functional enzyme and therefore could not be detected by the method of complementation. Only four yeast genes out of 20 examined were found to complement a defective function in *E. coli,* whereas in *Neurospora* only one gene, *qa-2*[+], has been able to complement a mutant in *E. coli.* A large number of other *Neurospora* genes tested were unable to complement *E. coli* mutants (S. R. Kushner, unpublished results, cited in Case, 1981; N. C. Mishra, unpublished results). In addition, the total number of eukaryotic genes for which complementation assays can be developed is limited. Therefore, immunological methods have been developed for the screening of the transformant colonies using radioactively labeled antibodies directed against purified fungal proteins (Clarke *et al.*, 1979). The immunological methods are very sensitive, specific, and effective in detecting even a small quantity of an antigen or a fragment of the antigen, retaining antigenic determinants. The immunological method can be used to screen transformant colonies on an agar plate. There are several immunological methods available (Erlich *et al.*, 1979; Broom and Gilbert, 1978; Clarke *et al.*, 1979; Young and Davis, 1983). All these methods represent variations of the same principle, i.e., specific antigen–antibody binding (involving radioiodinated ^{125}I-labeled antibody of *Staphylococcus aureus* protein A) and its visualization by autoradiography. A simple method devised by Clarke *et al.* (1979) and used to detect *E. coli* transformants carrying yeast hexokinase or phosphoglycerokinase genes is described here. This method is based on a covalent attachment of the antibody to CNBr-activated paper, followed by incubation of the paper with lysed colonies on agar plates. Colonies releasing antigen that can bind to the antibody are then detected by incubation with ^{125}I-labeled antibody and visualized by autoradiography. This method can easily detect 50–100 pg of antigen (yeast hexokinase or phosphoglycerokinase) when added to lysed colonies on a plate. By this method, 15 colonies out of a total of 6240 *E. coli* colonies showed a positive reaction with ^{125}I-labeled yeast hexokinase antibody (Clarke *et al.*, 1979).

4. Colony Hybridization

Grunstein and Hogness (1975) developed the method of colony hybridization for rapid screening of transformant colonies carrying a particular cloned DNA sequence by hybridization with a radioactive probe such as rRNA, tRNA, cDNA, or mRNA. The transformant colonies (bacterial or fungal) are grown on nitrocellulose filters lying on an agar plate; cells are lysed and the DNA is denatured *in situ* and then hybridized with ^{32}P-labeled DNA or RNA probes. The unhybridized probe is thoroughly washed and then the filter is autoradiographed to visualize the position of a colony containing the DNA sequence in question. This method has been modified to adapt for the screening of λ phage plaques (Benton and Davis, 1977) as well as yeast and *Neu-*

TABLE 2

Molecular Cloning of Genes

A. Fungal genes cloned in *E. coli* and detected by complementation of mutant host function

Organism	Gene	Linkage group	Comparable *E. coli* gene enzyme encoded	Reference
Yeast				
1.	*leu-2*	III	*leuB* is isopropylmalate dehydrogenase (EC 1.1.8.5)	Ratzkin and Carbon (1977)
2.	*his-3*	XV	*hisB* imidazoleglycerol phosphate dehydrogenase (EC 4.2.1.19)	Struhl *et al.* (1976); Ratzkin and Carbon (1977); Davis *et al.* (1979)
3.	*arg-4*	VIII	*argH* arginosuccinate lyase (EC 4.3.2.1)	Clarke and Carbon (1978)
4.	*trp-5*	VII	*trpA* tryptophan synthetase (EC 4.2.1.2.0)	Carbon *et al.* (1977); Walz *et al.* (1978)
5.	*trp-1*	VI	*trpC* N-(5'-phosphoribosyl) anthranilate isomerase	Clarke and Carbon (1978)
6.	*ura-3*	V	*pyrF* orotidine-5'-phosphate (OMP) decarboxylase (EC 41.1.2.3)	Petes *et al.* (1978)
7.	*ura-1*	XI	*pyrD* dihydroorotate dehydrogenase (EC 1.3.3.1)	Lauson and Jund (1982)
8.	*gal-1*	II	*galK* galactokinase (EC 2.7.1.6)	Citron *et al.* (1979)
9.	*his-5*	IX	*Hisc* histidinol-phosphate amino transferase (EC 1.1.1.23)	Harashima *et al.* (1981)
Neurospora				
1.	*qa-2*	VII	*aro-D* catabolic dehydrogenase (cDHQase) (EC 4.2.1.10)	Kushner *et al.* (1977); Vapnek *et al.* (1977)
2.	*trp-1*	IIIR	*TrpC* phosphoribosyl anthralate isomerase gene	Keesey and DeMoss (1982); Schechtman and Yanofsky (1983)
3.	*Pyr-4*	IIL	*PyrF* orotide-5'-phosphate carboxylase	Buxton and Radford (1983a,b)
4.	*nit-3*	IVR	Nitrate reductase	Smarrelli and Garrett (1982)

90

Fusarium

  1. Polygalactouronase activity expressed in *E. coli*, which allows *E. coli* to utilize agar as the sole carbon source (Martini *et al.*, 1978)

  2. Furadantin resistance (Martini *et al.*, 1978)

Mucor racemosus

  Suppression of *E. coli LeuB6* mutation (Cihlar and Sypherd, 1982)

B. Fungal genes cloned in *E. coli* and detected by immunological methods

  Yeast

    pgk; phosphoglycerkinase (EC 2.7.2.3) (Hitzeman *et al.*, 1979, 1980)

C. Fungal genes cloned in *E. coli* and detected by colony hybridization with labeled probes

  1. tRNA probe

    Yeast RNA genes (Petes *et al.*, 1978)

    Neurospora 5 S RNA gene and related pseudogene (Selker *et al.*, 1981)

  2. rRNA probe

    Yeast rRNA genes (Petes *et al.*, 1978)

    Neurospora rRNA genes (Cox and Penden, 1980; Free *et al.*, 1979)

  3. mRNA probe

    Yeast *pgk* (phosphoglycerokinase) (Hitzeman *et al.*, 1980)

    Yeast galactokinase (Rose *et al.*, 1981; Schell and Wilson 1979)

    Yeast glyceraldehyde-3-phosphate dehydrogenase (Holland and Holland, 1979)

    Yeast abundant class of mRNA (Petes *et al.*, 1978)

    Yeast H2a and H2b histone genes (Petes, 1980)

    Yeast enolase yeast ribosonal protein 52 gene (Petes, 1980)

    Aspergillus abundant class of mRNA (Zimmerman *et al.*, 1980)

    Aspergillus alcA (alcohol dehydrogenase) (Olsen *et al.*, 1982)

  4. cDNA probe

    Neurospora nit-5 (nitrate inducible genes) (G. Sorger, 1982, cited by Edwards, 1982)

  5. Synthetic oligonucleotide probe

    Yeast Cyc1 (isocytochrome 1) gene

91

(continued)

TABLE 2 (*Continued*)

Yeast cyc7 (iso-2-cytochrome-C) gene, Plaque hybridization to cloned cyc-1 gene (Petes, 1980)

Neurospora am (glutamate dehydrogenase) gene (Kinnaird and Fincham, 1982)

6. Acetate-inducible genes of Neurospora (in a charon-4 library) by differential plaque filter hybridization (Thomas, 1982)

D. Fungal gene cloned in fungi and detected by complementation or gene expression
   1. Yeast *his-4* (Hinnen *et al.*, 1979)
   2. Yeast *can-1* (Broach *et al.*, 1979), arginine permease
   3. Yeast *cdc-10* and *cdc-28* (Hsio and Carbon, 1979; Nasmyth and Reed, 1980)
   4. *Neurospora qa-3* (Schweizer *et al.*, 1981a), quinate dehydrogenase, also *qa-1* and *qa-4* (Schweizer *et al.*, 1981a)
   5. Neurospora *inl* gene (G. M. Szabo, personal communication, 1983)
   6. *Qa* gene cluster (Giles, 1982; Huiet, 1984)
   7. Yeast *phoE* (repressible) and *phoC* (constitutive) acid phosphatase (Hinnen and Meyhack, 1981)
   8. Yeast suppressor tRNA gene (Thomas and James, 1980a)
   9. Yeast gene for trichodermin resistance and ribosomal protein L3 (Fried and Warner, 1981)
   10. Yeast alcohol dehydrogenase genes (Williamson *et al.*, 1980)
   11. Yeast (*Saccharomyces cerevisiae*) *LEU-2*$^+$ gene cloned in *Schizosaccharomyces pombe* (Beach and Nurse, 1981)
   12. Yeast mating-type loci (HMLα, HMRa, MATa, and MATγ) (Strathern *et al.*, 1979)
   13. Yeast *his-5* (Harashima *et al.*, 1983)
   14. Yeast CHO1 (EC 2.7.8.8)-phosphatidyl serine synthetase (Letts and others, 1984); yeast Q-factor invertase fusion gene (Emr and others, 1983).
   15. Yeast (α-factor invertase) fusion gene (Emr *et al.*, 1983)

92

E. Prokaryotic genes cloned in yeast
1. *Cam^r* (chloramphenicol resistance)
   chloramphenicol acetyltransferase (Cohen *et al.*, 1980)
2. *Ac^r* (ampicillin resistance) (Chevallier *et al.*, 1980; Roggenkamp *et al.*, 1981)
3. *APH* (resistance to G418, aminoglycoside antibiotics) aminoglycoside phosphotransferase (Jimenez and Davies, 1980)
4. Nitrogen fixation gene cluster of *Klebsiella pneumoniae* (Zamir *et al.*, 1981)
5. β-Galactosidase gene of *E. coli* (Rose *et al.*, 1981; Guarente and Ptashne, 1981)
6. Kanamycin resistance gene of Tn601 (Hollenberg, 1981)
7. *E. coli* gene encoding the outer membrane protein II (Hollenberg, 1981)

F. Prokaryotic genes cloned in *Podospora*
1. Ampicillinase gene of *E. coli* plasmid pBR322 (Stahl *et al.*, 1981)

G. Higher eukaryotic genes cloned in yeast
1. *Dictyostelium* (repetitive) DNA controlling heat shock promoter function (Capello *et al.*, 1984)
2. *Drosophila* genomic DNA that complements yeast *ade-8* mutation (Henikoff *et al.*, 1981)
3. Rabbit globin gene (Beggs *et al.*, 1980)
4. Chicken ovalbumin gene (Mercereau-Puijalon *et al.*, 1980)
5. Human leukocyte interferon D (Hitzeman *et al.*, 1981)

H. Higher eukaryotic viral genes cloned in yeast
1. Gene for surface antigen of hepatitis B virus (HBSAg) (Valenzuela *et al.*, 1982; Miyanohara *et al.*, 1983)

93

rospora transformants (Hinnen *et al.*, 1978; Stohl and Lambowitz, 1983b). Transformants containing ribosomal DNA sequences as well as other conserved clone sequences can be detected by this method using heterologous DNA or RNA probes; this eliminates the need for preparing a specific probe. Transformants containing the yeast glyceraldehyde-3-phosphate dehydrogenase have been identified by this method of colony hybridization (Holland and Holland, 1979).

Radioactively labeled synthetic oligonucleotides can also be used as probes in the method of colony hybridization to detect a transformant containing a specific DNA sequence. A synthetic oligonucleotide can be made based on the amino acid sequence of a particular protein. An oligonucleotide (AGCACCTTTCTTAGC) complementary to bases 24–34 of the yeast cytochrome mRNA has been used to identify transformants containing yeast *iso-1* cytochrome *c* (Szostak *et al.*, 1979). A common technical problem of this method is the heavy background obscuring the desired spot. However, some of the problem can be overcome by the use of a lower concentration of probe (usually 10^5–10^6 cpm/ml) and by performing hybridization in 0.2% polyvinylpyrolidone, Ficoll, and bovine serum albumin as suggested by Szostak *et al.* (1979). Plaque hybridization using phage M13 (Messing *et al.*, 1977) is almost 100 times more sensitive than with λ phage. This property may greatly enhance the use of the synthetic oligonucleotides in the screening of a phage bank containing fungal DNA sequences.

Petes *et al.* (1978) have used the method of colony hybridization to detect the *E. coli* transformants carrying the DNA sequences for the yeast rRNA, tRNA, and certain mRNAs. These authors examined 2000 colonies. The largest class of colonies (201/2000) hybridized to a tRNA probe. It is expected that the arrangement of tRNA genes in the yeast genome will determine the proportion of colonies hybridizing to tRNA. The number of tRNA genes, calculated from the data of Schweitzer *et al.* (1969), is about 300. Thus the experimental observation that 10% of the colonies hybridized to tRNA suggests that the tRNA genes are scattered throughout the genome and not clustered (Petes *et al.*, 1978). This conclusion is in agreement with the previous findings of Hawthorne and Leupold (1974), Gilmore (1967), and Beckman *et al.* (1977), which showed that eight tyrosine-inserting ochre suppressors map at different positions in the yeast genome. Seventy-five colonies hybridized to rRNA; this frequency (3.7%) is in agreement with the predicted proportion of ribosomal DNA-containing colonies based on the earlier observation that 4.8% of the yeast genome is ribosomal DNA (Schweitzer *et al.*, 1969).

Surprisingly, these authors found a very small number (26/2000) of transformant colonies hybridizing to yeast mRNA. This observation does not imply that only a small proportion of the yeast genome is transcribed. Instead, it is a reflection of the abundant class of mRNA reported by Hereford and Rosbash (1977); these authors have estimated that the yeast mRNA is present in at least three classes of abundance: ~20 mRNA species are present at a frequency of 100 copies per cell, 400 species are present at a frequency of 10 copies per cell, and ~2500 mRNA species are present as a single copy in the cell. This conclusion is supported by the observation that the number of transformant colonies hybridizing to mRNA is significantly increased (up to 40%) when the amount of probe DNA was increased 10-fold. This is approximately the proportion of colonies that is expected to contain genes corresponding to the very abundant and moderately abundant classes of yeast mRNA.

5. Demonstration of the Physical Presence of Donor DNA in the Transformants by Southern Hybridization

The acquisition of prototrophy by recipient cells following treatment with donor DNA could be due to a reversion (back mutation) or due to the physical presence of the donor DNA and its expression in the recipient cells. Therefore, to establish the actual occurrence of a transformation event, it is important to demonstrate the presence of donor DNA in the recipient cells. In order to accomplish this, the recipient DNA is broken into fragments of appropriate size by digestion with site-specific restriction endonucleases and electrophoresed on an agarose gel. The DNA is visualized as bands after staining with ethidium bromide in UV light of long wave length. The DNA bands are then transferred to nitrocellulose filters, where they are hybridized to a radioactively labeled probe following the procedure of Southern (1975). This method is ultimately used to establish the physical presence of a donor DNA in transformants and to probe into the fate of the donor DNA in the transformants (i.e., whether or not the donor DNA has been integrated into the recipient genome at the homologous or heterologous sites).

D. FACTORS AFFECTING TRANSFORMATION FREQUENCY

The frequency of transformation in fungi is dependent on the nature of donor DNA, its purity, and a number of other factors described below. In Neurospora the frequency of transformation using total DNA

preparations as the donor DNA was very low, usually one to five transformants per million colonies analyzed (Mishra *et al.*, 1973; Kushner,
1978; Case *et al.*, 1979). A dramatic increase in transformation frequency was seen when chimeric plasmid DNA containing fungal DNA
inserts were used as the donor DNA. The frequency of transformation
of the *qa-2* locus of *Neurospora* using a hybrid plasmid was 30–100
transformants/µg DNA (Case *et al.*, 1979); this high frequency of
transformation of *Neurospora* has been further increased to 1000–
5000 transformants/µg DNA (Case, 1981). In yeast transformation
frequency usually varies depending upon the nature of the hybrid plasmid donor DNA. In yeast usually three kinds of hybrid plasmids are
used as donor DNA: (1) hybrid plasmids containing only yeast chromosomal DNA, which transforms yeast cells at low frequency (1–10
transformants/µg DNA); (2) hybrid plasmids (called RP7) containing a
1.4-kb yeast chromosomal fragment that contains the centromere-
linked *trp-1* gene and transforms yeast cells with an increased frequency (500–5000 transformants/µg DNA); and (3) hybrid plasmids
containing all or part of a yeast endogenous plasmid (2-µm circle) that
transforms yeast cells at the highest frequency (5000–20,000 transfor-
mants/µg DNA) (Beggs, 1978; Hinnen *et al.*, 1979; Struhl *et al.*, 1978;
Kingsman *et al.*, 1979; Gerbaud *et al.*, 1979).

1. Purity of DNA

The frequency of *Neurospora* transformants largely depends on the
purity of the hybrid plasmid used as the donor DNA. Impure DNA
preparations (usually containing RNA, which interferes in some manner during transformation) can reduce the frequency 100–1000-fold
(Case, 1981). In yeast, frequency of transformation is not much influenced when lysed bacterial cells are directly used as a source of donor
DNA during transformation (Kingsman *et al.*, 1979).

2. Nature of the Spheroplast

A good spheroplast preparation is a key to the success of transformation. The time required to digest the cell wall with the enzyme glusulase or zymolyase is a function of the age of the culture; a longer
incubation is required for the older cells. The spheroplast production
can be monitored microscopically as well as by determining the viable
cell count on standard medium (without sorbitol); spheroplasts are
osmotically sensitive and will not form colonies on regular medium
(without sorbitol). The problems of spheroplast production have been

recently reviewed (Pederby, 1979; Wiley, 1979; Kelbe *et al.*, 1983). Furthermore, Dhawale *et al.* (1984) have described a new method, using lithium acetate, which eliminates the need for spheroplasting of the recipient cells during fungal transformation.

3. *Nature of the Transforming DNA*

In *Neurospora,* large plasmids have been reported to mediate transformation at a higher frequency than small plasmids (Case, 1981). Also, circular DNA yields a higher frequency of transformation than linear DNA. In the *Neurospora* transformation system, a DNA preparation circularized by ligation serves as a more efficient transforming element than the linear molecule (Case, 1981). In yeast, however, if a hybrid plasmid is linearized by a cleavage in the yeast DNA (but not in the plasmid DNA) the transformation frequency is increased by a factor of 20; the mechanism for such an increase in the transformation by the linearization of a plasmid DNA is not yet understood.

4. *Conditions for Treatment of Spheroplasts with DNA*

There is no difference in the amount of plasmid DNA uptake by the spheroplasts up to 30 min; optimal transformation frequency can be reached after 5 min of treatment of the spheroplasts with the donor DNA. Exposure of the DNA-treated spheroplasts to PEG in the presence of $CaCl_2$ is critical for the success of transformation. The effect of PEG in the transformation of fungi is accomplished within 2–3 min (Case, 1981). This is also reported for mammalian cells, in which the effect of PEG is over within the first 3 min of treatment.

5. *Other Factors*

The other factors that must be considered are the ratio of the plasmid DNA to that of the spheroplasts and the density at which the spheroplasts are plated following the DNA treatment. The spheroplasts should be plated at a low density to avoid "Grigg's effect," a very well known phenomenon which occurs during the recovery of revertants in fungi (Grigg, 1958). If spheroplasts are plated at a high density the less frequent transformant cells may be killed by the debris of the dying nontransformant cells on a selective medium. Other factors that may be relevant but have not yet been studied in detail are the genetic factors involved in the uptake, processing, and integration of the donor DNA. Some of these aspects are considered later in Section XII.

V. Fate of Donor DNA Following Transformation

Mishra and Tatum (1973) and Mishra (1976) suggested that the donor DNA, after entry into the recipient cell, can establish itself as an autonomously replicating element or integrate into the host chromosome. Such integration can be effected at either the homologous or heterologous sites in the host genome. Theoretically, different kinds of transformants can occur in fungi depending on whether or not the donor DNA undergoes recombination. All these classes of transformants have been reported to occur in yeast and in *Neurospora* (Mishra and Tatum, 1973; Mishra, 1976; Hinnen *et al.*, 1978; Case *et al.*, 1979; Wootton *et al.*, 1980). The fate of the donor DNA can be monitored by detecting the predicted Southern hybridization pattern of the recipient DNA with a specific DNA probe (Fig. 2).

In order to monitor the fate of the donor DNA, the DNA is prepared from the recipient cells before and after transformation and digested with a known restriction enzyme, and then hybridized to a radioactively labeled probe. These probes include both the plasmid DNA and the hybrid plasmid DNA (containing a fungal DNA insert). The pattern of hybridization conclusively establishes the course of events that accompany the donor DNA following its entry into the recipient cells during the process of transformation. This is schematically presented in Fig. 2.

A. TRANSFORMATION WITHOUT RECOMBINATION

If the donor DNA does *not* undergo recombination and instead establishes itself as an autonomously replicating unit, then the Southern analysis will show a band of hybridization (with either probe) at a site characteristic of the migration of the plasmid molecule. This can be further verified by obtaining covalently closed circular DNA from the transformants and their subsequent hybridization to the radioactively labeled probe and the biological activity in a second round of the transformation experiment (Struhl *et al.*, 1979).

B. TRANSFORMATION WITH RECOMBINATION

If the donor DNA undergoes recombination with the recipient chromosome, either of the following patterns of hybridization can emerge, indicating a true specific event that has accompanied the process of genetic recombination.

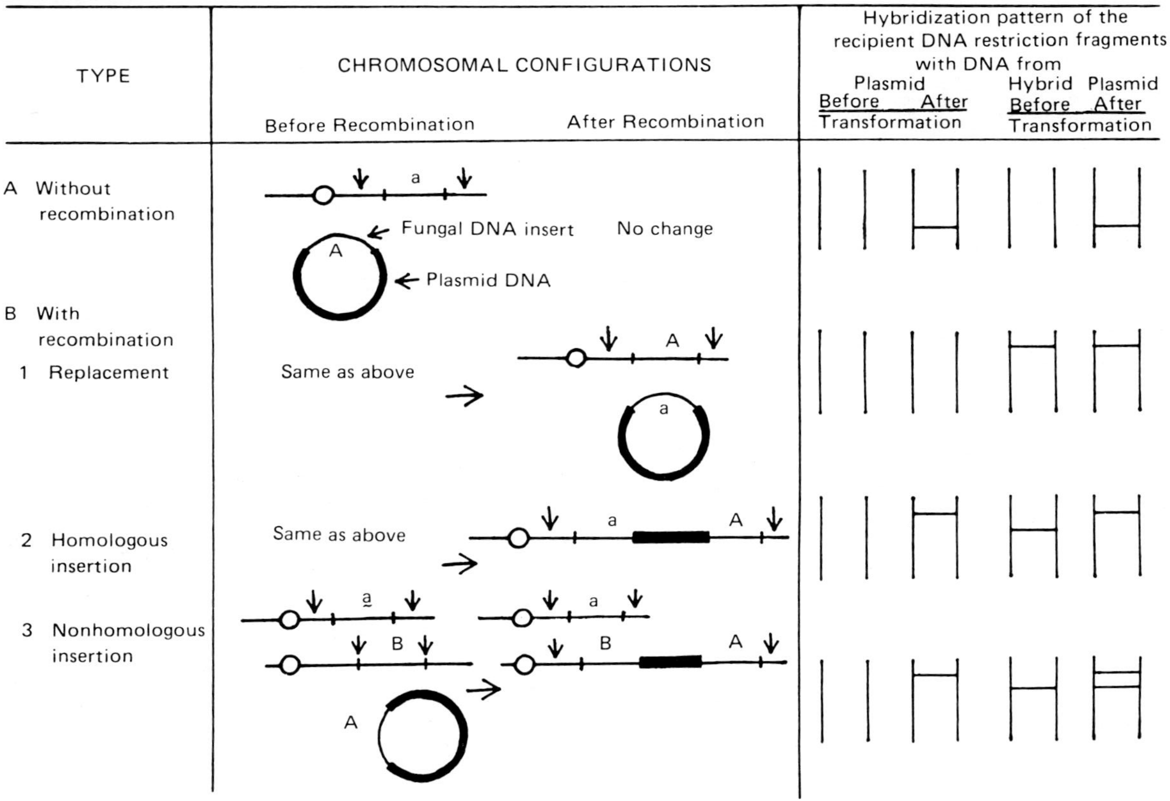

FIG. 2. Fate of donor DNA during fungal transformation: theoretically possible patterns of Southern hybridization between recipient and donor DNA(s) (Struhl *et al.*, 1979; Hicks *et al.*, 1979; Ilgen *et al.*, 1978; Stinchcomb *et al.*, 1979; Kingsman *et al.*, 1979; Mishra and Tatum, 1973; Mishra, 1976). (Adapted from Hicks *et al.*, 1978, with permission of authors and publisher.)

1. Replacement: This involves two crossover events between the host genome and the plasmid (as outlined in Fig. 2) such that the fungal DNA insert is exchanged for the homologous recipient DNA. Only one band of hybridization can be seen when the hybrid plasmid DNA is used as a probe. The plasmid DNA alone will not show any band of hybridization due to lack of homology. Also, the recipient DNA before and after transformation does not involve any structural change with respect to the outside restriction sites. Therefore, the recipient chromosomal DNA bands showing hybridization with the hybrid plasmid DNA will be of the same size (see Fig. 2) in either case (i.e., before or after transformation with the hybrid plasmid).

2. Homologous Insertion: The recombination may involve a single Campbell type of crossover leading to insertion of the hybrid plasmid at the homologous sites. In this case the size of the chromosomal fragment with respect to the outside cleavage will change; therefore the pattern of the recipient chromosomal DNA fragment with and without transformation will be distinguishable, as seen in Fig. 2.

3. Nonhomologous Insertion: The kind of recombination described above may result in insertion of the donor DNA (hybrid plasmid) at a nonhomologous site either on the same chromosome or on a different chromosome. This type of integration can be recognized by the appearance of an additional band of hybridization as seen in Fig. 2).

The recognition of these patterns of hybridization has been instrumental not only in understanding the nature of events following transformation but also in providing unambiguous evidence for gene transfer in higher eukaryotes by detecting the physical presence of donor DNA in the recipient cells. This fact suggests that the donor DNA (hybrid plasmid) essentially functions as an episome in the host cell. An episomal behavior of donor DNA was first described by Mishra (1976) in *Neurospora crassa*. In *Neurospora,* such as episome can be cured by treatment with DNA-intercalating drugs (Mishra, 1976). A theoretical consideration of the fate of the donor DNA has been undertaken elsewhere (Stinchcomb *et al.*, 1979; Hicks *et al.*, 1978; Mishra and Tatum, 1973; Mishra, 1976) and the different hybridization patterns have been actually visualized in both yeast and *Neurospora* (Hicks *et al.*, 1978; Stinchcomb *et al.*, 1979; Case *et al.*, 1979). The episomal behavior of the donor DNA implies the presence of genetic elements that control the replication of the donor plasmid in the fungal transformant. This realization led to the incorporation of these specific DNA sequences into the donor plasmid DNA in order to stabilize its replication. This was achieved by introducing small fungal DNA sequences known as *ars* (autonomously replicating sequences) into the donor plasmid DNA. The replication of the donor plasmid DNA can be ensured by a variety of means. These include addition of a portion of the 2-μm plasmid or a centromeric region or mitochondrial DNA or the ribosomal sequences of yeast. All these regions are supposed to include one or more DNA sequences involved in the replication of the yeast chromosome or its plasmid (Struhl *et al.*, 1979; Stinchcomb *et al.*, 1979; Kingsman *et al.*, 1979; Beggs, 1978; Zakian, 1981; Hyman *et al.*, 1982; Szostak and Wu, 1979) or of the *Neurospora* chromosome or of the *Neurospora* mitochondrial plasmid (Hughes *et al.*, 1983; Stohl and Lambowitz, 1983).

VI. Nature of Transformants as a Function of Donor DNA

The precise nature of the transformants is determined by the characteristics of the donor plasmid. Several classes of transformation systems have been identified both in yeast and in *Neurospora* as described below.

1. Transformation by hybrid plasmid alone (without any *ars*) occurs at low frequency (10–100/μg DNA) and is usually accompanied by integration of the donor plasmid into the fungal chromosome.

2. Hybrid plasmids containing parts of the endogenous fungal plasmid can transform at much higher frequency (5000–20,000/μg DNA). Such molecules replicate autonomously and usually 5–10 copies/cell are seen. These can integrate into the host chromosome and replicate as a part of the host chromosome.

3. Hybrid plasmids containing *ars* act as autonomously replicating genetic elements and transform at a very high frequency (500–5000/μg DNA). The presence of *ars* makes them somewhat mitotically stable and the transformation is usually not accompanied by recombination of the donor plasmids.

4. *Minichromosomes*: Carbon and his group have constructed donor plasmids that contain the entire centromeric region of a specific chromosome. The presence of a centromere makes the donor plasmid behave like a minichromosome. These minichromosomes are both mitotically and meiotically stable (Clarke and Carbon, 1980a; Hsio and Carbon, 1981). Dicentric minichromosomes in yeast were found to be unstable (Mann and Davis, 1983) and such behavior provides evi-

dence for the interaction of minichromosomes with the yeast cell spindle apparatus. Minichromosomes have not yet been constructed in any fungus other than yeast.

The natural occurrence of structures such as minichromosomes in yeast has been recently described (Strathern *et al.*, 1979). These authors have shown that in the yeast mutants involving mating type conversion, a circular chromosome is created by heterologous recombination. These circular chromosomes have been identified as 63-μm covalently closed circular DNA. A detailed discussion of the effect of the nature of vectors on the stability of yeast transformants has been presented by Hinnen and Meyhack (1981) and by Hollenberg (1981).

VII. Other Genetic Effects of Exogenous DNA during Transformation

The role of exogenous DNA in the production of mutants has been fully described in prokaryotes. Yoshikawa (1966) first reported the occurrence of auxotrophic mutations following genetic transformation in *Bacillus subtilis*. The molecular mechanism of the Mu phage-induced mutagenesis of *E. coli* has been described (Howe and Bade, 1975). Fink and collaborators (1978) have described the effect of the *ty-1* DNA sequence in the mutagenesis of the *his-4* locus of yeast. Mishra and others (1973) first noticed the effect of donor DNA in causing mutations during transformation of *Neurospora* recipient cells. It is assumed that these mutations were caused by insertions of donor DNA into the host genome. Mishra (1977a) reported the occurrence of morphological mutations in *Neurospora* following transformation. This included several bizarre morphological changes in the host cells, changes in osmotic sensitivity, and reduced fertility of the transformants (Mishra, 1977a). These changes induced by exogenous DNA during the transformation of *Neurospora* have been confirmed (Szabo *et al.*, 1978; Wootton *et al.*, 1980). It can be suggested that mutations in recipient cells affecting the structure of the cell wall may precede and facilitate the uptake of donor DNA by the recipient cell. Case and others (1979) have reported the occurrence of mutation in the *qa* gene cluster of *Neurospora* following insertion of plasmid DNA during the transformation. These authors have shown that transformation at the *qa-2* site was accompanied by a mutation at the adjoining *qa-4*$^{+}$ site. Szabo and Schablik (1982) have recently shown the occurrence of disomics for linkage group I and also for group VII of *Neurospora* following transformation of the *inl*⁻ locus on linkage group V. Most of these changes accompanying *Neurospora* transformants have also been reported to occur during transformation in yeast (Hicks *et al.*, 1978). Changes in cellular morphology, fertility, and ploidy are the most frequent aberrations that are seen to occur during yeast transformation. Polyploidy may result from cell fusion (and finally nuclear fusion) induced by the PEG added to facilitate the uptake of the donor DNA during transformation. The occurrence of polyploidy in yeast has been genetically analyzed (Hicks *et al.*, 1978); these authors have shown that a transformant is a diploid or triploid (depending on whether meiosis is normal or abnormal). This is consistent with the normal formation of a polyploid series in yeast. Most of the commercial yeasts belong to a polyploid series; however, this is not true in *Neurospora*. There seems to be a natural barrier to fusion of nuclei in vegetative cells of *Neurospora*. In yeast, the genetic transformation seems to be accompanied by changes in the morphology of the colony and in the mating ability of the yeast cell.

VIII. Applications of Transformation

Fungal transformation can be used for the following purposes: (1) the cloning of a desired sequence of DNA from a complex mixture of DNA molecules present in the eukaryotic cell and subsequent amplification in large amounts for biochemical characterization to ascertain its complete nucleotide sequence and to determine the role of different regions of nucleotide sequences as controlling elements, (2) the cloning of an artificially synthesized nucleotide sequence to examine its expression in eukaryotic cells, (3) the modification of a desired nucleotide sequence to examine its effect on the expression of a gene, and (4) the synthesis of a large number of peptides and chemicals or organisms of interest to science, medicine, commerce, and environmental preservation (Miller, 1979; Hitzeman *et al.*, 1981). Some of the applications of the fungal transformation system are discussed here.

A. GENE ISOLATION

The gene transfer system using a heterologous system provides a unique opportunity for molecular cloning of a desired DNA segment. A particular DNA sequence can be identified by different methods (see Section IV,C) and then enriched in *E. coli* or after reentry into an appropriate fungal host. A large number of the fungal genes have been cloned in this manner (see Table 2). Molecular cloning of fungal genes would be instrumental in understanding the structure and the organization of the fungal genome. The nucleotide sequences of a number of fungal genes are now available. The complete nucleotide sequence of the yeast iso-1-cytochrome *c,* yeast alcohol dehydrogenase, three isozymes of the glyceraldehyde-3-phosphate dehydrogenase, phosphoglycerokinase, *ars* (*ars-1* and *ars-2*), *trp-1,* and the actin genes and a partial nucleotide sequence of the promoter region of the *Neurospora qa* gene cluster and other genes are now available (Montgomery and Hall, 1978; Montgomery *et al.,* 1980; Faye *et al.,* 1981; Holland and Holland, 1980; Hitzeman *et al.,* 1980; Tschumper and Carbon, 1980, 1981; Ng and Abelson, 1980; Alton and Vapnek, 1981; Carlson and Botstein, 1982; Geever *et al.,* 1983; Rambosek and Kinsey, 1984; Woudt *et al.,* 1983; Arends and Sebald, 1984; Burke *et al.,* 1984; Viebrock *et al.,* 1983). The sequence analysis shows a remarkable absence of introns from the fungal genes except for the yeast actin gene (Ng and Abelson, 1980; Gallwitz and Sures, 1980) and *Neurospora am* (Kinnaird and Fincham, 1984) and histone (Woudt *et al.,* 1983) genes. The DNA sequence data also reveal several significant changes in the promoter region of the fungal genes as discussed later in this article.

B. Isolation and Characterization of the Eukaryotic Chromosomal Replicator

In bacteria DNA replication is controlled by a specific region located at a site called origin (*ori*). The origin of replication has been isolated from *E. coli* and its characterization will eventually provide a better insight into the mechanism of replication, perhaps to the same level of sophistication as our understanding of bacterial transcription and translation (see Kornberg, 1980, for a detailed discussion of replication). It seems that eukaryotic chromosomes are organized into a tandem repeat of replication units each with a site for the origin of replication.

This concept regarding the organization of eukaryotic chromosomes into the multiple units of replication has been recently examined by Davis and co-workers (Struhl *et al.,* 1979; Stinchcomb *et al.,* 1979). In order to isolate and characterize the eukaryotic origin of replication (or the autonomously replicating sequences, *ars*) the authors exploited the idea that the inclusion of *ars* in the donor plasmid DNA would increase the frequency of transformation since *ars* can confer the quality of autonomous replication on the hybrid plasmid in yeast transformants. In order to screen for the DNA segments containing *ars,* other eukaryotic DNA fragments ligated at specific sites in the yeast vector plasmid are then used as the donor DNA in a transformation experiment. The plasmids yielding high frequency of transformation are selected as the ones harboring *ars.*

Several other groups (Szostak and Wu, 1979; Chan and Tye, 1980; Hyman *et al.,* 1982) have independently used a similar method to isolate yeast *ars.* Szostak and Wu (1979) fragmented a yeast rDNA repeat unit into two segments, *Bgl*IIA and *Bgl*IIB (corresponding to the 5' and 3' end of the ribosomal RNA) by digestion with the restriction endonuclease *Bgl*II and then introduced the individual rDNA fragments into a hybrid pBR322 plasmid containing the yeast *leu-2*$^+$ gene. The chimeric plasmids were used to transform a *leu*$^-$ strain of yeast to prototrophy (i.e., *leu*$^+$). The purpose of this experiment was to select for plasmids containing *ars* of the yeast rDNA by their ability to transform yeast cells at high frequency. These authors found that the plasmids with the *Bgl*IIA fragment could increase the frequency of transformation only by 10-fold (100 colonies/μg DNA) as compared to a low transformation frequency (of 10 colonies/μg DNA) by a plasmid without an rDNA segment. The plasmids containing *Bgl*IIB rDNA segments increased the frequency of transformation 1000-fold (10,000 colonies/μg DNA). These authors also found that transformation by the plasmid with a *Bgl*IIA fragment was stable as compared to the ones caused by the plasmid with *Bgl*IIB fragments. The findings are consistent with the presence of a replication origin in the *Bgl*IIB fragment. It is known (Petes, 1980) that the yeast chromosome XII (between *gal-2* and *ura-4* loci) contains 140 tandemly repeated 9-kb units of DNA, each coding the 25 S, 18 S, 5.8 S and 5 S ribosomal RNAs; also contained in it is a 2-kb length of nontranscribed spacer. According to Szostak and Wu (1979), the origin of replication for yeast rDNA resides in this 2-kb nontranscribable spacer region. This origin of replication has not yet been characterized with respect to its nucleotide sequences.

Chan and Tye (1980) have also used high-frequency transformation as an indicator of the presence of *ars* in a hybrid plasmid. These authors have introduced *Sal*I-restricted yeast DNA into a chimeric pBR322 plasmid carrying a yeast *leu-2*$^+$ gene and selecting for *leu*$^-$ → *leu*$^+$ transformants at high frequency (500–4400 transformants/μg DNA). They selected 16 plasmids and of these 6 showed hybridization to the unique sequences of yeast DNA, whereas others showed multiple bands of hybridization suggesting a homology to families of sequences in the genome. Since the yeast genome has very little redundant DNA, their frequent occurrence in *ars* is thought to suggest the association of repetitive sequences with the ability to self-replicate. The facts that *Sal*I fragments (containing *ars*) were 8 kb and that only one out of four *Sal*I fragments acted as *ars* suggest the size of the yeast replication unit to be roughly 32 kb. This estimate (32 kb) of the size of the yeast replication unit is in good agreement with the size (36 kb) determined by electron microscopy (Beach *et al.*, 1980a). If these calculations are correct, yeast chromosomes contain about 450 replicons/haploid genome excluding the ones present in the ribosomal DNA (rDNA). Hyman *et al.* (1982) have demonstrated that the yeast mitochondrial DNA can function as *ars* when inserted into a vector.

1. Characteristics of ars

a. *Cis Effect.* *Ars* was found to have a cis effect on the replication of the plasmid that contained it. In cotransformation of yeast cells by two different plasmids, only the plasmid containing *ars* showed high (2000–5000 colonies/μg DNA) frequency of transformation. The second plasmid (without *ars*) showed a low (18 colonies/μg DNA) frequency of transformation (Stinchcomb *et al.*, 1979). This cis-acting property of *ars* confirms its nature as a control region (for replication) that does not encode the production of a diffusible material (Stinchcomb *et al.*, 1979).

b. *Mitotic Instability.* It has been shown that the transformants containing a plasmid with *ars* are lost if propagated on a nonselective medium for a long period. For example, *his-3*[+] transformants, when grown on medium with histidine for 10 generations, retain only 1–10% of the autonomously replicating plasmids. This loss is not due to a deficiency in the expression of the gene in question but is due to actual loss of the plasmid molecule as determined by the accompanying loss in the amount of supercoiled plasmid DNA. Most likely, the loss of the plasmid DNA results from mitotic instability of *ars*. However, several other explanations are plausible, such as (1) the *ars* may not contain all the information required for DNA replication or (2) the *ars* may compete with the host chromosomal replicator for limited initiator(s) or (3) the *ars* plasmid may replicate properly but fail to segregate during mitosis (Stinchcomb *et al.*, 1979, 1980).

c. *Other Characteristics.* Struhl *et al.* (1979) have utilized this ability of the *ars*-containing plasmid (to yield a high frequency of transformants) to isolate other eukaryotic *ars*. These authors ligated small DNA fragments from different eukaryotic organisms to the *E. coli* plasmid pBR322 and the chimeric plasmids showing a high frequency of transformants were isolated as the ones containing *ars*. In such experiments *ars* derived from *Neurospora* was the most effective, those from *Drosophila* and corn were the least effective. The *Neurospora ars* was as good as the yeast *ars* as determined by the growth rate of the transformants. In another study, a *Staphylococcus aureus* plasmid was shown to replicate in yeast cells; this finding casts doubt on the much-held view that replication signals in prokaryotes and eukaryotes are different (Goursot *et al.*, 1982).

The plasmid containing *ars* may integrate at different sites on the host chromosome. The *ars*-containing plasmids were also found to occur as dimers and trimers. All of them yielded transformants at the same frequency (i.e., 2000–5000 colonies/μg DNA). It has been determined that the *ars-1* associated with *trp-1*$^{+}$ is about 850 bp in length and is delimited with *Eco*RI and *Hin*dIII sites. When the radioactively labeled *ars-1* DNA was hybridized to a restriction endonuclease spectrum of yeast DNA it was found to yield only one band of hybridization (as visualized by autoradiography following Southern transfer). These data indicate that the *ars-1* does not bear any sequence homology with other *ars* present in the yeast chromosome, of which there must be at least 450 per haploid genome.

At present two yeast *ars* have been characterized; the *ars-1* associated with *trp-1*$^{+}$ and the *ars-2* associated with *arg-4*$^{+}$ (described by Hsio and Carbon, 1979). The *ars-1* was isolated independently by two different laboratories (Stinchcomb *et al.*, 1979; Kingsman *et al.*, 1979). Both *ars-1* and *ars-2* have been fully sequenced (Tschumper and Carbon, 1980, 1981). Both the *ars-1* and *ars-2* have been further characterized by subcloning of the yeast fragment containing these structures. The *ars-1* was found to be delimited by the *Eco*RI and *Hin*dIII sites and must contain a stretch of 838 bp for full activity; the region required for the *ars-1* activity has been localized at the 3′ end of the *trp-1* gene. By subcloning experiments the region delimited by *Pst*I and *Bgl*II is essential for the *ars-1* activity. Likewise the subcloning of *ars-2* shows that only a stretch of 100 bp is responsible for the *ars-2* activity; this region is delimited by a *Msp*I site on the left and the end point at the right [where a poly(dA) connector was added during the original isolation of the fragment containing *ars-2* (Tschumper and Carbon, 1981)]. The *ars-2* region contains an AT-rich region (18 bp long) of a dyad symmetry that also contains the canonical sequence of the TATA box; the latter is a characteristic sequence found in the eukaryotic promoter. Both the *ars-1* and *ars-2* contain a TATA box and

an AT-rich region of dyad symmetry; both of these structures may be
involved in the initiation of replication, particularly the TATA box if
the initiation is indeed RNA primed (Tschumper and Carbon, 1981).
ars-1 and ars-2 differ in significant ways; these differences are seen in
the structure, size, and efficiency of transformation. The ars-1 (838 bp)
is much larger in size than the ars-2 (100 bp); the ars-2 contains an
outflanking repeated sequence (δ); and the ars-2 containing plasmid is
almost four times more efficient in the transformation of yeast cells
than the same plasmid containing the ars-1. However, both ars-1 and
ars-2 were found to be equally unstable mitotically. ars-1 and ars-2
seem to belong to two different classes of ars, one containing only
unique sequences and the other containing repeats, as suggested by
Chan and Tye (1980). The association of dispersed repeated DNA sequences with ars raises the possibility of its role in acting as the origin
of replication in yeast and other fungi. This question has been investigated using the yeast transformation system. It has been shown that
the presence of dispersed repeat sequences (such as ty-1) does not confer the property of autonomous replication to a chimeric plasmid carrying it when introduced into the yeast cell by transformation (Struhl et
al., 1979; Scherer and Davis, 1980). Therefore the significance of the
association of δ sequences with certain yeast origins of replication is
not yet understood. If this has any role in DNA replication it would be
of interest to investigate the counterpart of such a structure in filamentous fungi, in which dispersed repeat sequences do not seem to
exist (Krumlauf and Marzluf 1979).

2. Significance of ars

The isolation and characterization of ars would provide a better insight into the mechanism of eukaryotic DNA replication and its control. It would be of interest to see the effect of other sequences as well
as the effect of mutagenesis on the functioning of ars. ars can be used
in in vitro systems (as suggested by Stinchcomb et al., 1979) to determine its role in DNA replication. It would also be of interest to isolate
ars from different mutants temperature sensitive (ts) for replication. It
would be also useful to isolate and characterize ars from other organisms to evaluate their role in the control of replication. It would be of
interest to isolate ars from fungal mitochondria (Hyman et al., 1982)
and their plasmids for comparative studies. In this direction the most
immediate effort should be directed to isolate ars of the yeast 2-μm
circle in order to determine why it is more effective than ars-1 and
ars-2. It is of interest to note here that the ars of the yeast 2-μm circle

requires an additional DNA sequence encoding a diffusible product (i.e., trans effect) for its proper functioning (Broach and Hartley, 1980). It is of interest to note that *E. coli* plasmids containing *Neurospora ars* can function in yeast (Stinchcomb *et al.*, 1980) but the same plasmids containing yeast *ars* (*ars-1* or *ars-2*) are not functional in *Neurospora* (N. C. Mishra, unpublished results).

C. Isolation and Characterization of Promoters

Transcription is one of the prerequisites for gene expression in all organisms. The mechanism of traditional control of gene expression in bacteria is very well understood because of the characterization of the different regions in a DNA segment responsible for active transcription. One region, called the promoter, is perhaps the most important since this binds with the RNA polymerase and facilitates transcription. The structure and function of the eukaryotic promoter have been extensively studied in recent years using the methods of molecular cloning (Corden *et al.*, 1980). These studies have established the presence of at least two canonical structures in the eukaryotic promoter and their role in the control of transcription (Corden *et al.*, 1980; Benoist and Chambon, 1981). These structures are identified as (1) a TATAATA (or Hogness) box and (2) a sequence of GGPyCAATCT; both located upstream 25 and 70 bp from the start of mRNA. However, a recent study shows that the DNA sequences further upstream (around 180 bp from the start of mRNA) may be essential for the *in vivo* transcription in higher eukaryotes. The characterizations of *Neurospora* and yeast promoter have been very useful in understanding the mechanism of transcription in fungi. The analysis of yeast promoter reveals the presence of a TATA box. Struhl (1981, 1982) has demonstrated the unequivocal role of the DNA sequences far upstream in promoting the *in vivo* transcription of yeast. However, the study of a *Neurospora* promoter region shows the absence of these canonical DNA sequences (see Fig. 3). The DNA sequence of a *Neurospora crassa* catabolic dehydroquinase gene promoter showed lack of the Hogness box (TATAATA box) as well as of another canonical sequence, GGPyCAATCT. Instead it was found to contain sequences rich in AT (almost 75%) in the location between −21 and −61 bp; also, the sense strand was rich in purine (9/10 A or G) in the region of 70 to 80 bp.

The structure of the *Neurospora* promoter does not exhibit any similarity with the yeast promoter either. However, both *Neurospora* and yeast promoters contain another canonical sequence, CACACA, in a

```
                    2  ,     3  ,
              1  ,    1    4  ,           4      2
GGCCTTTCGAATGGTCATCAGTGCAGTGACGAGCAATGTCACGGAAGCGGGGGACGGGTAATCGCTTATCCGCTCGTCGTGCAGACAAC
```

```
        3                                           A/T RICH (-G)
TTCGTCCGTGTATTAGAGATGGGAATGATGAGGGAACCGTGATTAAACAACAAAACATAAACACTTCAATTCAACCTTCTGGCCTGT
              -80          -68                               -32        -21
```

```
GAGTTGTTGGGTATAGTGCGGCGGCATCTTTCGGACGCATTCCCTGTTGCGCCCATCTCCCACAAGCCCATCGCACCCAACCAGAGGTA
        +1
```

		1											
		MET	ALA	SER	PRO	ARG	HIS	ILE	LEU	LEU	ILE	ASN	GLY	PRO
CCAAACACA		ATG	GCG	TCC	CCC	CGT	CAC	ATT	CTC	CTC	ATC	AAT	GGC	CCC

						20							
ASN	LEU	ASN	LEU	LEU	GLY	THR	ARG	GLU	PRO	GLN	SER	THR	ALA	GLN	SER
AAT	CTC	AAC	CTC	CTC	GGC	ACC	CGG	GAG	CCC	CAA	TCT	ACG	GCT	CAA	TCA

FIG. 3. Promoter region of *Neurospora* cDHQase (encoded by *qa-2*). (Adapted from Alton and Vapnek, 1981. From PROMOTERS: STRUCTURE AND FUNCTION, edited by Raymond L. Rodriguez and Micheal J. Chamberlin. © 1982 Praeger Publishers. Reprinted by permission of Praeger Publishers.)

position 15 bp upstream from the initiation codon (ATG). This may have a role in ribosome binding and the initiation of translation. The *Neurospora* message also contains another canonical sequence, GAGG, 11 nucleotides before the initiation codon (ATG). The (GAGG) is homologous to a sequence at the 3' end of 16 S rRNA of *E. coli* and could be a signal for the efficient translation of cDHQase in *E. coli* (Alton and Vapnek, 1981).

The *Neurospora* cDHQase structural gene contains another important sequence of GTGAC, which has been earlier identified as a catabolite receptor protein (CRP) binding sequence in *E. coli*. A number of *E. coli* catabolite-activated systems possess this sequence (Majors, 1975). It has been suggested that the cDHQase protein of *Neurospora* may bind with the *qa-1* gene product, perhaps in a manner analogous to the binding of the catabolite activator protein to the *lac* promoter region of *E. coli*. Such binding of the *qa-1* regulatory protein may control the expression of the *qa* gene cluster (Alton and Vapnek, 1981). However, these ideas remain to be confirmed by identifying the DNA sequences of the other promoter regions in the *qa* gene cluster and the sequencing of other promoters in the filamentous fungi.

Characterization of the promoter sequences from a variety of organisms can provide a better insight into the mechanism of transcription, which represents the first step in the gene expression of all organisms. An attempt has been made to isolate promoters from a group of evolutionarily divergent organisms by their ability to facilitate the expression of a gene conferring tetracycline (*tet*) resistance by a hybrid plasmid in *E. coli* (Neve *et al.*, 1978). These investigators first modified a hybrid plasmid pBR316 [containing the genes for resistance to ampicillin (*amp*) and to tetracycline (*tet*)] by inserting a chemically synthesized *Eco*RI linker at the promoter (RNA binding site) of the *tet* gene, thereby decreasing the efficiency of transcription of the *tet* gene and consequently reducing the level of resistance to tetracycline present in the growth medium. In a second step, they inserted small eukaryotic DNA sequences (obtained by restriction endonuclease digestion of the eukaryotic DNA) at the *Eco*RI site within the *tet* gene and then examined the hybrid plasmid for its ability to restore antibiotic resistance. Bacterial colonies not able to grow on plates containing 5 μg/ml of tetracycline were scored as *tet*-sensitive; those able to grow on plates containing 5 μg/ml of tetracycline but not higher drug concentrations were scored as containing a nonrecombinant plasmid, whereas those able to grow on plates containing tetracycline concentrations higher than 5 μg/ml were considered as harboring a recombinant plasmid with restored promoter activity. Thus the level of drug resistance was taken as a measure of the efficiency of the eukaryotic promoter inserted at the *Eco*RI site of the *tet* gene. In such an experiment, DNA sequences from *E. coli,* yeast, and *Euglena* chloroplasts showed promoter activity as high as 25–30% of the original plasmid promoter; DNA from *Neurospora* showed only 4.5% promoter ability. DNA segments from *Drosophila* (0.8%) and *Megasihia scalaris* (0.25%) showed negligible activity. The fact that the promoter efficiency of *Neurospora* DNA is in sharp contrast with that of yeast suggests that promoter sequences of *Neurospora* may have evolved further away from the prokaryotic DNA than those of yeast. This is further supported by the analysis of a *Neurospora* cDHQase structural gene promoter, which shows no similarity with the yeast or bacterial promoter or even with promoters of higher eukaryotes. However, no generalization can be made unless a large number of *Neurospora* and/or yeast promoters have been isolated and fully characterized for nucleotide sequence and *in vitro* and *in vivo* activities.

The promoter regions of several yeast structural genes are now known. All of them seem to contain the canonical sequence of a TATA box, besides which some of them contain an additional sequence, ACACACA, which has also been found in *Neurospora* cDHQase promoter.

Struhl (1981) has recently shown that a nucleotide sequence 113–155 bp upstream from the transcription start point is necessary for the wild-type expression of the yeast *his-3* gene and that the presence of the canonical TATA sequence alone is not enough for the wild-type expression of this gene. He has elucidated the role of different parts of the promoter region in the mRNA production and the expression of the *his-3* gene by deletion mapping. His approach included the deletion of a specific nucleotide sequence from the *his-3* gene before the transcribable region and the construction of an appropriate plasmid (containing the *his-3* deleted gene), which was used to transform auxotrophic (*his⁻*) yeast cells. The transformants were examined for their growth on minimal plates as well as on plates containing aminotriazole (a competitive inhibitor of yeast imidazoleglycerol phosphate dehydratase encoded by the *his-3⁺* gene). The transformants containing the *his-3* promoter deletions were found to possess three distinct classes of phenotypes. The first class, with deletions between 155 and 290 bp upstream from mRNA coding sequences, grew well at the wild-type rate on both minimal and aminotriazole plates. The second class, with deletions between 60 and 115 bp, grew at a reduced rate on minimal medium but not at all on a medium containing aminotriazole. The third class, having deletions of less than 45 bp from the mRNA coding region, did not grow at all on either medium. These data of Struhl (1981) and also of Lowry *et al.* (1983) clearly indicate the role of nucleotides (other than the TATA box) located upstream in controlling transcription. This is in agreement with findings regarding the role of the promoter in the control of eukaryotic transcription (Benoist and Chambon, 1981). The role of the distant nucleotide sequence must take into consideration the high organization of DNA (into nucleosomes) to be effective in controlling transcription. However, other plausible explanations have been suggested by Struhl (1981).

A comparison of the structure of several fungal genes shows the presence of several repeats of nucleotide sequences in the flanking region at both the 5′ and 3′ regions. Other important features include the lack of intervening sequences; only one gene (the yeast actin) out of 10 examined showed the presence of an intervening sequence (307 bp) at the 5′ end of the gene.

D. Isolation and Characterization of Centromeres and Telomeres

1. Centromere

Carbon and co-workers (Clarke and Carbon, 1980a; Hsio and Carbon, 1981) have utilized the effect of centromeric regions in stabilizing the *ars*-containing plasmid in yeast cells and in yielding high frequency of transformation as the criteria for the molecular cloning of yeast centromeres. Using this protocol they have isolated (from a yeast genomic library) colonies containing complete yeast centromere-3 (CEN-3) and centromere-11 (CEN-11) Clarke and Carbon, 1980a,b). The Giles group have reported the presence of the *Neurospora* centromeric region (CEN-7) in plasmids containing the *qa* gene cluster. They have been able to verify the presence of the centromere by comparing the structure of *Neurospora* inserts in different plasmids. These authors seem to have a *Neurospora* insert over 30 kb long carrying the *qa* cluster and the centromeric region (Schweizer *et al.*, 1981b). However, it is not certain if both sides of the centromere are included in their plasmid. In view of this fact, these authors have not been able to show the effect of a *Neurospora* centromere on the process of genetic transformation. The plasmids containing the yeast centromere have been isolated by a variety of methods such as chromosome walking (overlap hybridization), complementation of yeast mutation via transformation and immunological screening (Chinault and Carbon, 1979; Clarke and Carbon, 1980b; Clarke *et al.*, 1979). Yeast chromosome III has been isolated as a 1.6-kb segment of DNA (which is left of the centromere linked to the *cdc-10* locus of yeast). The presence of this DNA segment in a plasmid carrying a yeast chromosome replicator (*ars*) enables the plasmid to behave like a minichromosome both mitotically and meiotically. The yeast minichromosomes containing CEN-3 or CEN-11 or other centromere segregate as any ordinary chromosomes during meiotic division and show $2+:2-$ segregation. In this manner, it differs from the yeast *ars* as well as the yeast 2-μm plasmid fragments, which are not mitotically and meiotically stable even though their presence ensures plasmid maintenance (Clarke and Carbon, 1980a; Hsio and Carbon, 1981; Szostak and Wu, 1979). Clarke and Carbon (1980a) have shown that (1) a functional minichromosome requires both an *ars* as well as the CEN DNA segment, (2) the CEN DNA segment is unique in providing mitotic stability to the *ars*-containing plasmid, (3) the presence of the two chromosomal *ars* (i.e., *ars-1* and *ars-2*) or of the yeast endogenous plasmid *ars* is not enough to endow the plasmid containing them with the property of a functional chromosome. It is suggested that the centromeric DNA controls copy number and ensures proper meiotic segregation. The minichromosomes do not pair during meiotic division among themselves or with the normal yeast chromosomes, and (4) there is some evidence that suggests the diplodization of the CEN-3 plasmid during meiosis as revealed by tetrad analysis, and (5) centromeres do not play any role in the recognition of homologs during synapsis since homologous chromosomes with centromeres of different origins (i.e., nonhomologous) can effectively pair during meiosis (Clarke and Carbon, 1983; Zakian, 1983; Bloom and Carbon, 1982).

These characteristics of the minichromosome imply its capability to transfer stable characters to a yeast cell. It is obvious that further study of the minichromosome should be useful in understanding the organization of chromatin structure (Clarke and Carbon, 1980a) and its distribution during cell division. The determination of the complete nucleotide sequence of the fungal centromeric region can provide an insight to the mechanism of the behavior of the mini chromosome.

2. Telomeres

The roles of telomere in maintaining the linearity of eukaryotic chromosomes and their replication at the termini have been suggested earlier (McClintock, 1941, 1942; Orr-Weaver et al., 1981; Heumann, 1976; Watson, 1972). It has been further suggested that telomeres may contain certain special structural features (such as palindromes and single-strand interruptions at specific sites) which allow them to play such roles in maintaining the integrity of the chromosomal linear structure and its replication at the termini. Such roles of telomeres have been probed by extensive structural (such as restriction mapping) and functional (such as behavior of linear plasmid carrying cloned telomeres) analysis of the cloned telomeres (Szostak and Blackburn, 1982; Chan and Tye, 1983). These workers have cloned yeast as well as *Tetrahymena* telomeres using the fungal gene transfer system. The results of their investigation show the following: (1) Telomeres do contain palindromic structures and nicks at specific sites as predicted earlier (Heumann, 1976). (2) All telomeres possess a highly conserved structure which may be required to maintain the stability of linear chromosomes. All yeast telomeres share about 4 kb of homology at their termini. Furthermore, the yeast telomere can be substituted for the *Tetrahymena* telomere in a linear recombinant plasmid without any adverse effect in telomeric function, and (3) yeast telomeres contain several moderately repetitive *ARS*(s) (i.e., replication origin; see Chan and Tye, 1980). It is expected that comparison of such cloned telomeres from a variety of organisms *in vitro* or *in vivo* (using a transformation system) will provide information necessary to understand the topology and resolution of termini following DNA replication in the linear eukaryotic chromosomes. Such studies can also be exploited to understand the involvement of different enzymes during the replication and resolution of the termini of eukaryotic chromosomes.

E. Construction of an Artificial Chromosome

Eukaryotic chromosomes are highly organized structures and are composed of several components required for their maintenance and inheritance during cell division. The various structural components of eukaryotic chromosomes characterized are (1) *a structural gene,* which encodes for a protein, and is usually cloned by its ability to complement a defective function in the recipient cell, (2) *ars,* which is required for replication and is detectable by its ability to increase the frequency of transformation by the recombinant plasmid carrying it, (3) *a centromere,* which is required for chromosome movement during cell division and is clonable by its ability to endow stability to the recombinant plasmid carrying it, and (4) *a telomere,* which is required for the maintenance of the stability of a linear chromosome and is identifiable by its ability to maintain the linear structure of a recombinant plasmid during transformation. All of these components of eukaryotic chromosome have been individually cloned. Recently, two groups of workers have constructed artificial yeast chromosomes by the combined cloning of these structures into one recombinant plasmid, and they have examined its behavior during cell division after its introduction into a recipient cell using the fungal gene transfer system (Dani and Zakian, 1983; Murray and Szostak, 1983). In order to generate artificial chromosomes of different sizes, λ phage DNA segments of various lengths were added as fillers to the recombinant chromosomes. The results of their investigation show that (1) the stability of chromosomes during cell division is the function of the size of the artificial chromosomes; larger size chromosomes (50 kb or greater) were found to behave like the naturally occurring yeast chromosomes. Smaller chromosomes were highly unstable and were deleted at high frequency during cell division. (2) The recognition of the homologous chromosomes during meiosis is not mediated by the centromere or telomeres. Thus, DNA segments other than that of centromere and telomere must be involved in this important process (also see Clarke and Carbon, 1983). (c) The instability of an artificial chromosomes may be contributed to by factors such as the presence of foreign DNA (included as fillers in an artificial chromosome), improper spacing of the various structural components, and lack of optimal size, as discussed earlier by Zakian

(1983). Such construction of artificial chromosomes provides a means for the joint introduction of multiple genes controlling the structure of several proteins involved in the biosynthesis of antibiotics, hormones and other products of pharmaceutical importance. An added advantage of such a system would be to provide coordinate regulation of these genes in an artificial chromosome whose expression can be triggered in a controlled manner to maximize the yield of the gene products.

F. Determining the Function of Unknown Genes

In principle, the function of a complementing gene can be assigned to be the same as the defective gene which it complements. This has been the basis for the molecular cloning of a large number of fungal genes, presented in Table 2. Therefore, the fungal transformation system can be used in assigning function(s) to the genes by *in vivo* complementation and *in vitro* analysis of the gene product. For example, several genes whose functions are not known control the cell cycle of yeast. By transformation it is possible to isolate a DNA sequence that complements a particular cell cycle mutation. The hybrid plasmid containing the particular complementary DNA sequence can be analyzed for the specific gene product by *in vivo* translation in maxi- or minicells (Sancar *et al.*, 1979; Meagher *et al.*, 1977) or by *in vitro* translation (Yang and Zubay, 1978). The structure of the protein may yield a clue to its function. The DNA sequence of the complementing gene can be determined and may be used in computer-generated sequences of the encoded proteins. This information can be helpful in assigning a function to a gene for which there is no conventional way to determine its function. A number of fungal genes have been isolated and the assignment of their function is in progress; these include genes complementing *cdc-28* (Nasmyth and Reed, 1980) and *rad-52* mutants (R. K. Mortimer, personal communication) in yeast, and the *qa-1* gene of *Neurospora* Huiet, 1984); an attempt is being made in my laboratory to characterize the genes complementing *nuc* (nuclease) mutants of *Neurospora*.

G. Discovery of New Genes

Molecular cloning of a gene cluster can provide insight into the internal organization of the gene cluster and its flanking region. The Giles group has utilized a series of plasmids representing the entire *qa* gene cluster of *Neurospora*. As mentioned earlier, the *qa* gene cluster is known to contain four genes: *qa-1* encodes a regulatory protein, whereas *qa-2,-3,* and *-4* encode proteins involved in the catabolism of quinic acid. In their attempt to characterize the cloned *qa* cluster, these authors used Northern blots (Alwine *et al.,* 1977; Thomas, 1980) to identify the number of genes producing mRNA. In this elegant way, these authors were able to identify at least two more genes in the *qa* gene cluster (designated as *qa-x* and *qa-y*) that make mRNA. However, no translation product corresponding to these messages (i.e., *qa-x* and *qa-y* mRNA) have yet been identified. The genes *qa-x* and *qa-y* are under quinic acid control (Patel *et al.,* 1981) (see Fig. 4). A similar discovery of previously undetected acid phosphate gene (*pho-3*) has been reported in yeast (Rogers *et al.,* 1982).

H. DIRECTED MUTAGENESIS

A versatile hybrid plasmid (capable of replication in different organisms such as *E. coli,* yeast, and *Neurospora*) can be used for an understanding of certain basic processes by the method of directed mutagenesis. Several methods are available to induce an *in vitro* change in a gene (Carbon *et al.,* 1975; Shortle and Nathans, 1978); the *in vivo* effect of such *in vitro*-directed changes can be examined by the use of an appropriate transformation system. Such methods have proven quite successful in the understanding of the SV40 genome and its expression (Penden *et al.,* 1980; Weissman *et al.,* 1978; Mulligan and Berg, 1981). The deletion mapping of the yeast promoter function is one such example (Struhl, 1981). In another study, an *in vitro*-induced disruption in the yeast actin gene was found to have a lethal effect on the haploid yeast spore, thus supporting the conclusion that actin is an essential protein for the survival of yeast cells (Shortle *et al.,* 1982). Directed mutagenesis can be utilized to probe the function of certain genes that we do know at the present time. It can also be used to understand the mechanisms of DNA replication, repair, and recombination. The fact that several eukaryotic *ars* are now available and that the complete nucleotide sequences of yeast *ars-1* and *ars-2* are known provides an excellent basis for an additional *in vitro* modification of these *ars* for future study. These (*ars*) can be appropriately mutagenized and then used in transformation experiments to probe into the function of different parts of *ars* in its ability to initiate and to control chromosomal replication. In fungi, an approach can be made to understanding the role of different nuclease-deficient mutants in DNA repair. In such experiments, the wild-type and mutant strains (lacking a particular deoxyribonuclease) can be transformed with a versatile hybrid plasmid

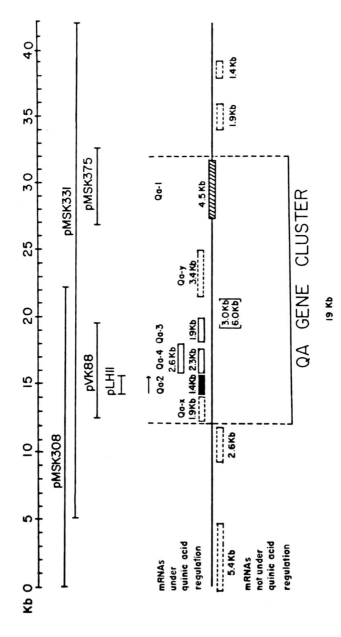

Fig. 4. Map of QA gene cluster in *Neurospora* as revealed by chromosome walking. (Adapted from Patel *et al.*, 1981, with permission of authors and publisher.)

(before and after mutagenesis of a particular gene in the plasmid). Later, the host cell can be examined for its ability to repair the damage in the plasmid and to facilitate its expression as monitored by the growth of the transformants on a selective plate. Methods of directed mutagenesis can also be used to promote a better expression of a gene or better yield of the gene products.

I. New Methods for Mapping Eukaryotic Chromosomes

A number of genes (for which no mutations or functions have been described) cannot be mapped by the classical methods of Mendelian genetics. However, these genes (available in the form of cloned DNA segments) can now be mapped based on the use of restriction fragment length polymorphisms (RFLPs) as genetic markers. This method first developed by Botstein *et al.* (1980) has been modified for the mapping of *Neurospora* genes (Metzenberg *et al.*, 1984). Thus, the methods of molecular cloning can be used to provide information regarding the chromosomal location of particular genes, their order, and physical mapping of chromosomal distances in terms of nucleotide base pairs. Some of these are described below.

1. Overlap Hybridization (Chromosome Walking)

A genomic library of a eukaryotic organism consists of an array of bacterial clones harboring chimeric plasmids (containing eukaryotic DNA segments as inserts). These eukaryotic inserts are usually in the form of overlapping stretches of DNA segments because of the manner in which they are initially prepared (by random shearing or partial digestion of eukaryotic DNA by a restriction endonuclease). Chinault and Carbon (1979) utilized this fact to develop methods to isolate DNA segments contiguous to previously cloned DNA. The method involves identifying clones with plasmids that cross-hybridize to known probes (as well as among themselves) and then identifying the common sequences by restriction mapping and cross-hybridization on a Southern blot (Southern, 1975). By the use of appropriate clones from a genomic library one can virtually "walk" on a chromosome arm in a particular direction from a fixed point (e.g., a centromere or a known gene) and reconstruct a stretch of DNA of 30 kb or longer. Using this approach of "chromosome walking" the physical maps of the third chromosome of yeast and of the seventh chromosome of *Neurospora* are available now. This method confirms the gene order established in these regions of yeast and of *Neurospora* chromosomes (Chinault and Carbon, 1979;

Clarke and Carbon, 1980a; Hsio and Carbon, 1979; Case, 1981). The order of the genes on chromosome III of yeast mapped by this method shows *his-4, leu-2, cdc-10,* the centromere, *pgk,* and a mating type locus, whereas in *Neurospora* a portion of chromosome VII shows the arrangement of genes in the *qa* cluster to be *qa-1, qa-3, qa-4,* and *qa-2* (Case, 1981) (see Fig. 4).

2. *Measurement of Genetic Map in Terms of Nucleotide Base Pairs*

In the case of yeast, a precise relationship between the genetic map distances (as measured by recombination frequencies) and the actual physical distances in terms of base pairs of nucleotides has been established (Clarke and Carbon, 1980a; Nasmyth and Reed, 1980). These authors have determined that a distance of 8 cM (centimorgan) between *leu-2* and *cdc-10* encompasses 25 kb, suggesting that 1 cM equals 3 kb. However, Nasmyth and Reed have determined that a genetic distance of 7 cM between yeast *cdc-28* and *try-1* genes includes only 12.25 kb, suggesting 1.75 kb/cM. A similar determination has been made by Hicks and his group (Strathern *et al.,* 1979) from the study of a ring chromosome in yeast; these authors have found 1 cM to be equal to 2.7 kb; this is in agreement with the finding of Clarke and Carbon (1980a). However, such estimates in humans present a sharp contrast to that made in yeast; in human 1 cM has been found to correspond to 1000 kb (Ruddle, 1981). Any disparity between physical and genetic distances may be due to their location in different parts of a chromosome since the recombination frequencies are usually reduced near or around the centromere. Such disparity may also result from the differences in the regulatory mechanisms controlling the frequency of recombination in various groups of organisms.

3. *Insertional Mapping*

Szostak and Wu (1979) have described a new method of chromosome mapping in yeast (see also Falco and Botstein, 1983). These authors have constructed a recombinant plasmid containing yeast *leu-2*[+] and a part of the yeast ribosomal DNA. This chimeric plasmid was used to transform a yeast *leu-2*[−] strain to *leu-2*[+]. In some of these transformants, the *leu-2*[+] gene was found to be located on chromosome XII among the ribosomal RNA genes. The new location of *leu-2*[+] was unambiguously determined by the predicted restriction pattern, which is distinguishable from its location on chromosome III where the yeast *leu-2* gene is naturally located. The new location of the *leu-2* gene on chromosome XII was due to the fact that the transforming chimeric plasmid contained ribosomal RNA genes. Thus this method can be used to map genes for which no mutations are known or whose phenotype is difficult to score or which are represented by highly or moderately repeated copies. This method is analogous to the use of Tn elements for gene mapping in bacteria (Kleckner, 1977).

In this new method, a chimeric plasmid with a yeast $leu-2^+$ gene along with the gene in question will be used to transform a yeast $leu-2^-$ strain to prototrophy and then the $leu-2^+$ transformant will be analyzed for a new site of insertion to reveal the location of the gene in question. A general method can be developed using other yeast selectable markers.

4. Chromosomal Location of Genes

Recombinant plasmids containing a fungal gene can be hybridized to total DNA fragments or to individual chromosomes to reveal their chromosomal location. Hicks and Fink (1977) have used a plasmid containing the yeast $leu-2^+$ gene to determine its location on chromosome III; these authors showed a proportional relation in the intensity of hybridization of the $leu-2^+$ plasmid with the DNA preparations from different yeast strains aneuploid for chromosome III. Szostak and Wu (1979) have determined the location of the gene for ribosomal RNA on chromosome XII of yeast. Likewise, *Neurospora* rDNA has been located mainly toward one end of chromosome II (linkage group V) apparently as a cluster of 110–140 repeats of 8.6 kb each; thus 17–30% of *Neurospora* chromosome II appears to consist of rDNA sequences (Perkins and Barry, 1977; Cox and Penden, 1979).

5. Replacement of a Resident Chromosomal Gene

The fungal gene transfer system has been used to replace a wild type gene by an *in vitro*-modified mutant gene at a new location in the same chromosome. This method of gene replacement can be used for linkage analysis as well as for determining the role of *in vitro* modification in the expression of a gene. The basis for such gene replacement has been discussed earlier (see Section V,B,1 and Fig. 2). Using this technique, Scherer and Davis (1979) have successfully replaced the *his-3* on yeast chromosome IV by a *his-3* mutant gene (deleted for 150 bp *in vitro*) following a transformation of an *ura-3* recipient by a chimeric plasmid containing the $ura-3^+$ and the deleted *his-3* genes. In such experiments the $ura-3^+$ gene was used just as a marker to select the transformants. The results of this elegant experiment clearly established the role of the fungal gene gransfer system in the study of directed mutagenesis and other genetic manipulation.

J. Insertion Sequences

McClintock (1951) described genetic elements that caused gene instability and chromosomal rearrangement; however, the molecular nature of these genetic elements remained elusive until the development of recombinant DNA technology. Genetic elements capable of transposition and chromosomal rearrangement (called *insertion sequences* or *transposons*) have been described from both prokaryotes and eukaryotes (Kleckner, 1977; Finnegan *et al.*, 1977; Scherer and Davis, 1980; Shapiro *et al.*, 1977; Roeder *et al.*, 1980). The prokaryotic transposons carry genes encoding enzymes responsible for the transposition (Simon *et al.*, 1980), whereas the eukaryotic transposons seem to lack this property and differ in other significant ways (Roeder *et al.*, 1980; Scherer and Davis, 1980). In order to understand the nature of eukaryotic transposons, it would be of interest to mention that eukaryotes contain dispersed repeated gene families, at least some of which have now been characterized from *Drosophila* (Finnegan *et al.*, 1977), such as copia and sequence 412 as well as from yeast, such as *ty-1* sequence (Cameron *et al.*, 1979). Studies using the yeast transformation system and other methods of molecular genetics have now established that some of these dispersed repeated DNA sequences are indeed transposons. Thus, these studies clarify the relationship between two important observations in classical genetics and molecular biology (i.e., the phenomenon of gene instability and the occurrence of dispersed repeated DNA sequences) and link them together by showing that some of these dispersed repeated sequences are indeed transposons. This correlation has been possible in yeast only. The dispersed repeated DNA sequences of yeast were isolated as the *Eco*RI fragments of the chromosomal DNA from two different yeast strains (B596 and S288C) cloned in λgt phage and later identified as transposons. Their identification as transposons was based on the observations that (1) the restriction fragments containing *ty-1* elements varied in different yeast strains, (2) the *Sup 4* region in two strains differed in the presence or absence of a *ty-1* element, and (3) the culture of the strain S288C was found to acquire different-sized fragments containing *ty-1* when propagated for a month (Cameron *et al.*, 1979). The yeast *ty-1* elements have been very well characterized (Cameron *et al.*, 1979;

Scherer and Davis, 1980; Roeder *et al.*, 1980; Gafner and Philippsen, 1980; Farrabaugh and Fink, 1980; Kingsman *et al.*, 1981). The *ty-1* elements have a large effect on gene expression and may facilitate adaptation of a particular cell in a new environment (Cameron *et al.*, 1979); like the *Drosophila* copia and 412 sequences, the yeast *ty-1* may contribute to the poly(A) RNA pool (Cameron *et al.*, 1979). Certain retroviruses (MMLV) possess a structure similar to the transposons (Shoemaker *et al.*, 1980; Shiba and Saigo, 1983). Transposition seems to occur widely in both prokaryotes and eukaryotes. Detailed analyses of transposable elements have provided excellent means for development of new vectors for the transfer of genes to a particular target in eukaryotes (Rubin and Spradling, 1982).

The *ty-1* element contains a 5300-bp DNA sequence flanked on both sides by a direct repeat of another repeated DNA sequence (300 bp in length) called the δ region. The yeast genome contains at least 35 copies of *ty-1* and 100 copies of the δ sequences occur mostly in association with *ty-1* or as direct repeats of independently dispersed sequences in the yeast genome. The presence of δ usually indicates that *ty-1* has been excised out leaving behind the δ sequence. Two *ty-1* elements (*Ty-912* and *Ty-917*) are known to cause mutation of *his-4*$^+$ to *his-4*$^-$ in yeast. Yeast *his-4*$^-$ mutations (*his-4-912* and *his-4-917*) caused by insertion of *ty-1* transposons have been fully characterized. It has been shown that the mutation *his-4*$^+$ → *his-4*$^-$ is caused by insertion of *ty-1* in the proximal end of the *his-4*$^+$ gene whereas the reversion of *his-4*$^-$ to *his-4*$^+$ is caused by the excision (perfect or imperfect), inversion, or translocation of the inserted *ty-1* element. The *his-4-912* mutation results from the insertion of a *ty-1* element into the 5' noncoding region of the yeast *his-4*$^+$ gene and the majority of *his-4*$^+$ revertants result from excision of the *ty-1* element by a recombination event between the flanking δ regions. During this process, a δ region is left behind in the revertant. The excision of the yeast transposon *ty-1* is similar to that of the bacterial transposon *Tn-9* (MacHattie and Jackowski, 1977). However, the *ty-1* shows several significant differences from the prokaryotic transposon. The excision of the *ty-1* from the yeast *his-4* mutant can lead to complete or partial wild-type function or to a new phenotype. Unlike the prokaryotic insertion elements, the controlling elements of eukaryotes exist as a two-component system like as *spm* (suppressor–mutator) of corn (McClintock, 1951). Two classes of *spm* have been described in yeast (Roeder *et al.*, 1980): *spm-1* suppresses the *his*$^-$ phenotype of both *his-4-912* and *his-4-917,* whereas *spm-2* and *spm-3* suppress *his-4-917* but not *his-4-912. spm-2* and *spm-3,* however, increased the frequency of reversion of *his-4-912* to *his-4*$^+$. Thus *spm-2* and *spm-3* may control factors that promote re-combination between homologous δ regions to excise the internal *ty-1* sequence leaving behind a single copy of δ. These yeast revertants (*his-4*) carrying a δ region display a new phenotype of cold sensitivity for histidine requirement during growth. *Spm-1* suppresses the cold sensitivity of all δ-containing revertants. Therefore it is suggested that *spm-1* acts on a δ region (Roeder *et al.,* 1980). Likewise *spm-2*$^-$ and *spm-3*$^-$ suppress the cold sensitivity of the *his-4-912* revertant carrying the δ sequence. The molecular nature of *spm*$^-$ as well as the mechanism of their action remain to be elucidated. It is expected that the nature of *spm* would be explored soon utilizing the yeast transfor-mation system (Roeder *et al.,* 1980).

These studies have been significant in defining the role of the dis-persed repeated DNA sequence as transposons. Their characterization provides a better understanding of the molecular basis of gene in-stability and chromosomal rearrangements that could not be elucidated by classical genetic analysis. The dispersed repeated sequences like *ty-1* do not possess any role in replication since the plasmid containing *ty-1* was incapable of autonomous replication in a yeast transformation system (Scherer and Davis, 1980). The *ty-1* may be associated with the unlinked insertion of *leu-2*$^+$ genes during yeast transformation (Hin-nen *et al.,* 1978). The *ty-1* may also play a role in the mating-type interconversion in yeast by facilitating intrachromatid recombination (Strathern *et al.,* 1979). Diverse effects of *ty*-mediated gene expression in yeast have been recently described (Errede *et al.,* 1980; Ciriacy and Williamson, 1981). It has been shown to cause over-production of a mating-type locus signal or of the isocytochrome *c* gene product or constitutive production of yeast alcohol dehydrogenase II (an enzyme that is glucose reversible). No transposons have yet been characterized from other fungi; in *Neurospora,* certain position effects regarding the expression of a morphological mutation *rg,* defective for phos-phoglucomutase (Mishra and Threlkeld, 1967; Mishra and Tatum, 1970), may involve a transposition of a *ty-1*-like (Mishra, 1967, also unpublished results) genetic element.

K. RECOMBINATION

Studies in fungi have been instrumental in providing the molecular mechanism of genetic recombination (Holliday, 1964; Hotchkiss, 1977); however, the biochemistry of recombination as well as its control is only poorly understood in eukaryotes (Stahl, 1980). It is believed that a better understanding of the molecular basis of recombination will be made in the near future through the use of fungal transformation systems. It has already been used to provide evidence for (1) an intrachromosomal gene conversion in yeast (Klein and Petes, 1981), (2) recombination between an autonomously replicating plasmid and yeast chromosomal DNA (Falco *et al.*, 1983; Jayaram and Broach, 1983), and (3) the different mechanisms of gene conversion during mitosis and meiosis in yeast (Jackson and Fink, 1981).

L. BIOTECHNOLOGY

The fungal transformation system can be utilized for the production of animal proteins and many pharmaceuticals. This possibility for biotechnology has been strengthened by the demonstration of abundant production of a human interferon and of human hepatitis B virus surface antigen in yeast cells transformed with a chimeric plasmid containing the gene for the human interferon (Hitzeman *et al.*, 1981) or for the hepatitis B virus surface antigen (Valenzuela *et al.*, 1982; Miyanohara *et al.*, 1983). It seems that most of the human genes (or other eukaryotic genes) that are intronless (or are originally cloned from a cDNA) if attached to a yeast promoter region (such as that of ADH1) can be easily expressed in a yeast transformant; thus this discovery overcomes the major obstacle in biotechnical features that favor application of yeast. Other important considerations in the genetic engineering of yeast and its application in biotechnology are the facts that (1) the biochemical steps in the production of glycoproteins and the secretion of proteins in yeast are similar to those in animal cells, (2) the fermentation behavior of yeast can be easily manipulated, and (3) the foreign proteins (such as human interferon and hepatitis virus B surface antigen) produced in yeast transformants are biologically active and made in amounts commercially profitable for the manufacture of vaccine. These discoveries offer immense possibilities for control of human diseases.

IX. Mobilization and Expression of Genes in Various Groups of Organisms

Molecular cloning of genes from different organisms is now possible in fungi. Also, using a shuttle vector it is possible to transfer fungal

genes into *E. coli* and back to fungal cells using the appropriate transformation systems. These systems provide an opportunity to examine
the expression of genes in heterologous environments. These include
(1) expression of fungal genes in a bacterium (*E. coli*), (2) expression of
bacterial (*E. coli*) genes in fungi, and (3) expression of eukaryotic genes
in fungi. A discussion of the different heterologous systems for gene
expression is presented here.

A. Expression of Fungal Genes in Bacteria

The expression of a fungal gene in bacteria depends on the accuracy
of the transcription and translation by the bacterial RNA polymerases
and the protein synthesizing system. Walz and others (1978) have
shown the role of a bacterial insertion sequence (*IS-2*) in acting as a
promoter for the yeast tryptophan synthetase (*trp-5*) gene in *E. coli*.
These investigators have shown that the insertion sequence was transferred to the hybrid plasmid from a F plasmid (and not from the *E. coli*
chromosome) by *recA*-independent recombination. This observation
provides an ingenious method to facilitate the expression of eukaryotic
genes and should be used in future for initiation of fungal gene transcription in bacteria. It has been further shown that the expression of
the yeast *ura-1* gene is initiated from a bacterial promoter in *E. coli*
but from a yeast promoter in yeast (Lauson and Jund, 1982). The
composite plasmid (with the *ura-1* gene) was found to carry a 0.9-kb
segment.

A number of fungal genes have been expressed in *E. coli*. The fact
that their expression is not influenced by the orientation of the fungal
DNA sequences in the vector plasmid suggests that the fungal genes
are transcribed from an internal promoter (of the fungal gene) and not
from the bacterial promoter. The expression of these fungal genes necessarily implies that *E. coli* RNA polymerases and the protein-syn-
thesizing system are able to recognize the appropriate control signals
for an accurate transcription and translation of the fungal message. It
is known that the intervening sequences present in an eukaryotic message must be spliced out before translation; therefore, the fact that
fungal transcripts lack an intervening sequence must be the major
factor in facilitating their accurate translation in the bacterial cell. A
number of the fungal genes that have been found to complement bacterial mutants are described in Table 2. The enzymes produced by the
fungal genes in bacteria were found to possess the physicochemical
properties of the enzymes encoded by the fungal genes.

It is of interest to mention here that the expression of a fungal gene in bacteria is an exception rather than a rule. Also it is plausible that the fungal genes may require some mutation before their expression in *E. coli* as demonstrated for yeast *arg-3* gene (Crabeel *et al.*, 1981) or for *his-5* gene (Harashima *et al.*, 1981). The results of this study pave new pathways for the manipulation of fungal gene expression. A better understanding of the bacterial transcriptional and translational machinery has provided a basis for their manipulation to promote expression of eukaryotic genes. Some of these have recently been discussed (Guarente *et al.*, 1980). These include (1) an addition of a bacterial promoter sequence (particularly the Shine–Delgarno sequence) by gene fusion, (2) a use of cDNA in cloning to avoid the processing of intervening sequences, and (3) the easy secretion of eukaryotic protein to prevent its digestion in the bacterial cell.

B. Expression of Bacterial Genes in Fungi

The fact that a bacterial plasmid containing the fungal *ars* can replicate in fungal cells provides an opportunity for examination of the expression of bacterial genes in fungi. A number of bacterial genes have been found to be expressed in yeast; these include the genes for resistance to ampicillin, chloramphenicol, kanamycin, and a 2-deoxystreptamine (G148) antibiotic.

The yeast cells harboring the hybrid plasmid containing an ampicillin gene were found to produce mRNA for β-lactamase as well as the functional enzyme (Chevallier and Aigel, 1979; Chevallier *et al.*, 1980), which was immunologically similar to *E. coli* β-lactamase. These authors also found that in yeast, the β-lactamase is produced as a preprotein, which is converted to the mature enzyme. This was confirmed by an *in vitro* study in which the yeast cell-free extract was found to process the bacterial preprotein into mature β-lactamase. The ampicillin gene-specific mRNA was found to be polyadenylated and had a half-life of 20 min, which is much longer than the half-life of this mRNA in *E. coli*; it is suggested that these changes in the mRNA are linked to the host metabolism, which is discussed later in this paragraph. However, according to Hollenberg (1979) the activity of the β-lactamase per unit cell mass is 25–100 times lower in yeast as compared to that found in *E. coli*; this low activity may, however, result from inefficient processing of the preprotein. It is not certain whether the β-lactamase gene is transcribed from a yeast promoter since the

hybrid plasmid containing the plasmid gene is fused to a DNA sequence of the yeast endogenous plasmid. Chevallier *et al.* (1980) have
suggested that the long half-life of β-lactamase mRNA may be due to
mRNAs that are not translatable. This suggestion is compatible with
the observation of Lasson and Lacroute (1979) that mRNAs engaged in
protein synthesis are more readily destroyed by RNase than messages
that are not being translated. Quite recently it was reported that the
bacterial gene for resistance to ampicillin was expressed in *Podospora*
as well (Stahl *et al.,* 1981).

Cohen and others (1980) have constructed a chimeric plasmid containing the *E. coli* R factor-derived chloramphenicol resistance (*cam*^r)
gene, which conferred resistance to a very high level of chloramphenicol (500 μg/ml) on the yeast cells. These authors have also shown that
the yeast cells carrying the chimeric plasmid produce an active chloramphenicol acetyl transferase enzyme, which is responsible for the
antibiotic resistance. Hollenberge (1979) has shown the expression of a
gene controlling kanamycin resistance in yeast transformants.

Jimenez and Davies (1980) have reported the expression of a transposable element carrying a gene for resistance to aminoglycoside antibiotic G418 in yeast cells. The yeast transformants carrying the chimeric
plasmid were resistant to a high drug concentration (1 mg/ml). The yeast
transformants were also found to contain high levels of the detoxification
enzyme, aminoglycoside phosphotransferase, which phosphorylates and
inactivates the antibiotic G418. In all cases of antibiotic resistance
conferred by chimeric plasmids to yeast cells, drug resistance has always been correlated to the presence of other markers present in the
plasmid and was found to be expressed only under conditions in which
plasmids were maintained in the tranformed cells. The introduction of
an antibiotic resistance gene into yeast offers distinct possibilities in
the area of molecular cloning. The utilization of these genes as selectable markers of a vector in the eukaryotic cells may lead to the development of yeast and other fungi as eukaryotic cloning systems.

A chimeric plasmid containing the *Neurospora qa* genes and the
genes for chloramphenicol and G418 antibiotic resistance has been
constructed. However, the *Neurospora* cells transformed with this
plasmid expressed only the *Neurospora qa* genes (Case, 1981). The
resistance to antibiotic G418 could not be examined in this fungus,
since *Neurospora* mutates readily to confer drug resistance. Therefore,
it is difficult to determine whether the newly acquired G418 resistance
in the *Neurospora* transformants was due to the presence of a chimeric
plasmid or due to a mutation of a *Neurospora* gene. Other factors such

as the integration of the donor DNA (or at least part of it) into the host
chromosome may inhibit the expression of the (integrated) plasmid
DNA in *Neurospora* transformants (Case, 1981).

The expression of a bacterial gene in yeast has also been examined
by the gene fusion technique. This method is based on the observation
that substitution of amino acids up to 30 in the N-terminal region of
the β-galactosidase enzyme by a variety of other proteins results in a
hybrid protein with substantial β-galactosidase activity. Thus the expression of the gene fused on the 5' end (N terminus of β-galactosidase)
of the *lac-z* gene can be monitored by measuring the activity of the β-
galactosidase (Mark *et al.*, 1981). This method has been utilized to
measure promoter activity, its location, and regulation for two yeast
genes fused to the *lacZ* gene. The yeast genes studied by this method
include *ura-3* (Rose *et al.*, 1981) and iso-1-cytochrome (*cyc-1*) (Guarente and Ptashne, 1981). These studies were facilitated by the fact
that (1) β-galactosidase activity can be detected on plates containing
glucose and a chromogenic substrate x-gal (5-bromo-4-chloro-3-indo-
lylgalactoside), which makes the colonies (containing β-galactosidase)
turn blue, (2) regulation of the *ura-3* gene is known both in *E. coli* and
yeast, and (3) the DNA sequence of yeast *cyc-1* is completely known
and appropriate plasmids are available to examine the expression of
these genes in both *E. coli* and yeast by appropriate transformation
systems. Some of the fusion plasmids were found to express β-galactosidase in yeast transformants and the expression of β-galactosidase
was under uracil regulation in yeast. Only a small set of the fusion
plasmids carrying a deletion in the 3' end of the *ura-3* gene expressed
β-galactosidase activity in yeast, whereas virtually all the fusion plasmids containing a complete deletion of the *ura-3* gene (including the
flanking regions) produced β-galactosidase activity in *E. coli*. These
data suggest that there are many sequences that can cause initiation
of translation in *E. coli*. But the fact that most of the deletions in the
ura-3 gene did not express β-galactosidase activity in yeast suggests
that expression in yeast requires additional or different signals. The
gene fusion technique has been used in another study (Guarente and
Ptashne, 1981) to monitor the expression of *E. coli* β-galactosidase
gene in yeast cells. The yeast transformants carrying the chimeric
plasmid were found to synthesize a chimeric protein with cytochrome *c*
at the N terminus and a functional β-galactosidase at the C terminus.
The expression of β-galactosidase activity by the plasmid containing
the *cyc* gene (with its entire promoter region of 1100 bp) was found to
be controlled in the manner seen for the regulation of cytochrome *c*

(i.e., reduction of synthesis in cells grown on glucose). The expression of β-galactisodase in another plasmid carrying a deletion between 700 and 250 nucleotides upstream from the start of the *cyc* gene was negligible (0.3%). In another transformant carrying a plasmid with a deletion between 1100 and 300 bp, the expression of β-galactosidase was restored in yeast cells grown on the glucose medium but not on the raffinose medium (16%); however, the enzyme expression was highly increased in *E. coli* cells transformed with this plasmid. Another chimeric plasmid, in which the β-galactosidase gene was fused to the sea urchin histone H2A gene, was unable to produce active β-galactosidase in either of the host organisms (i.e., yeast or *E. coli*). These data are illustrated in Fig. 5.

These data lead to the following conclusions: (1) The region beyond 300 nucleotides is responsible for the enzyme activity in the presence of raffinose or deletions in this region make the expression of the gene insensitive to glucose. There are several other possibilities as discussed by these authors (Guarente and Ptashne, 1981). (2) The *cyc-1* promoter lies within a segment 250–300 bp upstream; this has recently been confirmed by Smith *et al.* (1979). (3) A sea urchin gene cannot be recognized as a promoter in yeast cells.

An ingenious method has been used to transfer the bacterial *nif* (nitrogen fixation) gene cluster to yeast (Zamir *et al.*, 1981). These authors transformed a *his⁻* yeast strain with a mixture of two plasmids; one plasmid (pYehis) contained the yeast *his⁺* gene and another (pWK220) contained the *nif* gene cluster from *Klebsiella pneumoniae*; then the *his⁺* transformants were selected on minimal medium. Among 87 *his⁺* transformants, 2 were found to contain the *nif* gene cluster. The restriction and the Southern hybridization pattern of the transformant DNA were found to be compatible with tandemly duplicated pWK200 sequences in the yeast chromosome. Thus, the entire *nif* gene cluster (46 kb) was found to be integrated into the yeast chromosome. The tetrad analysis showing 1:1 segregation confirmed the chromosomal location of the *nif* gene cluster. During mitosis, one of the *nif* regions was found to be lost; however, the remaining *nif* gene cluster was found to be stable even after 40 generations of growth on a nonselective medium. It was also shown that DNA extracted from the yeast transformant containing the *nif* gene cluster, when used to transform *E. coli*, yielded transformants capable of growth in a nitrogen-free medium and able to reduce acetylene. In preliminary experiments the *nif* gene cluster has not yet been expressed in the yeast transformants; however, the stable yeast transformants offer a good

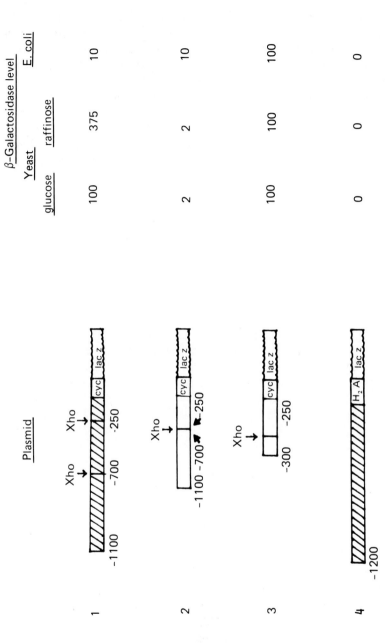

Fig. 5. Effect of deletion of promoter region on the expression of β-galactosidase in yeast/*E. coli* transformants. (Adapted from Guarente and Ptashne, 1981, with permission of authors and publisher.)

opportunity for the examination of transcription and translation of the
nif gene cluster (encoding the enzymes nitrogenase and nitrogen reductase). The tandem integration of the *nif* gene cluster suggests that
very large DNA segments such as bacteriophages can be introduced
into the yeast genome. This opens the possibility of cloning λ phage in
its entirety into yeast and of developing new genetic systems that can
be used to probe into several regulatory control mechanisms. Such a
system will have the advantage of the wealth of biochemical and genetic knowledge that we already possess about both yeast and λ bacteriophage. The outcome of these experiments opens the way for the
transfer of any gene or set of genes into yeast, which can be appropriately modified so that the introduced gene(s) are able to express themselves in the host cells. Such a gene or gene cluster can be easily
introduced as an additional (mini)chromosome by combining a yeast
centromeric region with a vector carrying the gene cluster (Clarke and
Carbon, 1980a; see also pp. 115–116).

It is of interest to mention that the plasmid DNA isolated from
certain phytopathogenic fungi (*Fusarium oxysporum* f.sp. *lycopersici*
and *Rhizoctonia solani*) was found to confer on *E. coli* the ability to
utilize agar as a carbon source. Phytopathogenic fungi are known to
degrade complex carbohydrates, therefore, the acquisition of the ability to grow on agar as sole carbon source by an *E. coli* transformant is
compatible with the possible role of the plasmid in phytopathogenesis.
The *E. coli* transformants were also resistant to the antibiotic furandantin (Martini *et al.,* 1978; Grimaldi *et al.,* 1978). Such a transformation system also provides a novel method to assign a function to a
particular DNA sequence (such as the furandantin resistance determined by the fungal plasmid). The discovery of such a system can offer
immense possibilities in the construction of new vectors and new
transformation systems in fungi (particularly filamentous fungi in
which efficient vectors like those of yeast are not yet known). This
aspect is further discussed in Section X.

C. EXPRESSION OF EUKARYOTIC GENES IN FUNGI

1. Expression of Fungal Genes

The fungal transformation system (particularly in yeast) provides
an excellent opportunity for the molecular cloning of eukaryotic genes
that cannot be selected by complementation in *E. coli*. A number of
yeast and *Neurospora* genes that could not be detected by complemen-

tation in *E. coli* were, however, found to be expressed in fungi on reintroduction. The genes detected by expression in fungal cells are listed in Table 2. Stinchcomb *et al.* (1980) have shown that DNA segments from *Neurospora* and other eukaryotes can function as *ars* in yeast. The chimeric plasmid containing *Neurospora ars* was found to grow as an autonomously replicating plasmid in yeast cells. The yeast transformants with *Neurospora ars* have a doubling time of 4.3 hr, which is very similar to that of yeast. Higher eukaryote (corn and fruit fly) DNAs were not as effective *ars* as *Neurospora* DNA (Stinchcomb *et al.*, 1980). These data support the evolutionary closeness of the *Neurospora ars* to yeast.

Huiet (cited in Case, 1981) has constructed a chimeric plasmid containing the *Neurospora qa-2$^+$* and the yeast *leu-2$^+$*, *his-3$^+$*, *ura-3$^+$* or *trp-1$^+$* genes to transform comparable mutants in *Neurospora*. In these experiments *qa-2$^+$* transformants were obtained but no yeast gene was found to be expressed. Mishra had used yeast chimeric plasmids containing either yeast *ars-1* or *ars-2* and yeast *trp-1, leu-2,* and *arg-4* genes to transform corresponding *Neurospora* mutants, but no expression of yeast genes was observed in *Neurospora* (N. C. Mishra, unpublished results). In such experiments the chimeric plasmids containing *ars* were used to optimize the possibility of the propagation of the chimeric plasmid in *Neurospora*. This approach was based on the observation that a yeast chimeric plasmid containing the *Neurospora ars* was functional in yeast. However, this expectation was not realized since these experiments did not yield any transformants. The failure of the intergeneric transformation could be ascribed to any of the following reasons: (1) a yeast *ars* is nonfunctional in *Neurospora,* (2) the yeast DNA could not be integrated into the *Neurospora* chromosome in order to be functional, or (3) the yeast genes cannot complement *Neurospora* mutations. Additional experiments must be performed to resolve these possibilities.

The intergeneric transfer of genes in *Neurospora* and yeast can be achieved by devising a vector that functions as an autonomously replicating plasmid in *Neurospora*. This can be further facilitated by the method of cotransformation by two different plasmids containing DNA sequences homologous to the host cells or by adding a centromeric region to a versatile plasmid (see Section X for the discussion of a versatile plasmid) that can function equally well in both *Neurospora* and yeast. A versatile plasmid containing a *Neurospora* centromere may be able to fit well in the cytoskeleton of *Neurospora* cytoplasm and might help in its inclusion into the *Neurospora* nucleus. This is particularly important since *Neurospora* chromosomes are never exposed to cytoplasm (Barry, 1972). No intergeneric transformations among other fungi have been attempted (except for PEG-mediated cell fusion and subsequent gene transfer). The filamentous fungi, showing a great degree of differentiation, offer several interesting possibilities to examine the expression of a phase-specific gene following transformation. Therefore, it should be useful in analysing cell division, morphogenesis, and cytodifferentiation.

2. *Expression of Higher Eukaryotic Genes in Fungi*

Although in *Neurospora* the expression of a higher eukaryotic genome has not yet been attempted such experiments have been quite successful in yeast. The yeast transformation system provides several advantages in examining the expression of other eukaryotic genes; this is based on the facts that (1) yeast, being a eukaryote, would have a similarity in transcription and translation systems, (2) recombinant plasmids are available that can be used to compare the expression of the same genes in yeast as well as in *E. coli,* and (3) a variety of plasmids are available that can be maintained in multicopy as autonomously replicating units or exist as an independent chromosome or integrate into the host chromosome and can be used to examine their effect on gene expression. The expression of higher eukaryotic genes in yeast may be hindered by its presumed difficulty in processing the foreign intervening sequences (see Beggs *et al.*, 1980; Langford *et al.*, 1983). However, this problem can be overcome by the use of cDNA or of intronless genes of eukaryotes (Hitzeman *et al.*,1981).

The structural gene for β-galactosidase (*lac-4*) from a related yeast, *Kluyveromyces lactis,* have been transferred to *Saccharomyces cerevisiae* (baker's yeast) via the construction of a composite plasmid and yeast transformation system (Dickson, 1980). All transformants expressed *lac-4* β-galactosidase activity; however, the yeast transformants were unable to grow on lactose, probably because the lactose does not enter the yeast cell (as determined by the lack of uptake of [^{14}C]lactose). The enzyme activity was, however, shown in the cell-free extract. Also, the enzyme produced was characteristic of the donor DNA (i.e., *K. lactis*). Transformants carrying multicopies of the transforming plasmid possessed a much higher level (25 ×) of β-galactosidase activity than transformants in which a β-galactosidase gene was integrated into the yeast chromosome (the exact site of integration is not established but it is assumed to be at the *leu-2* locus on chromosome III). As expected, the transformants containing the autonomously replicating plasmid were unstable, whereas the one in which the donor DNA was integrated into the yeast chromosome were stable. These experiments clearly indicate that other eukaryotic genes can be stably transferred and expressed in yeast.

It seems unlikely that prokaryotic cells will provide an adequate environment for the physiological expression of eukaryotic genes. Higher eukaryotic mRNA differs from the prokaryotic message in terms of processing and splicing of the transcription product. In this context, yeast seems to provide a natural environment for the expression of eukaryotic messages; yeast mRNA are capped and polyadenylated like other eukaryotic messages and seems to possess splicing mechanisms for tRNAs and at least for one structural gene (actin). In view of these advantages, Beggs and others have transformed yeast cells with a composite plasmid containing a complete rabbit β-globin gene with two intervening sequences and extended flanking regions (Beggs et al., 1980). The investigators found that the yeast transformants produced β-globin-specific transcripts (800–900 nucleotides long). However these lacked 20–40 nucleotides at the 5' end and did not undergo splicing of the intervening sequences. Thus yeast cells fail to effect splicing of the primary β-globin gene. It is quite possible that yeast lacks appropriate splicing enzymes. Alternatively, the transcript and the appropriate splicing enzyme(s) may be spatially separated or the splicing enzyme may be unable to act on a transcript that lacks nucleotides at the 5' end. A complete transcript can generate secondary and tertiary structures essential for splicing (Beggs et al., 1980). Thus at least in this case, the yeast cells seem inadequate for the expression of the higher eukaryotic genes, which at present must be studied in systems using animal viruses as vectors (Mulligan and Berg, 1981).

Several clever methods have been devised for the expression of eukaryotic messages in E. coli; one of these is based on the fusion of a eukaryotic gene to the lacZ promoter. A similar approach has been undertaken to examine the expression of a cDNA sequence (coding for chicken ovalbumin) in yeast cells. A chimeric plasmid containing a cDNA copy of chicken ovalbumin mRNA was fused to the beginning of the E. coli β-galactosidase gene (so that in the hybrid protein the seventh amino acid of β-galactosidase was connected to the fifth amino acid of ovalbumin). This was ligated to a hybrid plasmid containing pBR322 and yeast 2-μm plasmid. The selective marker used for the yeast transformation was the ura-3 gene. The production of chicken ovalbumin by the yeast transformants was determined by radioimmunoassay. The ratio of ^{125}I-labeled ovalbumin bound to antibody in the presence and absence of competing material was used to indicate the nature of ovalbumin produced in the *E. coli* and the yeast transformants. In both cases, the ovalbumins produced were similar but appeared to differ from the native chicken ovalbumin. These observations suggested that some of the antigenic determinants present in the native protein were missing from the ovalbumin produced in the yeast or the *E. coli* transformants. This difference may result from the substitution at the N terminus (by the galactosidase amino acid sequence) or be due to the absence of posttranslational modifications (glycosylation, phosphorylation, and acetylation) that occur in the chicken (Mercereau-Puijalon *et al.*, 1980). It would be of interest to investigate whether or not a yeast cell is capable of performing any of these modifications. It has been estimated that a yeast cell synthesizes 1000–5000 molecules of ovalbumin. The higher yield of animal protein by the yeast cell confirms the validity of a suggestion that yeast would provide a good system for the manufacture of animal (including human) proteins (Mercereau-Puijalon *et al.*, 1980).

An important discovery regarding the expression of eukaryotic genes in fungi has been made by Henikoff and his collaborators (1981; Henikoff and Furlong, 1983). These investigators have described the complementation of a defective yeast function by a *Drosophila* genomic DNA sequence. A chimeric plasmid (pYF) was constructed that contained plasmid pBR322, a yeast endogenous plasmid (2-μm plasmid), the yeast chromosomal *leu-2*$^+$ gene, and a *Drosophila* genomic DNA sequence (as expected, such a plasmid was capable of replication both in *E. coli* and yeast). This chimeric plasmid was used to transform a yeast double mutant (*leu-2, ade-8*) strain. It was found that yeast *leu-2*$^+$ transformants thus obtained were also *ade-8*$^+$ prototrophs. An analysis of the transcripts from the yeast cell transformed by the chimeric plasmid (pYF) showed that a 0.8-kb poly(A)-containing RNA was indeed specific to a *Drosophila* DNA sequence and that the *Drosophila* DNA sequence alone was responsible for the complementation of the defective yeast function (*ade-8*). The *Drosophila* DNA sequence that complemented the yeast *ade-8* mutation was shown by *in situ* hybridization to be localized on the chrommomeric band (No. 27) of the left arm of the *Drosophila* chromosome II. This is in agreement with the published map location of the Drosophila *ade-8* gene. The Drosophila gene of *ade-8* has been shown to complete a *Schizosaccharomyces pombe ade-5* mutation which affects the structure of GAR transformylase (Henikoff and Furlong, 1983).

3. Expression of a Human Gene in Yeast

The question of the expression of a gene of heterologous origin has been very adequately dealt with during the successful cloning of a human leukocyte interferon (LeIF-D) gene in yeast (Hitzeman et al., 1981). These investigators have pointed out certain important features required for the expression of eukaryotic genes in yeast. These include (1) the use of an intronless gene in order to circumvent the problems of splicing, (2) a similarity in the codon preference in the genetic dictionary such that amino acids are coded by the same triplets as during the synthesis of the abundant class of yeast proteins, (3) a lack of the signal peptide sequence so that the foreign protein is retained inside the cell and avoids being digested during transport via an unfamiliar fungal secretory pathway, and (4) use of an A-rich sequence in the untranslated leader region (i.e., the sequence preceding the initiator ATG). Such sequences are predominantly present in yeast; the ADH1 promoter region that was fused to the human interferon gene contains a sequence of AATTCATG; leader sequences containing G are not expressed in yeast. Based on these considerations, Hitzeman and collaborators (1981) chose to examine the expression of a human interferon gene, which lacks an intron and shows a biased use of arginine codons. The human gene was fused to the yeast ADH1 promoter region in a chimeric plasmid capable of autonomous replication and selection in both yeast and E. coli. The yeast cells transformed by this chimeric plasmid were found to synthesize large amounts of biologically active human interferon. The human interferon produced by the yeast transformant was found to accumulate in the cytosol (1×10^6 molecules/ cell). The results of these studies pave the way for the genetic engineering of fungi, particularly yeast, for the production of animal proteins and pharmaceuticals. As pointed out by these investigators, yeast provides a system of choice for the production of glycoproteins since the addition of the inner core of carbohydrates into the polypeptide and the secretory pathways for the transport in yeast are similar to those found in animal cells (Novick et al., 1980; Ballou et al., 1980).

4. Expression of a Eukaryotic Viral Gene

In the past, study of certain human viruses (such as hepatitis B) has proven very difficult because of the fact that they cannot be propagated in systems other than their primary hosts. It is therefore of interest to examine their expression in fungal cells, particularly in view of the fact that human genes could be expressed in yeast cells. Two groups of workers have independently demonstrated the production and assembly of immunogenic particles in the yeast transformants by a chimeric plasmid carrying the DNA sequence coding for HBV surface antigen (Valenzuela *et al.*, 1982; Miyanohara *et al.*, 1983). Such expression of a human viral gene in yeast cells opens the possibility for production of vaccine and control of human viral diseases by immunization. Also the production and assembly of HBV surface antigen into 20- to 22-nm particles offers an effective approach for the analysis of organelle biogenesis in a higher organism (Valenzuela *et al.*, 1982). The expression of the herpes simplex virus thymidine kinase (HSV-ITK[+]) gene has been recently demonstrated in yeast cells (McNeil and Friesen, 1981).

X. Construction of a Versatile Vector

Beggs (1978) created the first genetic shuttle vector by adding the yeast 2-μm plasmid to the *E. coli* plasmid. This composite plasmid was capable of propagation both in yeast and in *E. coli* and thus provided a vehicle for shuttling genes back and forth between these two organisms, which are the ideal representatives of prokaryotes and eukaryotes. Later on, it was found by different groups of workers that only a portion of yeast 2-μm plasmid can confer the ability on an *E. coli* plasmid to propagate in yeast cells (Hicks *et al.*, 1978; Gerbaud *et al.*, 1979). As mentioned earlier, the yeast chromosomal DNA sequences carrying the origin of replication (called *ars* or the autonomously replicating sequence) can also give to an *E. coli* plasmid similar ability to replicate as an autonomous element in yeast (Kingsman *et al.*, 1979; Struhl *et al.*, 1979; Hsio and Carbon, 1979; Szostak and Wu, 1979). It is of interest to mention here that the plasmid DNA recovered from the *E. coli–*yeast*–E. coli* shuttle remains unchanged, as shown by restriction mapping (Hsio and Carbon, 1979). It has also been shown that any mutation that increases the efficiency of expression of a shuttle plasmid in *E. coli* does not affect its expression in yeast (Hsio and Carbon, 1979). A shuttle vector can be actually made to function as a minichromosome by addition of a centromeric region to the plasmid carrying *ars* (Clarke and Carbon, 1980a). Another genetic shuttle has also been constructed that can function in *E. coli* as well as in mammalian cells; this has been accomplished by combining *E. coli* plasmid with an animal viral DNA (SV40) (Mulligan and Berg, 1981). This shuttle vector has been used for the mobilization of genes from *E. coli* to mammalian cells and vice versa. The development of shuttle vectors has been very useful in the understanding of the physiological expression of a gene in homologous and heterologous environments. However, as yet no vector is available for the mobilization of genes among yeast, filamentous fungi, and *E. coli.* As discussed earlier (see Section IX), yeast chimeric plasmids carrying *ars* are unable to replicate in *Neurospora.*

A covalently closed circular plasmid DNA (90 kb in circumference) has been obtained by the CsCl ethidium bromide gradient centrifugation of a DNA preparation from a phytopathogenic fungus, *Fusarium oxysporum (sp. Lycopersici)* (Grimaldi *et al.*, 1978). This plasmid was found to propagate in *E. coli,* and it confers on the host cell the ability to utilize agar as the sole carbon source (by producing the enzyme polygalacturonase) and also a new resistance to the antibiotic furadantin.

Another plasmid has been obtained from *Podospora anserina* (Stahl *et al.*, 1980). This plasmid is very well characterized with respect to its restriction maps and its ability to confer senescence on a young culture of *Podospora* by transformation (Tudzynski and Esser, 1979; Tudzynski *et al.*, 1980). A composite plasmid containing pBR322 and the *Podospora* plasmid was also found to confer senescence on *Podospora* by transformation; the *Podospora* transformants were also found to express the ampicillin gene of pBR322 (Stahl *et al.*, 1981). Furthermore, both senescence and ampicillin resistance were expressed if the hybrid plasmid contained (instead of the *Podospora* plasmid) a *Sal* fragment (5.05 kb) isolated from juvenile mitochondrial DNA, which is only partially homologous to the plasmid (Kuck *et al.*, 1981; Stahl, 1981).

The work of the Esser group as discussed above provides an excellent example of how a shuttle vector can be developed for filamentous fungi. Some of these plasmids or their combinations can be engineered to provide a versatile vehicle for filamentous fungi. Indeed, this expectation has been fulfilled by Stohl and Lambowitz (1983) by construction of a chimeric plasmid containing a segment of *Neurospora* mitochondrial plasmid DNA with *ars* function. The chimeric plasmid developed by these workers can replicate both in *E. coli* and *Neurospora* and thus is indeed a shuttle vector (Stohl and Lambowitz, 1983).

XI. Biochemistry of Gene Transfer

Unlike bacterial transformation, the processes involved in the uptake of donor DNA and its subsequent processing and integration into resident chromosomes during fungal transformation are not yet

known. In *Neurospora*, DNA uptake during transformation is facilitated by the use of certain cell wall-deficient mutants as recipients.
The hindrance offered by the fungal cell wall can be overcome by
treatment with an appropriate enzyme that generates spheroplasts.
There are certain indications that the uptake of donor DNA may be
genetically controlled (Kingsman *et al.*, 1979).

Enzymes involved in the recombination of donor DNA would necessarily include deoxyribonuclease, DNA polymerase, and ligase. The
role of these enzymes in *in vivo* and *in vitro* recombination has been
very well elucidated (see Kornberg, 1980; Stahl, 1980). However, no
gene similar to the *recA* gene of *E. coli* has yet been identified in yeast,
Neurospora, or other fungi. It is of interest to mention that Mishra
(1982) has recently demonstrated the lack of a *recA*-like gene from the
higher eukaryotes. The role of a nucleaseless mutation (*nuc-1*) in the
DNA uptake and in the transformation of *Neurospora* has been described (Mishra, 1979b). The DNA uptake by the *nuc-1* mutants of
Neurospora was only about 40% that of the wild-type strain. Schablik
and Szabo (1981) have reported the presence of a protein in young
cultures of *Neurospora* that stimulates DNA uptake. Further purification and characterization of this DNA uptake stimulating factor may
provide a better insight into the mechanism of DNA uptake during the
transformation of fungi.

XII. Gene Transfer in Different Groups of Fungi

A. *Neurospora*

1. *Transformation with Total DNA Preparations*

Mishra *et al.* (1973) first reported the occurrence of the DNA-mediated genetic changes in *Neurospora*. These authors chose to examine
transformation of the inositol (*inl*) locus of *Neurospora crassa* because
of the fact that no spontaneous reversion of a particular allele (*89601*)
of *inl⁻* had been reported earlier (Giles, 1951; E. L. Tatum, unpublished results); also, the induced reversion frequency of this *inl⁻*
allele (*89601*) was very low (Giles, 1951). In view of the fact that no
deletion mutants were known, it was thought that a low reversion
frequency of the *inl* locus could be distinguishable from its transformation frequency under appropriate experimental conditions (Mishra *et
al.*, 1973). As mentioned earlier these authors showed that the effect of
the wild-type donor DNA preparation (called allo-DNA) was specific

for the transformation of the *inl* locus of *Neurospora* since the mutant's own DNA (called iso-DNA) was ineffective in increasing the reversion frequency of the *inl*⁻ allele. The allo-DNA was found to increase the reversion of the *inl*⁻ allele to *inl*⁺ by 30–40-fold as compared to the reversion frequency of this allele in the absence of any DNA treatment or in the presence of iso-DNA treatment. The allo-DNA-mediated transformation frequency initially seen at the inositol locus of *Neurospora* was roughly $1/10^6$. This value is comparable to that reported by others (Case *et al.*, 1979; Wootton *et al.*, 1980) using total allo-DNA preparations for transformation of *Neurospora*. The transformation frequency at the *rg-1* (phosphoglucomutase) locus of *Neurospora* was unusually high, 15/3000 colonies as compared to that of *inl* ($1/10^6$ colonies) (Mishra *et al.*, 1973). This high frequency of transformation could not be explained at that time; however, it seems that the high frequency of transformation might have resulted from the fact that *rg-1* is very close to the centromere, and the transforming DNA might have contained the centromere of *Neurospora*. The inclusion of the centromeric region is now established as a factor in increasing the frequency of transformation in yeast (Clarke and Carbon, 1980a). This possibility should be reexamined in *Neurospora,* using *rg-1* as well as other centromere linked loci such as *his-2*. Mishra *et al.* (1973; and unpublished results) have inferred that the complex cell wall of the *Neurospora* wild-type strain is indeed a factor hindering DNA uptake and genetic transformation. This conclusion has been based on the following facts: (1) wild-type strains of *Neurospora* cannot be transformed unless the cell wall is digested (G. Szabo, personal communication; N. C. Mishra, unpublished results), (2) only cell wall-deficient mutants such as *rg* or *os-1* carrying *inl*⁻ were found to be transformable by the original workers (Mishra *et al.*, 1973), and (3) Wootton and others (1980) have shown that the conidia of a *inl*⁻ *os-1* strain of *Neurospora* when germinated in medium of high osmolarity are more readily transformable.

The fact that in all of these transformation experiments, the recipient contained the *inl*⁻ mutation has been interpreted to suggest the role of *inl*⁻ (because of the possible role of a cell wall impairment in the DNA uptake). This fact remains to be established despite strong correlation between presence of the *inl*⁻ mutation and transformation by total allo-DNA preparations of *Neurospora*. Even though Case *et al.* (1979) used an *inl*⁻-containing mutant as recipient in their earlier experiments, they have used other recipient strains without *inl*⁻ showing high frequency of transformation, suggesting that *inl*⁻ is not required for the transformation. However, in these experiments the authors have used a cloned *Neurospora* DNA segment as source of donor DNA, therefore, this cannot exclude the possibility of some role played by the inositol locus itself in facilitating the donor DNA uptake.

The frequency of Neurospora transformation (as measured at the *inl* locus) by a total allo-DNA preparation can be increased 5–10 times by using a young (22–30 hr) culture of *Neurospora* as recipient (Mishra, 1977b; Aradi *et al.*, 1978). Mishra has shown that the DNA uptake by the recipient cell is linear up to 1 hr of the DNA treatment of a young culture. In an old culture (30 hr or more), this linearity of donor DNA uptake with period of treatment is lacking. Both $CaCl_2$ and ATP were found to increase the DNA uptake by young *Neurospora* mycelium (Mishra, 1977b, 1979b). The stimulating effect of PEG as reported by others (Schablik *et al.*, 1979) was not found to be a significant factor for DNA uptake (Mishra, 1977a, 1979a). Actively growing cultures of Neurospora were found to take up a high amount of DNA (6 µg of DNA/mg dry weight of mycelium) (Mishra, 1977a). Schablik and Szabo (1981) have described a protein factor that stimulates DNA uptake by the *Neurospora* mycelium. It is suggested that this factor may endow the fungal cell with some kind of competence for transformation.

In the absence of evidence for a physical integration of the donor DNA into the recipient genome, Mishra *et al.* (1973) called this effect of allo-DNA-mediated genetic changes "DNA-induced reversion" and *not transformation,* even though other compelling evidence (obtained from genetic analysis of the allo-DNA-induced revertants) suggested the mechanism involved a process of transformation. The experiments carried out by Szabo and his collaborators could not provide the crucial evidence required to demonstrate that transformation was indeed the mechanism behind the allo-DNA-induced reversion of *Neurospora* (Schablik *et al.*, 1977; Szabo *et al.*, 1978; Szabo and Schablik, 1982).

The critical evidence for gene transfer in *Neurospora* was provided by Mishra (1979a,b) who demonstrated that the nature of the transformants (as well as the transformation frequency) was determined by the kinds of donor DNA used. In his experiments Mishra (1979a,b) used different kinds of donor DNA prepared from three different strains of *Neurospora*; these were (1) an allo-DNA preparation from the wild-type strain (which did not require *inl* for growth at any temperature), (2) another allo-DNA preparation from a temperature-sensitive mutant (*inl⁻ -ts*) strain of *Neurospora* (which required inositol for growth at 37°C but not at 25°C), and finally (3) an iso-DNA preparation from the mutant strain (89601) (which required inositol for growth at any temperature). In all transformation experiments mutant 89601 was used as recipient. In these experiments the nature of transformants was shown to be a function of the donor DNA. The allo-DNA from the wild-type strain produced wild-type (inl^+) transformants; that is, these did not require inositol for growth at any temperature, whereas the treatment with allo-DNA from the temperature sensitive strain (ts-inl^-) produced temperature-sensitive transformants (i.e., ts-inl^-). The treatment with iso-DNA failed to cause any transformation. These data unambiguously demonstrated that the nature of the transformants was a function of the donor DNA and therefore the mechanism involved was indeed transformation and not DNA-induced reversion (Mishra, 1979a,b). Mishra could not use the method of recombinant DNA technology to demonstrate transformation at the inositol locus since no inositol mutant (inl^-) of *Escherichia coli* is known to this date and therefore it is not possible to clone the *Neurospora* inositol gene in *E. coli*. In addition, there are certain difficulties in the expression of other *Neurospora* genes in *E. coli*. So far, only one *Neurospora* gene, *qa-2*, has been expressed in *E. coli* (Case, 1981). However, the *Neurospora* inositol gene can now be cloned in yeast since appropriate yeast inl^- mutants are available (Letts *et al.*, 1983) provided the *Neurospora* inl^+ gene can complement the yeast inl^- mutation.

2. Genetics of Transformants

Mishra and others have shown that the transformants obtained after allo-DNA treatment differed from the revertants (obtained without any DNA treatment or after iso-DNA treatment) in the mode of the inheritance of the newly acquired character of inositol independence (inl^+) (Mishra *et al.*, 1973; Mishra and Tatum, 1973; Mishra, 1976; Szabo and Schablik, 1982). These authors reported a typical Mendelian segregation of 1:1 for the inl^- and inl^+ alleles in crosses between inl^+ revertant × inl^- strains. Among the allo-DNA mediated transformants there were two classes: Class I, which showed a Mendelian segregation, and Class II, which showed a nonMendelian segregation of the transformed character in the progeny of a cross between transformant (inl^+) × inl^-. At least some of the Class II transformants were found not to be heterokaryotic for inl^+ and inl^- alleles (Mishra and Tatum, 1973). The non-Mendelian transmission (of the Class II transformants) can be explained by the assumption that the donor DNA establishes itself as a plasmid in the nucleus of the transformants and that this plasmid behaves as an episome. The episomal nature of the donor DNA is supported by the tetrad analysis of the transformants (Mishra and Tatum, 1973; Szabo and Schablik, 1982) and further elucidated by the use of certain DNA-intercalating drugs to rid the transformed strains of the plasmid. In such experiments Mishra (1976) was successful in showing that only Class II transformants were capable of loss of the *inl*+ phenotype (become inositol requirers, *inl*−) upon treatment with ethidium bromide or acridine dyes, whereas the Class I transformants (which carried the *inl*+ genetic information in their chromosomes like any other chromosomal gene) showed no loss of the *inl*+ character on treatment with these drugs. The *inl*− isolates obtained after drug treatment of the transformants (*inl*+) were found to be allelic to the original *inl*− (89601) strain (Mishra, 1976).

Szabo and Schablik (1982) have found some evidence for partial disomy as the basis for the non-Mendelian transmission of the inositol independence among some allo-DNA mediated transformants of *Neurospora*. These authors have also found that the allo-DNA mediated transformants showed a significant increase in the number of aberrant tetrads (6:2, 2:6, 5:3, 3:5) in a cross with an *inl*− strain as compared to a *inl*+ revertant (obtained without any DNA treatment) when crossed to the same *inl*− strain. Furthermore, these authors have found some differences in the specific activity and immunological properties of the enzyme inositol 1-phosphate synthetase (encoded by the *inl*+ allele) among the *inl*+ transformants, the wild-type strain, and the *inl*+ revertant (Zsindely *et al.*, 1979). Using the techniques of DNA hybridization Szabo and co-workers found no difference in the number of copies of the *inl* gene among the known wild-type strains, the transformants, and the revertant strains (Feher *et al.*, 1979). However, they found that the transformants' DNA has a significant decrease in the T_m, suggesting a 1.2–1.7% mismatch (Feher *et al.*, 1979).

Szabo and Schablik (1982) have suggested the occurrence of partial disomy as the basis for the non-Mendelian transmission of the inositol independence among some allo-DNA-mediated transformants of *Neurospora*. However, this interpretation seems unwarranted, particularly in view of the fact that the occurrence of autonomously replicating plasmids among *Neurospora* transformants has been confirmed by Southern blot analysis (Grant *et al.*, 1984). The data of Grant *et al.* (1984) clearly support the original view of Mishra and Tatum (1973) that donor DNA may establish as an autonomously replicating plasmid following transformation of *Neurospora*.

3. Transformation with Cloned DNA

As soon as the recombinant DNA technology became available Giles and his group constructed a chimeric plasmid containing the *Neurospora* gene for quinic acid metabolism. These authors were able to show the expression of the *Neurospora qa-2*[+] gene in *E. coli* (Alton *et al.*, 1978; Hautala *et al.*, 1977; Kushner *et al.*, 1977). The chimeric plasmid was found to complement an *E. coli aro-D* mutation and the enzyme encoded by *Neurospora* DNA in the chimeric plasmid possessed both the physicochemical and immunological characteristics of the *Neurospora* catabolite dehydroquinase enzyme (encoded by the *Neurospora qa-2*[+]). Later Giles and his group (Case *et al.*, 1979) used this chimeric plasmid to transform *Neurospora qa-2*[-] mutant to *qa-2*[+] following the methods of yeast transformation (Hinnen *et al.*, 1978). Giles and co-workers have used the *Neurospora qa* gene cluster extensively for the characterization of a transformation system in this filamentous fungus. This genetic region of *Neurospora* has been fully characterized both biochemically and genetically and can be used for the understanding of the genetic control mechanism (Case and Giles, 1975, 1976). The *qa* cluster consists of four genes, *qa-1*, *qa-2*, *qa-3*, and *qa-4*; of these *qa-1* has a regulatory function and encodes a protein that controls the expression of the other three genes, *qa-2*, *qa-3*, and *qa-4*. The different biochemical reactions catalized by the *qa* gene cluster are described below (see Fig. 6).

With the use of a chimeric plasmid containing the *qa-2*[+] region of *Neurospora*, Giles and others provided conclusive evidence for the presence of donor DNA in the transformation of this filamentous fungus (Case *et al.*, 1979). These authors showed the integration of the donor DNA at the homologous as well as heterologous sites. They did not find any evidence for the nonintegrative type of transformants described earlier in *Neurospora* (Mishra and Tatum, 1973; Mishra, 1976) and in yeast (Struhl *et al.*, 1976). The absence of the nonintegrative type of transformants can be explained by the absence of the *Neurospora* origin of replication in the chimeric plasmid used for the transformation of the *Neurospora qa-2* mutation. Using a recombinant plasmid, the Giles group has shown the successful transformation of other mutations in the *Neurospora qa* gene cluster (Schweizer *et al.*, 1981a; Case, 1981). These authors have also discovered the presence of two other genes in the *qa* gene cluster that are under quinic acid control (Patel *et al.*, 1981). The Giles group has recently constructed several cosmid recombinant vectors which include larger segments

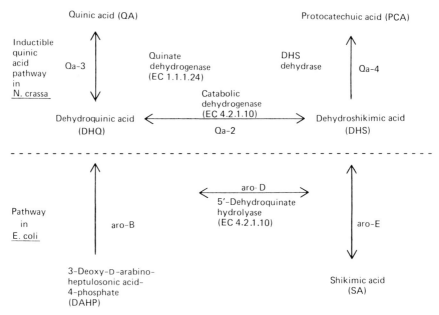

F<small>IG</small>. 6. Quinic acid pathways in *Neurospora* and *E. coli*. (Adapted from Kushner *et al.*, 1977, with permission of authors and publisher.)

(30–40 kb) of the *Neurospora* DNA. The recombinant cosmids have been used for the transformation of *Neurospora* (Schweizer *et al.*, 1981a; Case, 1981). A number of features in the genome of *Neurospora* has been elucidated by the transformation system developed by the Giles group (Case *et al.*, 1979): these include (1) a correlation between the linkage map and the physical map of actual base sequences (encoding genes) in the *qa* gene cluster, (2) discovery of new genes (*qa-x* and *qa-y*) that are under quinic acid control; it is suggested that at least one of these genes (*qa-x* and *qa-y*) may code for a permease (Patel *et al.*, 1981), and (3) a characterization of the promoter of the *qa-2* gene (Alton and Vapnek, 1981). These studies have provided enough basis for the thinking that the use of a *Neurospora* transformation system will be crucial for the isolation and characterization of each component of the *qa* gene cluster, which will in turn clarify the molecular basis of the regulatory control mechanisms in this lower eukaryote.

Molecular cloning of *Neurospora* ribosomal genes has been recently described by two different groups (Free *et al.*, 1979; Cox and Penden, 1980); these groups of workers have used pBR322 or phage to construct a genomic library of *Neurospora,* and the *E. coli* colonies harboring recombinant plasmids with *Neurospora* rDNA were identified by the method of a colony hybridization. The data obtained by these authors prove clustered organization of the ribosomal gene in *Neurospora*; this is in agreement with the earlier chromosomal localization of ribosomal genes on the tip of the left arm of the second *Neurospora* chromosome. Such an organization of the *Neurospora* ribosomal genes is supported by other recent studies (Krumlauf and Marzluf, 1979) and is comparable to the organization of yeast ribosomal genes (Cox and Penden, 1980). A list of *Neurospora* genes cloned by the method of transformation is listed in Table 2. The different kinds of vectors used for the molecular cloning of *Neurospora* genes are included in Table 3. A *Neurospora* genomic library has been recently constructed in our laboratory (Schablik *et al.,* 1983) and elsewhere (Schechtman and Yanofsky, 1983).

B. Yeast

Hinnen *et al.* (1978) first described the transformation of yeast by a recombinant plasmid carrying the yeast *leu-2*$^+$ gene, which was earlier cloned in *E. coli* and was found to complement the *E. coli leuB* mutation (Ratzkin and Carbon, 1977; Hicks and Fink, 1977). Soon a number of yeast genes were cloned in *E. coli* and some of them were found to complement corresponding mutations in *E. coli.* The yeast genes that have been cloned either in *E. coli* or in yeast are listed in Table 2. A variety of hybrid vectors have been used during the transformation of yeast; these are described in Table 3. The general features of these vectors are that they include (1) an *E. coli* plasmid such as *ColE* or its derivative pBR322 that allows its replication in *E. coli*; (2) a yeast origin of replication (called *ars*) or a part of the yeast endogenous plasmid (i.e., the 2-μm plasmid), which also allows the replication of the chimeric plasmid in yeast cell, and (3) a yeast chromosomal segment carrying a selectable marker such as *leu-2*$^+$ and/or *ura-3*$^+$ that facilitates ready detection of the transformants (harboring composite plasmid) on minimal medium when *leu-2*$^-$ or *ura-3*$^-$ cells are used as recipients during transformation. Such a vector can be made to act as a naturally occurring yeast chromosome by introducing a particular centromeric region into the vector (Hsio and Carbon, 1981; Clarke and Carbon, 1980a). At present the centromeres III, IV, and XI have been introduced into the composite yeast plasmid. Such a vector carrying the yeast centromeric region has been called a minichromosome; it

TABLE 3

Characteristics of Some Cloning Vectors of Fungi

Kinds of vectors	Designation of vectors	Components and other characteristics (and references)
		Yeast
Hybrid plasmid	pYE Leu-10	ColE1, yeast chromosome III DNA segment containing $leu\text{-}2^+$ locus (β-iso-propylmalate dehydrogenase), complements E. coli leuB and yeast $leu\text{-}2^-$ mutations (Ratzkin and Carbon, 1977; Hinnen et al., 1978)
	pYE Leu-11 -12 -17	ColE1 and yeast chromosomal DNA segments, specific suppressor of E. coli leu-B6 mutation (Ratzkin and Carbon, 1977)
	pYE his-3	ColE1 and yeast chromosomal DNA fragment containing $his\text{-}3^+$ (im-idazoleglycerol-phosphate dehydrogenase), complements E. coli hisB and yeast his-3 mutation (Ratzkin and Carbon, 1977; Hinnen et al., 1979)
	pYE arg-1 -2 -3	ColE1 and yeast chromosomal fragment containing arg-4 gene (arginosuccinate lyase, EC 4.3.2.1), which complements E. coli argH mutations (Clarke and Carbon, 1978)
	pYE his-4 (pst-1)	pBR313 containing entire $his\text{-}4^+$ region of yeast (a 20-kb Bam-HI fragment of yeast chromosomal DNA) (Hinnen et al., 1979; Fink et al., 1978)
	YIP 3000	pBR313 with 24-kb BamHI fragment of yeast chromosomal DNA (Jackson and Fink, 1981)
	pYE57E2	ColE1 containing yeast pgk (phosphoglycerate kinase) region (13 kb in length) (Hitzeman et al., 1980)
	pgap 49 pgap 492 pgap 63	pSF2124 with yeast glyceraldehyde-3-phosphate dehydrogenase locus (Holland and Holland, 1980)
		pSF2124 with yeast glyceraldehyde-3-phosphate dehydrogenase locus (Holland and Holland, 1980)
	pJD102	pBR322 and yeast chromosomal DNA segment (4.7 kb) containing galac-tokinase gene (Citron et al., 1979)
	pBR322-6 -8	pBR322 and 1.8-kb HindIII yeast chromosomal DNA containing $ura\text{-}3^+$ gene (coding for OMP decarboxylase) (Chevallier et al., 1980; Bach et al., 1979)

148

Plasmid	Description
PTY-39 G 18	pCR1, yeast 2-μm plasmid DNA and *ura-3+* gene (Gerbaud et al., 1979)
pRB-45	pBR322, yeast *leu-2+* gene and *E. coli* β-galactosidase gene fused to yeast *ura-3+* gene (Rose et al., 1981)
pLG669-z	pBR322 (pYcyc1) with *E. coli lac-2* gene fused to yeast *cyc-1* gene (Guarente and Ptashne, 1981)
pJDB-36 -41 -71 -45 -219	pMBJ, 2-μm yeast plasmid and yeast chromosomal DNA containing *leu-2+* gene (Beggs, 1978) (autonomously replicating)
YEp13	pBR322, yeast *leu-2+* gene, yeast 2-μm plasmid and yeast *Can1+* gene (coding arginine permease) (Broach et al., 1979)
pMP78-1	pBR325 and yeast 2-μm DNA (capable of synthesizing active β-lactamase gene in yeast cells) (Roggenkamp et al., 1981)
pYTE1	pBR322, yeast 4.1-kb DNA segment containing *SUP-4A* gene and *Eco*RI fragment of yeast 2-μm plasmid (Thomas and James, 1980a)
pYT11-Leu-2	pBR325, yeast 2-μm DNA plasmid and *leu-2* structural gene (Cohen et al., 1980), pBR325 carries gene for chloramphenicol resistance, which is expressed in yeast
pOMP2	pBR322, yeast 2-μm plasmid, yeast *ura-3+* gene, and cDNA copy of chicken ovalbumin mRNA fused to the beginning of *E. coli* β-galactosidase gene (Mercereau-Puijalon et al., 1980)
pYE	pBR322, one yeast 2-μm DNA, and a *Drosophila* genomic DNA fragment that complements yeast *ade-8* mutation (Henikoff et al., 1981)
YRp7 (57kb)	pBR322 with 14-kb yeast chromosomal DNA, which contains yeast *trp-1+* gene and *ars-1* (Struhl et al., 1979; Stinchcomb et al., 1980) autonomously replicating
pYE (arg4)	ColE1 and yeast chromosomal DNA segment (12 kb) with *arg-4+* region and *ars-2* (Hsio and Carbon, 1979)
psz26	pBR322, yeast rDNA *Bgl*IIA fragment, and yeast *leu-2+* gene

(continued)

149

TABLE 3 *(Continued)*

Kinds of vectors	Designation of vectors	Components and other characteristics (and references)
	pszl (8.7 kb)	pBR322, yeast rDNA *Bgl*IIB fragment, and yeast *leu-2*+ gene (Szostak and Wu, 1979), autonomously replicating
	pCH group I	YRp7 plasmid with yeast centromere regions [pCH-4 and pCH-25 contain yeast centromere 3 (CEN-3)] (Clarke and Carbon, 1980a)
	YI plac4 YRp71ac4 pWK220	yRp7 plasmid containing *lac-4*+ (β-galactosidase) gene of *K. lactis* (Dickson, 1980) Yeast shuttle plasmid with *Klebsiella penumoniae nif* gene cluster and yeast *His-4*+ gene (Zamir *et al.*, 1981)
Plasmids with a series of overlapping yeast DNA inserts	pYE11G4	ColE1 and overlapping region of chromosome III containing *leu-2* locus (Chinault and Carbon, 1979)
	pYe (cdc10)	ColE1 containing centromeric region of chromosome III with *cdc-10* locus (Clarke and Carbon, 1980a)
	pYE98 F4T pYE65H3T pYe101C3T pYe(CEN3)11 pYe(CEN3)21 pYE(Cen3)41	Contain overlapping region of chromosome III (right of the centromere 3 and beyond *cdc-10* locus) (Clarke and Carbon, 1980a)
λ phage vector	λgt1-SC401 SC466 SC481	Contain yeast chromosomal DNA *Eco*RI fragments SC481 and contain *gal1*, *gal7*, and *gal10* gene cluster of yeast chromosome DNA *Hind*III fragments (St. John and Davis, 1979)
	λgt-λBCYC	Contains *cyc1* gene (Cameron *et al.*, 1979)
	λgtSC-his-3 λgt-4SC2601his-3	Contain *Eco*RI fragment of yeast chromosomal DNA with *his-3* gene (IGP dehydratase EC 4.2.1.19). Complements *E. coli hisB* mutation (Struhl *et al.*, 1976; Struhl and Davis, 1977)
	λgt-SCS1 λgt-SCB1	Contains *Eco*RI fragments (12.5 kb) of yeast DNA from strain S288C Contains *Eco*RI fragments (14 kb) of yeast DNA from strain B596 (Cameron *et al.*, 1979)

150

Podospora

Vector type	Plasmid	Description	Neurospora DNA insert size (kb)	Restriction	Strain used
Hybrid plasmid vector	psp 17	pBR322 and *Podospora* plasmid inserted in the tetracycline gene (Stahl et al., 1980)			
	pKP482	pBR322 and *Sal*IV fragment of mtDNA of *Podospora* (Kuck et al., 1981)			

Neurospora

Vector type	Plasmid	Description	Neurospora DNA insert size (kb)	Restriction	Strain used
Hybrid plasmid vector	pVK57	pBR322 and a 2.7-kb *Neurospora* chromosomal DNA ($qa\text{-}1^c$ $qa\text{-}2^+$) inserted at the *Hind*III site (Alton et al., 1978)			
Hybrid plasmid vector	pVK88	pBR322 and a 6.8-kb *Neurospora* chromosomal DNA ($qa\text{-}1^c$ $qa\text{-}2^+$) (Alton et al., 1978)			
Cosmid vector		*Neurospora* rDNA fragments inserted in pBR322 (Free et al., 1979)			
		Containing *Neurospora qa* gene(s) (Schweizer et al., 1981a; Case, 1981)			
	pMSK 308		22.2	*Eco*RI	74A ($qa\text{-}1^-$)
	331		36.5	*Eco*RI	74A $qa\text{-}1$
	332		13.0	*Eco*RI	74A $qa\text{-}1^+$
	334		72.4	*Eco*RI	74A $qa\text{-}1^+$
	335		36.5	*Eco*RI	74A $qa\text{-}1^+$
	pMSK 374	Subclone of 335 (*Pst*I)			
	363	Subclone of 335 (*Pst*I)			
	347	Subclone of 331 *Eco*RI			
	364	Subclone of 335 *Hind*III			
	370	Subclone of 335 *Eco*RI			
	375	Subclone of 335 *Bam*HI			
	376	Subclone of 335 *Bam*HI			
	pES 155	Contains *Neurospora qa* region			
λ phage vector	λ641	Contains *Hind*III fragments of *Neurospora* chromosomal DNA which hybridizes to *Xenopus* rDNA (Cox and Penden, 1980)			
		Catcheside has cloned *Neurospora qa-2* gene in λ phage vector (cited in Case, 1981)			
		Sorger has cloned *Neurospora* nitrate-inducible gene in λ charon ua vector (cited by Edwards, 1982)			

behaves stably during mitosis and meiosis, like any of the seventeen naturally occurring chromosomes of the yeast cell. In yeast, a cotransfer of genetic markers has been shown to occur when recipient cells were treated with two separate hybrid plasmids carrying yeast or other genes. A recombination of markers has been shown to occur at high frequency if the plasmids used for cotransformation possess a significant nucleotide sequence homology. The genetic analysis of the yeast transformant reveals the presence of all possible kinds of transformants, as depicted in Fig. 2. The occurrence of different kinds of transformants suggests an episomic behavior of the chimeric plasmid; such episomic behavior of yeast vectors is comparable to that of the *Neurospora* plasmid described earlier (Mishra, 1976; Hughes *et al.*, 1983; Stohl and Lambowitz, 1983; Stohl *et al.*, 1984).

The yeast plasmid, particularly the one carrying the yeast origin of replication (*ars*) or the centromeric region, seems to reside within the nucleus and not in the cytoplasm. This is again comparable to the location of the *Neurospora* plasmid (Mishra and Tatum, 1973; Mishra, 1976; Stohl and Lambowitz, 1983). The non-Mendelian transmission of the yeast plasmids most probably stems from a deficiency in their ability to replicate in yeast cells or from an unequal distribution during cell division.

The degree of instability of the hybrid plasmid in yeast transformants can be readily estimated by determining the ratio of prototrophic colonies present in cultures grown on minimal (selective) medium and on a complete (nonselective) medium (Struhl *et al.*, 1976). The yeast vectors are essentially shuttle vectors and the latter property is not impaired even if the plasmids are transferred back and forth from yeast to *E. coli* or vice versa (Hsio and Carbon, 1979). Sometimes the vector does pick up certain mutations that affects its expression. It has been shown that a plasmid carrying the *ura-3* [+] gene produces at least 30 times as much enzyme in yeast as in *E. coli* (Chevallier *et al.*, 1980).

At present among the different transformation systems (both prokaryotic and eukaryotic), the yeast transformation system is the most thoroughly characterized and has emerged as the major cloning system for the study of several basic or applied problems in molecular genetics.

C. *Aspergillus*

A genomic library of *Aspergillus nidulans* has been constructed, and the clones carrying DNA segments of the abundant class of mRNA present during conidiogenesis have been identified by colony hybridization (Zimmerman et al., 1980).

Methods for gene transfer in *Aspergillus nidulans* have been developed by different groups of workers (Ballance et al., 1983; Tilburn et al., 1983; Haynes et al., 1983; Yelton et al., 1984). These workers have demonstrated the complementation of a defective *Aspergillus* function by a recombinant plasmid carrying either *Aspergillus* or *Neurospora* DNA segments. The genes that have been cloned using the *Aspergillus* transformation system are *trp-C* (phosphoribosyl anthranilate isomerase) by Yelton et al. (1984), *pyr-4* (orotidine-5′-phosphate decarboxylase) by Ballance et al. (1983), and *admS* (acetamidase) by Tilburn et al. (1983). These *Aspergillus* transformants showed mitotic stability, suggesting the integration of the donor DNA segments into the resident *Aspergillus* chromosomes. Such integration of donor DNA into the recipient chromosome was confirmed (Yelton et al., 1984; Ballance et al., 1983) using the methods of Southern blot hybridization (Southern, 1975). However, unlike in yeast (Beggs, 1978) or in *Neurospora* (Stohl and Lambowitz, 1983), attempts to generate a chimeric plasmid capable of autonomous replication in *Aspergillus nidulans* by addition of *Aspergillus* ribosomal repeats or mitochondrial origin of replication have not yet been successful (Tilburn et al., 1983). Development of a gene transfer system in *Aspergillus* would be useful in understanding the molecular basis of a morphogenesis such as conidial development (Yelton et al., 1983; Timberlake and Bernard, 1981; Timberlake, 1980a; Zimmerman et al. 1980). It is of interest to note here that the yeast genes (such as *trp-2, his-3, ura-3* and *Leu-2*) were not expressed in *Neurospora* (Case, 1981; Mishra, unpublished data). However, *Neurospora pyr-4* gene was expressed in yeast (Ballance et al., 1983) and an *Aspergillus* gene has been found to complement yeast OTCase (ornithine carbamoyl transferase) gene (Berse et al., 1983). Also, a *Neurospora* gene (*pyr-4,* orotidine-5′-phosphate decarboxylase) has been found to express in *Aspergillus nidulans* (Ballance et al., 1983). Thus, it appears that among filamentous fungi *Aspergillus* is more readily amenable to transformation by genes from heterologous systems, and the development of a gene transfer system in *Aspergillus* could provide a rapid means for genetic manipulation of several industrial fungi (including *Aspergillus niger*).

D. *Podospora* AND OTHER FUNGI

A transformation system using recombinant plasmids has been described in *Podospora anserina*. Stahl et al. (1980) have isolated and characterized a plasmid from the mitochondria of this fungus that has been shown to cause senescence (or onset of early aging) in this fungus (Esser and Tudzynski, 1980). Tudzynski and others (1980) have used this plasmid or a recombinant plasmid (containing pBR322 and the *Podospora* plasmid DNA) to transform a strain of *Podospora*. The *Podospora* recipient used carried two mutations (*gr, viv*) that prevented senescence in the recipient. The transformants were detected by their senescent phenotype. The presence of the hybrid plasmid in the *Podospora* transformants was shown by the production of β-lactamase encoded in the pBR322 component of the hybrid plasmid. Further, the hybrid plasmid isolated from *Podospora* transformants was recovered by transformation of *E. coli* (Kuck *et al.*, 1981; Stahl *et al.*, 1981).

A high frequency transformation system for the fission yeast *Schizosaccharomyces pombe* has been recently developed (Beach and Nurse, 1981; Sakaguchi and Yamamoto, 1982). These authors have shown that the chimeric plasmid carrying the *ura-1* locus does not multiply as a monomer but assumes a polymeric (10–12-mer) form in yeast. It has been shown that existence of polymeric arrangement is essential for the establishment of high-frequency transformation in fission yeast (Sakaguchi and Yamamoto, 1982).

A transformation system for the slime mold *Dictyostelium discoideum* has been described; this system utilizes a hybrid plasmid with *ars* and a transposon conferring resistance to antibiotic G418 (Peterhirth *et al.*, 1982). It is believed that this vector would be useful in analysis of several problems in developmental genetics.

A genomic bank of *Mucor racemosus* is available (Cihlar and Sypherd, 1980) and a chimeric plasmid containing two *Hin*dIII fragments (3 and 1.7 kb) of the *Mucor* genomic DNA was found to complement a defective *leuB6* allele of *E. coli*; this complementation of a defective bacterial function was apparently by some kind of suppression (Ratzkin and Carbon, 1977); neither fragment alone was functional in this respect. An extensive analysis of this system may reveal the molecular nature of *leuB6* allele of *E. coli*, which is required for the understanding of the similar suppression of *E. coli* mutant genes by yeast DNA (Ratzkin and Carbon, 1977). A gene transfer system for *Mucor* is expected to be developed soon.

A fungal gene encoding cutinase has been cloned and characterized from *Fusarium solani* (Soliday *et al.*, 1984). Because cutinase is involved in the penetration of plants by pathogenic fungi, an understanding of the mechanism of action of cutinase by its structural analysis should provide a molecular basis for understanding phytophathogenesis and a means of protection of plants by fungal attack.

XIII. Discussion

Conclusive evidence for gene transfer in fungi is made available by the methods of molecular genetics (molecular cloning, Southern hybridization, and donor DNA specific gene expression) (Hinnen *et al.*, 1978; Case *et al.*, 1979; Mishra, 1979a,b; Stahl *et al.*, 1981). The methodology of gene transfer provides a powerful tool for molecular probing of the organization and expression of the eukaryotic genome and for directed changes in higher organisms. The development of a versatile vector and of new methods that can manipulate the expression of cloned genes in a heterologous environment will facilitate the introduction of new characters into host organisms. Addition of a centromeric region has been used to stabilize the new genome in a recipient cell. These developments can be utilized for understanding of the basic mechanisms of all living systems, particularly the eukaryotes, such as the process of DNA replication, repair, recombination, and control of gene expression. The data reviewed here indicate that the gene transfer studies in fungi have already started giving answers to many of these questions in molecular biology and genetics of eukaryotes.

A detailed study of *ars* and its use in *in vitro* systems will provide a better understanding of the mechanism of eukaryotic DNA replication and its control. It would be of interest to isolate mammalian *ars* using fungal gene transfer systems. A comparison of *ars* from X chromosomes before and after inactivation may provide an insight into the molecular mechanism of X-chromosome inactivation in mammals. Mouse *ars* has been recently isolated using the yeast transformation system (Beach *et al.*, 1980b); its characterization could provide a better understanding of mammalian chromosomal replication. It would be of interest to investigate the nature of *ars* in mammalian cancer cells using the fungal transformation system in order to probe into the possibility that cancerous transformation in mammals may be accompanied by a change in *ars*.

The gene transfer system can be used to study DNA repair by isolating the genes involved in DNA repair and by identifying their gene product by *in vitro* and *in vivo* translation in a mini- or maxicell (Sancar *et al.*, 1979). The cloning of the *rad-1* and *rad-3* (Higgins *et al.*, 1983a,b) and *rad-52* (R. K. Mortimer, personal communication) genes of yeast as well as of several nuclease (*nuc-1–nuc-7*) genes of *Neurospora* (Mishra and Forsthoefel, 1983) are some of the examples of measures taken to understand DNA repair mechanisms.

A better insight into the mechanism of DNA recombination will stem from the study of recombination between two chimeric plasmids as well as between a plasmid DNA and the chromosomal DNA. Eventually an *in vitro* system will be developed to characterize the different proteins (enzymes) that facilitate DNA recombination *in vivo*. The ability to construct chimeric plasmids of choice and their functional analysis by complementation in fungal transformation can be used to develop systems for the study of mechanisms of recombination *in vitro*. Chimeric plasmids with appropriate markers (such as a characteristic restriction pattern and/or appropriate genetic mutations) can be used *in vitro* to assay for factors (enzymes and proteins such as *recA* protein) that promote recombination between chimeric plasmids; the resulting recombinant plasmid can be analyzed by its new restriction pattern and by transformation assay. This approach would be analogous to the study of factors involved in the replication of the viral genome (Kornberg, 1980; Alberts *et al.*, 1980). It seems almost certain now that the fungal endogenous plasmids or their mutants (most likely the yeast 2-μm plasmid) will be commissioned to probe into the mechanism of eukaryotic DNA replication in a manner analogous to the use of the bacterial viruses (G4, φX174) to probe into the mechanism of prokaryotic DNA replication. Furthermore, the fungal gene transfer system can be used to clone the eukaryotic DNA polymerase genes; it is possible that an extensive characterization of the eukaryotic DNA polymerases will provide a better understanding of the DNA replication mechanisms. It has been recently shown that both *Neurospora* and yeast DNA polymerases are sensitive to aphidicolin (Mishra, 1984; Sugino *et al.*, 1981), a known inhibitor of several eukaryotic DNA polymerases. This fact can be utilized to screen for possible DNA polymerase mutants of fungi by selecting for mutants resistant to the drug. However, there is a major obstacle in this approach since fungi are not sensitive to aphidicolin *in vivo* due to the poor uptake of the drug by the complex fungal cell wall. This problem can be easily overcome first by selecting a drug-permeable mutant (or by the use of *slime* or other cell wall-deficient mutants of *Neurospora*) that can then be used in a second step to select for DNA polymerase mutants by their drug resistance *in vivo*. These mutants should be further examined for the resistance of DNA polymerase *in vitro* to aphidicolin. DNA polymerase mutants of yeast and/or *Neurospora* can then be used for the molecular cloning and extensive characterization of the DNA polymerase genes and enzymes. Such studies in *Neurospora* can be aided by the fact that its DNA polymerase has been characterized (Elassouli and Mishra, 1978, 1981) and aphidicolin-resistant (*aph^r*) mutants with altered DNA polymerase have been isolated (Mishra, 1984).

Genes for the same enzyme from a number of organisms have been cloned; a comparison of their structure (particularly the promoter region) by sequence analysis and expression in a heterologous gene transfer system should provide a better insight into the nature of the regulatory elements involved in the expression of these genes (Buxton and Radford, 1983).

Isolation and characterization of eukaryotic promoters using fungal transformation will provide a full understanding of the control mechanisms involved in eukaryotic gene expression. The fungal gene transfer system can and will be used to analyze the different kinds of repeated DNA sequences found in higher organisms and to probe into their physiological role. Significant progress in this direction has already been made by assigning the function of the transposons to a class of dispersed repeated DNA sequences in yeast. The data discussed in this article suggest a distinct possibility for the isolation and characterization of different parts of a gene [such as promoters, leader sequences, introns, exons, flanking sequences, caps, ribosome-binding sites, and poly(A) sequences]; such analysis will be facilitated by determining their complete nucleotide sequence (Maxam and Gilbert, 1977; Sanger *et al.*, 1977) and by functional analysis of the mutant sequences by complementation in an appropriate transformation system. The progress in this area may culminate in understanding of the total DNA sequences in the chromosomes of yeast as a model system for eukaryotes; such analysis is probable now and can lead into several computer-generated speculations regarding the nature and organization of the higher eukaryotic genome. This newly discovered methodology of fungal gene transfer will allow understanding of the complex processes such as cell division, mating-type differentiation, and other cytodifferentiation (including organelle biogenesis) in fungi and in other higher eukaryotes. A start in this direction has already been made by Osley and Hereford (1982), who have isolated a DNA segment from yeast capable of *ars* function and regulation of cell cycle-specific gene expression.

At the present time, gene transfer by molecular cloning has been characterized in only four fungal species (yeast, *Neurospora, Aspergillus,* and *Podospora*). Attempts are under way to make the gene transfer systems of *Neurospora* and *Podospora* reach the sophistication of the yeast system in terms of the availability of different kinds of vectors and ability to get expression of any gene cloned in yeast (Grant *et al.,* 1984; Stohl and Lambowitz, 1983). Efforts must be directed to develop gene transfer systems for these two filamentous fungi to provide similar capability to clone any eukaryotic gene with its expression.

It would be highly desirable to construct a versatile vector that can shuttle gene in different groups of fungi. However, at the present, this seems difficult for two reasons: first, it would require additions of several *ars*(s) from different fungi unless a particular *ars* is found that readily function in different groups of fungi. Second, most fungi are resistant to antibiotics, therefore other selective markers must be developed for the detection of transformants. It is possible that screening for new antibiotic resistance may solve this problem. Alternatively, resistance to certain antimetabolites (*para*-fluorphenylalanine, methyltryptophan, aminopterin, or inhibitors of cell wall synthesis) or ability to degrade and grow on complex carbohydrates may be developed as possible selectable markers.

It would be further necessary to develop an autonomously replicating vector, perhaps one with a centromere in order to permit it to exist as a minichromosome. The development of a versatile vector for gene transfer in fungi will be very useful in analyzing certain features characteristic of a particular fungus in homologous and heterologous systems. For example, *Neurospora* and *Podospora* do not tolerate even a partial diploidy (since the partial diploids break down readily). This phenomenon can be examined by comparing the mitotic and meiotic behavior of different kinds of artificially constructed minichromosomes (some with homology to *Neurospora* chromosomes and others without any homology). With the use of a versatile vector, some minichromosomes can be introduced into yeast and/or *Aspergillus* and examined for their behavior during mitosis and meiosis. Filamentous fungi also provide systems for the study of nuclear–limited effects (i.e., phenomena of gene expression in a diploid but *not* in a heterokaryon) (Pontecorvo, 1958). Filamentous fungi also provide a system for the study of differentiation and development, which can be analyzed by the new methodology of gene transfer. Study of filamentous fungi can provide a better insight to DNA sequence organization and their role in the control of gene expression (Britten and Kohne, 1968) since fungi in general seem to possess few repeated DNA sequences (Krumlauf and Marzluf, 1979). A study of the mating type and the heterokaryon incompatibility factors in *Neurospora* can provide an insight into the mechanism of interactions at the cellular level that could be reminiscent of a primitive immunological interaction. A molecular analysis of the *Neurospora* clock mutants (Feldman *et al.*, 1979) can provide a better understanding of the basic mechanism underlying the phenomenon of biorhythms or the biological clock. A similar analysis of the *Neurospora* spore-killer mutants (Turner, 1977) is expected to yield a better understanding of the phenomenon of segregation distortion, which has been very well described in higher organisms (Bennett, 1975; Silver and Artzt, 1981), but its molecular analysis in mammals is too complex. Methods of "chromosome walking" (described in Section VIII) can be successfully used for the isolation and molecular cloning of complex genes like the alleles of clock or spore-killer mutants in *Neurospora*.

It would be very useful to develop gene transfer systems for *Aspergillus*, *Sordaria*, *Ustilago,* and *Schizophyllum* because of their well-established genetics. The development of a gene transfer system in these fungi would be useful in understanding the mechanism of genetic recombination and its control, particularly in view of the fact that several appropriate genetic systems for the study of recombination are already available in fungi (Catcheside, 1974; Stahl, 1980). It is of interest to mention that such studies *via* gene transfer have already suggested distinct mechanisms for gene conversion during mitosis and meiosis in yeast. A study of the phytopathogenic fungi would provide an understanding of the mechanism of the host–pathogen relationship at the molecular level; such an understanding can lead to a distinct possibility of genetic engineering of plants for disease control.

Genetic engineering of fungi can also lead to the production of better or newer antibiotics, favorable changes in the nature of soil for agriculture, biodegradation of new chemical products, and better management of our environment. In nature, fungi are the only source of enzymes responsible for the degradation of major wood components such as lignin and hemicellulose (Griffin, 1981). Molecular cloning of these important fungal genes (encoding enzymes for digestion of lignin and hemicellulose) is crucial for the development of biomass conversion as a supplementary energy source. These are some of the possibilities that new methodology of gene transfer promises for us in the future (Erickson, 1978; Johnson and Burnett, 1978; Vournakis and Elander, 1983; Abelson, 1983). Therefore, it is important to develop a newer and better gene transfer system in different groups of fungi by developing better vector systems and by examining their genetic behavior. This can be facilitated to a great extent by developing strains that are readily transformable since studies with yeast suggest that transformability itself could be genetically controlled. These studies will not only keep biologists busy enjoying a specific scientific pursuit but will indeed provide a better understanding of eukaryotic genetic organization and expression that can and will form the basis for their possible genetic manipulation.

ACKNOWLEDGMENTS

The author wishes to thank Drs. Alton, Case, Clarke, Giles, and Tschumper for making available preprints of their publications; and Drs. Karl Esser, Rollin D. Hotchkiss, George Marzluf, Gabor Szabo, Marcella Schablik, and Paul Tudzynski for their critical reading of the manuscript. The author also wishes to thank Drs. E. W. Caspari, R. D. Hotchkiss, M. Gabor Hotchkiss, and S. F. H. Threlkeld for their encouragement and Ms. Debra Chavis for her utmost care and cooperation during the typing of the manuscript. The data presented from my laboratory were supported by a grant from the U.S. Department of Energy (Contract No. EP78–5–09–1071).

REFERENCES

Abelson, P. H. (1983). Biotechnology: An overview. *Science* **219,** 611–613.
Alberts, B. M., Barry, J., Bedinger, P., Burke, R. C., Hibner, U., Liu, C. C., and Sheridan, R. (1980). Studies of replication mechanism with the T4 bacteriophage *in vitro* system. *In* "ICN-UCLA Symposia on Molecular and Cellular Biology" (B. M. Alberts and C. F. Fox, (eds.), Vol. 19, pp. 449–473. Academic Press, New York.
Alton, N. K., and Vapnek, D. (1981). Characterization of the promoter and identification of the 5′ end of the catabolic dehydroquinase mRNA from *Neurospora crassa. In* "Promoters: Structure and Function" (M. J. Chamberlain and R. L. Rodriquez, eds.), pp. 345–362. Praeger, New York.
Alton, N., Hautala, A., Giles, H., Kushner, R., and Vapnek, D. (1978). Transcription and translation in *E. coli* of hybrid plasmid containing the catabolic dehydroquinase gene from *Neurospora crassa. Gene* **4,** 241–259.
Alwine, J. C., Kemp, D. S., and Stark, D. S. (1977). Method for detection of specific RNA in agarose gels by transfer to diazobenzyloxymethyl paper and hybridization with DNA probes. *Proc. Natl. Acad. Sci. U.S.A.* **74,** 5350–5355.
Aradi, J., Schablik, M., Zsindely, A., Csongor, J., and Szabo, G. (1978). Factors influencing the uptake of DNA by *Neurospora crassa. Acta Biochim. Biophys. Acad. Sci. Hung.* **13,** 259–267.

Arends, H., and Sebald, W. (1984). Nucleotide sequence of the cloned messenger RNA and gene of the ADP-ATP carrier from *Neurospora crassa*. *EMBO J.* **3**, 377–382.

Avery, O. T., McLeod, C. M., and McCarty, M. (1944). Studies on the nature of the substance inducing transformation of pneumococcal types. Induction of transformation by a deoxyribonucleic acid fraction isolated from pneumococcus type III. *J. Exp. Med.* **79**, 137–158.

Bach, M., Lacronte, F., and Botstein, D. (1979). Evidence for transcriptional regulation of orotidine-5'-phosphate decaryboxylase in yeast by hybridization of mRNA to the yeast structural gene cloned in *Escherichia coli. Proc. Natl. Acad. Sci. U.S.A.* **76**, 386–390.

Ballance, D. J., Buxton, F. P., and Turner, G. (1983). Transformation of *Aspergillus nidulans* by the orotidine-3'-phosphate decarboxylase gene of *Neurospora crassa. Biochem. Biophys. Res. Commun.* **112**, 284–289.

Ballou, L., Cohen, R. E., and Ballou, C. E. (1980). *Saccharomyces cerevisiae* mutants that make mannoprotein with a truncated carbohydrate outer chain. *J. Biol. Chem.* **255**, 5986–5991.

Barry, E. F. (1972). Meiotic chromosome behavior of an inverted insertional translocation in *Neurospora. Genetics* **71**, 53–62.

Beach, D., and Nurse, P. (1981). High frequency of transformation of the fission yeast *Schizosaccharomyces pombe. Nature (London)* **290**, 140–142.

Beach, D., Piper, M., and Shall, S. (1980a). Isolation of chromosomal origins of replication in yeast. *Nature (London)* **284**, 185–187.

Beach, D., Piper, M., and Shall, S. (1980b). Isolation of eukaryote DNA replication origins. *In* "ICN-UCLA Symposia on Molecular and Cellular Biology" (B. Alberts, C. F. Fox, and F. J. Stusser, eds.), Vol. 19, pp. 358–368. Academic Press, New York.

Beadle, G. W., and Tatum, E. L. (1941). Genetic control of biochemical reactions in *Neurospora. Proc. Natl. Acad. Sci. U.S.A.* **27**, 499–506.

Beckman, J. S., Johnson, P. F., and Abelson, J. (1977). Cloning of yeast tRNA genes in *E. coli. Science* **196**, 205–208.

Beggs, D. (1978). Transformation of yeast by a replicating hybrid plasmid. *Nature (London)* **275**, 104–108.

Beggs, J. D., VandenBerg, J., VanDoyen, A., and Weissman, C. (1980). Abnormal expression of chromosomal rabbit β-globin gene in *Saccharomyces cerevisiae. Nature (London)* **283**, 835–840.

Bennett, D. (1975). The T-locus of the mouse *Cell* **6**, 441–454.

Benoist, C., and Chambon, P. (1981). Early promoter of simian virus 40. *Nature (London)* **29**, 306–310.

Benton, W. D., and Davis, R. W. (1977). Screening λgt recombinant clones by hybridization to single plaque *in situ. Science* **196**, 180–182.

Berse, B., Dmochowska, A., Skrzypek, M., Weglenski, P. Bates, M. A., and Weiss, R. L. (1983). Cloning and characterization of ornithine carbomyltransferase gene from *Aspergillus nidulans. Gene* **25**, 109–117.

Bicknell, J. N., and Douglas, H. C. (1970). Nucleic acid homologous among species of *Saccharomyces. J. Bacteriol.* **101**, 505–512.

Bloom, K. S., and Carbon, J. (1982). Yeast centromere DNA is an unique and highly ordered structure in chromosomes and small circular minichromosomes. *Cell* **29**, 305–317.

Bolivar, F. (1978). Molecular cloning vectors: Derivatives of plasmid pBR322. *In* "Genet ic Engineering" (H. W. Boyer and S. Nicosia, eds.), pp. 59–63. Elsevier/North-Holland, Amsterdam.

Bolivar, F., and Backman, K. (1979). Plasmids of *Escherichia coli* as cloning vectors. *In* "Methods in Enzymology" (R. Wu, ed.), Vol. 68, pp. 245–267. Academic Press, New York.

Bolivar, R., Rodriquez, R. L., Greene, P. J., Betlach, M. C., Heyneker, H., and Boyer, H. W. (1977). Constitution and characterization of new cloning vehicles. II. A multipurpose cloning system. *Gene* **2**, 95–113.

Botstein, D., White, R. L., Skolnick, M., and Davis, R. W. (1980). Construction of a genetic linkage map in man using restriction fragment length polymorphisms. *Am. J. Human Genet.* **32**, 314–331.

Britten, R. J., and Davidson, E. (1971). Repetitive and nonrepetitive DNA sequences and a speculation on the origin of evolutionary novelty. *Q. Rev. Biol.* **46**, 111–133.

Britten, R. J., and Kohne, D. E. (1968). Repeated sequences in DNA. *Science* **161**, 529–540.

Broach, J. R., and Hartley, J. (1980). Replication functions associated with the yeast plasmid, 2μ circle. *In* "ICN-UCLA Symposia in Molecular and Cellular Biology" (B. Alberts, C. F. Fox, and F. J. Stusser, eds.), Vol. 19, pp. 387–388. Academic Press, New York.

Broach, J. R., Strathern, J. N., and Hicks, J. B. (1979). Transformation in yeast: Development of a hybrid cloning vector and isolation of the CAN gene. *Gene* **8**, 121–133.

Broom, S., and Gilbert, W. (1978). Immunological screening method to detect specific translation products. *Proc. Natl. Acad. Sci. U.S.A.* **75**, 2746–2749.

Burke, J. M., Breitenberger, C., Heckman, J. E., Dujon, B., and Rajbhandary, U. L. (1984). Cytochrome *b* gene of *Neurospora crassa* mitochondria. Partial sequence and location of introns at sites different from those in *Saccharomyces cerevisiae* and *Aspergillus nidulans. J. Biol. Chem.* **259**, 504–511.

Buxton, F. P., and Radford, A. (1983a). Cloning of the structural gene for orotidine 5′ phosphate carboxylase by expression in *Escherichia coli. Mol. Gen. Genet.* **190**, 403–405.

Buxton, P. P., and Radford, A. (1983b). Cloning of the *Pyr-4* gene of *Neurspora. Heredity* **50**, 210–211.

Cairns, J. (1963). The chromosomes of *Escherichia coli. Cold Spring Harbor Symp. Quant. Biol.* **28**, 43–46.

Cameron, J. R., Philippsen, P., and Davis, R. W. (1977). Analysis of chromosomal integration and deletions of yeast. *Plasmids* **5**, 1429–1448.

Cameron, J. R., Loh, E. Y., and Davis, R. W. (1979). Evidence for transcription of dispersed repetitive DNA families in yeast. *Cell* **16**, 739–751

Cappelo, J., Zuker, C., and Lodish, H. F. (1984). Repetitive Dictyostelium heat shock promoter functions in *Saccharomyces cerevisiae. Mol. Cell. Biol.* **4**, 591–598.

Carbon, J., Ratzkin, B., Clarke, L., and Richardson, D. (1977). The expression of cloned eukaryotic DNA in prokaryotes. *Brookhaven Symp. Biol.* **29**, 277–296.

Carlson, M., and Botstein, D. (1982). Two differentially regulated mRNAs with different 5′ ends encode secreted and intracellular forms of yeast invertase. *Cell* **28**, 145–154.

Case, M. E. (1981). Transformation of *Neurospora crassa* utilizing recombinant plasmid DNA. *In* "Genetic Engineering of Microorganisms for Chemicals" (A. Hollander, R. D. DeMoss, S. Kaplan, J. Konisky, D. Savage, and R. S. Wolfe, eds.), pp. 87–100. Plenum, New York.

Case, M. E., and Giles, N. H. (1975). Genetic evidence on the organization and action of the qa-1 gene product: a protein regulating the induction of three enzymes in quinate catobolism in *Neurospora crassa*. *Proc. Natl. Acad. Sci. U.S.A.* **72**, 553–557.

Case, M. E., and Giles, N. H. (1976). Gene order in the qa gene cluster of *Neurospora crassa*. *Mol. Gen. Genet.* **147**, 83–89.

Case, E., Schweizer, Kushner, R., and Giles, N. H. (1979). Efficient transformation of *Neurospora crassa* by utilizing hybrid plasmid DNA. *Proc. Natl. Acad. Sci. U.S.A.* **76**, 5259–5263.

Catcheside, D. G. (1974). Fungal Genetics. *Annu. Rev. Genet.* **8**, 279–300.

Chan, C. S. M., and Tye, B. (1980). Autonomously replicating sequences in *Saccharomyces cerevisiae*. *Proc. Natl. Acad. Sci. U.S.A.* **77**, 6329–6333.

Chan, C. S. M., and Tye, B. (1983). Organization of DNA sequences and replication origins at yeast centromeres. *Cell* **33**, 563–573.

Chevallier, M., and Aigel, M. (1979). Qualitative detection of penicillase produced by yeast strains carrying chimeric yeast–coli–plasmids. *FEBS Lett.* **180**, 179–180.

Chevallier, M., Bloch, J., and Lacrente, F. (1980). Transcriptional and translational expression of a chimeric bacteria-yeast plasmid in yeasts. *Gene* **11**, 11–19.

Chinault, A., and Carbon, J. (1979). Overlap hybridization screening: Isolation and characterization of overlapping DNA fragments surrounding the *leu2* gene on yeast chromosome III. *Gene* **5**, 11–126.

Cihlar, R. L., and Sypherd, P. S. (1980). The organization of the ribosomal RNA genes in the fungus *Mucor racemosus*. *Nucleic Acids Res.* **8**, 793–804.

Cihlar, R. L., and Sypherd, P. S. (1982). Complementation of the LeuB6 allele of *E. coli* by cloned DNA from *Mucor racemosus*. *J. Bacteriol.* **151**, 521–523.

Ciriacy, M., and Williamson, V. M. (1981). Analysis affecting ty-mediated gene expression in yeast. *Mol. Gen. Genet.* **182**, 159–163.

Citron, B. A., Feiss, M., and Donelson, J. E. (1979). Expression of the yeast galactokinase gene in *Escherichia coli*. *Gene* **6**, 251–264.

Clarke, L., and Carbon, J. (1975). Biochemical construction and selection of hybrid plasmids containing specific segments of the *Escherichia coli* genome. *Proc. Natl. Acad. Sci. U.S.A.* **72**, 4261–4365.

Clarke, L., and Carbon, J. (1978). Functional expression of cloned yeast DNA in *Escherichia coli*: Specific complementation of argininosuccinate lyase *(argH)*. *J. Mol. Biol.* **120**, 517–532.

Clarke, L., and Carbon, J. (1980a). Isolation of a yeast centromere and construction of functional small circular chromosomes. *Nature (London)* **287**, 504–509.

Clarke, L., and Carbon, J. (1980b). Isolation of the centromere linked CDC10 gene by complementation in yeast. *Proc. Natl. Acad. Sci. U.S.A.* **77**, 2173–2177.

Clarke, L., and Carbon, J. (1983). Genomic substitution of centromeres in *Saccharomyces cerevisiae*. *Nature (London)* **305**, 23–28.

Clarke, L., Hitzeman, R., and Carbon, J. (1979). Selection of specific clones from colony banks by screening with radioactive antibody. *In* "Methods in Enzymology" (R. Wu, ed.), Vol. 68, pp. 436–442. Academic Press, New York.

Cohen, J. D., Eccleshall, T. R., Needleman, R. B., Federoff, H., Buchferer, B. A., and Marmur, J. (1980). Functional expression in yeast at the *Escherichia coli* plasmid gene coding for chloramphenicol acetyltransferase. *Proc. Natl. Acad. Sci. U.S.A.* **77**, 1078–1082.

Cohen, S. N., and Chang, A. C. Y. (1974). A method for selective cloning of eukaryotic

DNA fragments in *Escherichia coli* by repeated transformation. *Mol. Gen. Genet.* **134**, 133–141.

Cohen, S. N., Chang, A. C. Y., and Hsu, L. (1972). Nonchromosomal antibiotic resistance in bacteria: genetic transformation of *Escherichia coli* by R-factor DNA. *Proc. Natl. Acad. Sci. U.S.A.* **69**, 2110–2114.

Collins, J., and Hohn, B. (1978). Cosmid: A type of plasmid gene cloning vector that is packageable *in vitro* in bacteriophage heads. *Proc. Natl. Acad. Sci. U.S.A.* **75**, 4242–4246.

Collins, R. A., Stuhl, L. L., Cole, M. D., and Lambowitz, A. M. (1981). Characterization of a novel plasmid DNA found in mitochondria of *N. crassa. Cell* **24**, 443–452.

Corden, J., Wasylyk, B., Buchwalder, A., Sassone-Corsi, P., Kedinger, C., and Chambon, P. (1980). Promoter sequences of eukaryotic protein-coding genes. *Science* **209**, 1406–1414.

Cox, R. A., and Penden, K. (1980). A study of the organization of the ribosomal ribonucleic acid gene cluster of *Neurospora crassa* by means of restriction endonuclease analysis and cloning in bacteriophage lambda. *Mol. Gen. Genet.* **174**, 17–24.

Crabeel, M., Messenguy, F., Lacroute, F., and Glausdorff, N. (1981). Cloning arg-3, the gene for ornithine transcarbamoyl transferase from *Saccharomyces cerevisiae*: Expression in *E. coli* requires secondary mutations; production of plasmid β-lactamase in yeast. *Proc. Natl. Acad. Sci. U.S.A.* **78**, 5026–5030.

Cryer, D. F., Eccleshall, R., and Marmur, J. (1975). Isolation of yeast DNA. *Methods Cell Biol.* **12**, 39–44.

Dani, G. M., and Zakian, V. A. (1983). Mitotic and meiotic stability of linear plasmids in yeast. *Proc. Natl. Acad. Sci. U.S.A.* **80**, 3406–3410.

Davis, R., St. John, T., Struhl, K., Stinchcomb, D., Scherer, S., and McDonell, M. (1979). Structural and functional analysis of the *His3* gene and galactose inducible sequences in yeast. *In* "ICN-UCLA Symposia on Molecular and Cellular Biology" (R. Axel, T. Maniatis, and C. R. Fox, eds.), Vol. 14, pp. 51–55.

Dhawale, S. S., Paietta, J. V., and Marzluf, G. A. (1984). A new rapid and efficient transformation procedure for *Neurospora. Curr Genet.* **8**, 77–79.

Dickson, R. C. (1980). Expression of a foreign eukaryotic gene *Saccharomyces cerevisiae*: β-galactosidase from *Kluyveromyces lactis. Gene* **10**, 347–356.

Edwards, D. (1982). Session on molecular genetics. *Neurospora Newslett.* **29**, 7–8.

Elander, P. R. (1980). New genetic approach to industrially important fungi. *Biotechnol. Bioeng.* **22**, Suppl. 1, 49–61.

Elassouli, S. M., and Mishra, N. C. (1978). *Neurospora* DNA polymerases. *Naturwissenschaften* **65**, 63–64.

Elassouli, S. M., and Mishra, N. C. (1981). Properties of *Neurospora* DNA polymerases. *FEMS Microbiol.* **13**, 181–185.

Emr, S. D., Schekman, R., Flessel, M. C., and Thorner, J. (1983). An MFα-1-suc-2 (λ factor-invertase) gene fusion for the study of protein localization and gene expression in yeast. *Proc. Natl. Acad. Sci. U.S.A.* **80**, 7080–7084.

Erickson, R. J. (1978). The potential of genetic engineering technologies in the production of industrially important enzymes. *In* "Genetic Engineering" (H. W. Boyer and S. Nicosia, eds.), pp. 209–212. Elsevier/North Holland, Amsterdam.

Erlich, H. A., Cohen, S. N., and McDevitt, H. O. (1979). Immunological detection and characterization of products translated from cloned DNA fragments. *In* "Methods of Enzymology" (R. Wu, ed.), Vol. 68, pp. 443–454. Academic Press, New York.

Errede, B., Cardillo, T. S., Sherman, F., Dubois, E., Deschamps, J., and Waime, J. M.

(1980). Mating signals control expression of mutations resulting from insertion of a transposable repetitive element adjacent to diverse yeast genes. *Cell* **25**, 437–436.

Esser, K., and Kuenen, R. (1967). "Genetics of Fungi." Springer Verlag, Berlin and New York.

Esser, K., and Tudzynski, P. (1980). "Senescence in Fungi: Senescence in Plants," pp. 67–84. CRC Press, Boca Raton, Florida.

Esser, K., Tudzynski, P., Stahl, U., and Kuck, U. (1980). A model to explain senescence in the filamentous fungi *Podospora anserina* by the action of plasmid-like DNA. *Mol. Gen. Genet.* **178**, 213–216.

Esser, K., Kuck, U., Stahl, U., and Tudzynski, P. (1983). Cloning vectors of mitochondrial origin for eukaryotes, a new concept in genetic engineering. *Curr. Genet.* **7**, 239–243.

Falco, S. C., and Botstein, D. (1983). A rapid chromosome-mapping method for cloned DNA fragments of yeast DNA. *Genetics* **105**, 857–872.

Falco, S. C., Rose, M., and Botstein, D. (1983). Homologous recombination between episomal plasmids and chromosomes in yeast. *Genetics* **105**, 843–856.

Farabaugh, P. J., and Fink, G. R. (1980). Insertion of the eukaryotic transposable element *ty-1* creates a 5-base pair duplication. *Nature (London)* **286**, 352–356.

Faye, G., Leung, D. W., Tatchell, K., Hall, B. D., and Smith, M. (1981). Deletion mapping of sequences essential for *in vivo* transcription of the iso-1-cytochrome C gene. *Proc. Natl Acad. Sci. U.S.A.* **78**, 2258–2262.

Feher, Z., Schablik, M., and Szabo, G. (1979). Hybridization of DNA from *Neurospora crassa* strains may indicate base sequence alterations as a conseqeunce of genetic transformation. *Acta Biol. Acad. Sci. Hung.* **30**, 387–392.

Feldman, J. F., Gardner, G., and Denison, R. (1979). Genetic analysis of circadian clock of *Neurospora. In* "Biological Rhythms and Their Central Mechanisms" (M. Suda, O. Hayashi, and H. Nakagawa, eds.), pp. 57–66. Elsevier/North Holland, Amsterdam.

Fincham, J. R. S., Day, P. R., and Radford, A. (1978). "Fungal Genetics," 4th Ed. Blackwell, Oxford.

Fink, G. R., Hicks, J. B., and Hinnen, A. (1978). Yeast as a host for hybrid DNA. *In* "Genetic Engineering" (H. W. Boyer and S. Nicosia, eds.), pp. 163–171 Elsevier/North-Holland, Amsterdam.

Finnegan, D. J., Rubin, G. M., Young, M. W., and Hogness, D. S. (1977). Repeated gene families in *Drosophila melanogaster. Cold Spring Harbor Symp. Quant. Biol.* **42**, 1953–1063.

Free, S. J., Ric, P. W., and Metzenberg, R. W. (1979). Arrangement of the genes coding for ribosomal ribonucleic acids in *Neurospora crassa. J. Bacteriol.* **137**, 1219–1225.

Fried, H. M., and Warner, J. R. (1981). Cloning of yeast gene for trichodermin resistance and ribosomal protein L3. *Proc. Natl. Acad. Sci. U.S.A.* **78**, 238–242.

Gafner, J., and Philippsen, P. (1980). The yeast transposon Tyl generates duplications of target DNA on insertion. *Nature (London)* **286**, 414–418.

Gallwitz, D., and Sures, I. (1980). Structure of a split yeast gene: Complete nucleotide sequence of the actin gene in *Saccharomyces cerevisiae. Proc. Natl. Acad. Sci. U.S.A.* **77**, 2546–2550.

Geever, R. F., Case, M. E., Tyler, B. M., Buxton, F., and Giles, N. H. (1983). Point mutations and DNA rearrangements 5′ to the inducible qa-2 gene of *Neurospora* allow activator protein independent transcription. *Proc. Natl. Acad. Sci. U.S.A.* **80**, 7298–7302.

Gerbaud, C., Fournier, P., Blanc, H., Aigle, M., Heslot, H., and Guerineua, M. (1979).

High frequency of yeast transformation by plasmids carrying part or entire 2-μm yeast plasmids. *Gene* **5**, 233–253.

Giles, N. H. (1951). Studies on the mechanism of reversion in biochemical mutants of *Neurospora crassa*. *Cold Spring Harbor Symp. Quant. Biol.* **16**, 283–313.

Giles, N. H. (1982). Summary of session on regulation. *Neurospora Newslett.* **29**, 9.

Gilmore, R. A. (1967). Super-suppressors in *Saccharomyces cerevisiae*. *Genetics* **56**, 637–658.

Glover, D. M. (1980). Genetic engineering cloning DNA. *In* "Outline Studies in Biology" (W. J. Brammer and M. Edinin, eds.), pp. 1–79. Chapman & Hall, New York.

Gorbstein, C. (1980). "Double Image of Double Helix." Freeman, San Francisco.

Goursot, R., Goze, A., Niaudet, B., and Ehrlich, S. D. (1982). Plasmids from *Staphylococcus aureus* replicate in yeast. *Nature (London)* **298**, 488–490.

Grant, D. M., Lambowitz, A. M., Rombosek, J. A., and Kinsey, J. A. (1984). Transformation of Neurospora with recombinant plasmids containing the cloned glutamate dehydrogenase (*am*) gene. Evidence for autonomous replication. *Mol. Cell. Biol.* **4**, 2041–2051.

Grigg, G. W. (1958). Competitive suppression and the detection of mutations in microbial populations. *Aust. J. Biol. Sci.* **11**, 69–84.

Griffin, D. H. (1981). "Fungal Physiology." Wiley, New York.

Grimaldi, G., Guardiola, J., and Martini, G. (1978). Fungal extrachromosomal DNA and its maintenance and expression in *E. coli* K-12. *Trends Biochem. Sci.* **3**, 248–253.

Grunstein, M., and Hogness, D. S. (1975). Colony hybridization: a method for the isolation of cloned DNAs that contain a specific gene. *Proc. Natl. Acad. Sci. U.S.A.* **72**, 3961–3964.

Guardiola, J., Grimaldi, G., Costantio, P., Micheli, G., and Cervone, F. (1982). Loss of nitrofuran resistance in *Fusarium oxysporum* is correlated with loss of a 46.7 kb circular DNA molecule. *J. Gen. Microbiol.* **128**, 2235–2242.

Guarente, L., and Ptashne, M. (1981). Fusion of *Escherichia coli lacZ* to the cytochrome C gene of *Saccharomyces cerevisiae*. *Proc. Natl. Acad. Sci. U.S.A.* **78**, 2199–2203.

Guarente, L., Roberts, T. M., and Ptashne, M. (1980). A technique for expressing eukaryotic genes in bacteria. *Science* **209**, 1428–1430.

Gunge, N. (1983). Yeast DNA plasmids. *Ann. Rev. Microbiol.* **37**(253), 76.

Harashima, S., Sidhu, R. S., Toh-e, A., and Oshima, Y. (1983). Cloning of *His5* gene of *Saccharomyces cerevisiae* by yeast transformation. *Gene* **16**, 335–431.

Hautala, J. A., Conner, B. H., Jacobson, J. W., Patel, G. L., and Giles, N. H. (1977). Isolation and characterization of nuclei from *Neurospora crassa*. *J. Bacteriol.* **130**, 704–713.

Hawthorne, D. C., and Leupold, U. (1974). Suppressor mutation in yeast. *Curr. Top. Microbiol. Immunol.* **64**, 1–47.

Haynes, M. J., Corrick, C. M., and King, J. A. (1983). Isolation of genomic clones containing the *amds* gene of *Asperigillus nidulans* and their use in the analysis of structure and regulatory mutations. *Mol. Cell. Biol.* **3**, 1430–1439.

Henikoff, S., and Furlong, C. E. (1983). Sequence of a Drosophila DNA segment that functions in *Saccharomyces corivisiae* and its regulation by a yeast promoter. *Nucleic Acids Res.* **11**, 789–800.

Henikoff, S., Tachell, B., Hall, D., and Nasmyth, A. (1981). Isolation of a gene from *Drosophila* by complementation in yeast. *Nature (London)* **289**, 33–37.

Hereford, L. M., and Rosbash, M. (1977). Number and distribution of polyadenylated RNA sequences in yeast. *Cell* **10**, 453–462.

Heumann, J. M. (1976). A model for replication of the ends of linear chromosomes. *Nucleic Acids Res.* **3**, 3167–3171.

Hicks, J., and Fink, G. R. (1977). Identification of chromosomal location of yeast DNA from hybrid plasmid pYeleu10. *Nature (London)* **269**, 265–267.

Hicks, J. B., Hinnen, A., and Fink, G. R. (1978). Properties of yeast transformation. *Cold Spring Harbor Symp. Quant. Biol.* **43**, 1305–1313.

Higgins, D. R., Prakash, S., Reynolds, P., and Prakash, L. (1983a). Molecular cloning and characterization of the RAD1 gene of *Saccharomyces cerevisiae*. *Gene* **26**, 119–126.

Higgins, D. R., Prakash, S., Reynolds, P., Polakowska, R., Weber, S. and Prakash, L. (1983b). Isolation and characterization of RAD-3 gene of *Saccharomyces cerevisiae* and inviability of *rad-3* deletion mutation. *Proc. Natl. Acad. Sci. U.S.A.* **80**, 5680–5684.

Hinnen, A., and Meyhack, B. (1981). Vectors for cloning in yeast. *Curr. Top. Microbiol. Immunol.* **96**, 101–117.

Hinnen, A., Hicks, J. B., and Fink, G. R. (1978). Transformation of yeast. *Proc. Natl. Acad. Sci. U.S.A.* **75**, 1929–1933.

Hinnen, A., Farabaugh, P., Ilgen, C., and Fink, G. R. (1979). *In* "ICN-UCLA Symposia on Molecular and Cellular Biology" (R. Axel, T. Maniatis, and C. R. Fox, eds.), Vol. 14, pp. 43–50. Academic Press, New York.

Hitzeman, R. A., Chinault, A. C., Kingsman, A. J., and Carbon, J. (1979). Detection of *Escherichia coli* clones containing specific yeast genes by immunological screening. *In* "ICN-UCLA Symposia on Molecular and Cellular Biology" (R. Axel, T. Maniatis, and C. R. Fox, eds.), Vol. 14, pp. 57–68. Academic Press, New York.

Hitzeman, R. A., Clarke, L., and Carbon, J. (1980). Isolation and characterization of the yeast 3-phosphoglycerokinase gene (PGK) by an immunological screening technique. *J. Biol. Chem.* **255**, 12073–12080.

Hitzeman, R. A., Hogie, F. E., Levine, H. L., Goeddel, D. V., Ammerer, G., and Hall, B. D. (1981). Expression of a human gene for interferon in yeast. *Nature (London)* **293**, 717–720.

Holland, J. P., and Holland, M. J. (1979). The primary structure of a glyceraldehyde-3-phosphate dehydrogenase gene from *Saccharomyces cerevisiae*. *J. Biol. Chem.* **254**, 9839–9845.

Holland, J. P., and Holland, M. J. (1980). Structural comparison of two nontandemly repeated yeast glyceraldehyde-3-phosphate dehydrogenase. *J. Biol. Chem.* **255**, 2596–2605.

Hollenberg, C. P. (1979). *In* "Proceedings of the Symposium on Plasmids of Medical, Environmental and Commercial Importance" (eds. K. N. Timmis, and A. Puhler, eds.). Elsevier/North-Holland, Amsterdam.

Hollenberg, C. P. (1981). Cloning with 2 μm DNA vectors in *Saccharomyces cerevisiae*. *Curr. Top. Microbiol. Immunol.* **96**, 119–144.

Holliday, R. (1964). A mechanism for gene conversion in fungi. *Genet. Res.* **25**, 282–304.

Hotchkiss, R. D. (1955). The biological role of the deoxypentose nucleic acids. *In* "Progress in Nucleic Acids Research" (E. Chargaff and J. N. Davidson, eds.), Vol. 2, pp. 435–473. Academic Press, New York.

Hotchkiss, R. D. (1977). Concluding remarks. *In* "Modern Trends in Bacterial Transformation and Transfection" (A. Portales, R. Lopez, and M. Espinosa, eds.), pp. 321–329. North Holland, New York.

Howe, M. D., and Bade, E. G. (1975). Molecular biology of bacteriophage Mu *Science* **190**, 624–632.

Hsio, C., and Carbon, J. (1979). High-frequency transformation of yeast by plasmids containing the cloned yeast *arg4* gene. *Proc. Natl. Acad. Sci. U.S.A.* **76**, 3829–3833.

Hsio, C., and Carbon, J. (1981). Direct selection procedure for the isolation of functional centromeric DNA. *Proc. Natl. Acad. Sci. U.S.A.* **78**, 3760–3764.

Hudspeth, M. E. S., Timberlake, W. E., and Goldberg, R. B. (1977). DNA seqeunce organization in the water mold *Achyla*. *Proc. Natl. Acad. Sci. U.S.A.* **74**, 4332–4336.

Hughes, K., Case, M. E., Geever, R., Vapnick, D., and Giles, N. H. (1983). Chimeric plasmid that replicate autonomously in both *Escherichia coli* and *Neurospora crassa*. *Proc. Natl. Acad. Sci. U.S.A.* **80**, 1050–1057.

Huiet, L. (1984). Molecular analysis of the *Neurospora qa-1* regulatory region indicates that two interacting genes control *qa* gene expression. *Proc. Natl. Acad. Sci. U.S.A.* **81**, 1179–1178.

Hyman, B. C., Cramer, J. H., and Round, R. H. (1982). Properties of a *Saccharomyces cerevisiae* mtDNA segment conferring high frequency yeast transformation. *Proc. Natl. Acad. Sci. U.S.A.* **79**, 1578–1582.

Ilgen, C., Farabough, P. J., Hinnen, A., Walsh, J. M., and Fink, G. R. (1979). Transformation of yeast. *In* "Genetic Engineering" (J. Setlow and A. Hollander, eds.), Vol. 1, pp. 117–132. Plenum, New York.

Ish-Horowicz, D., and Burke, J. F. (1981). Rapid and efficient cosmid cloning. *Gene* **9**, 2989–2999.

Jackson, J. A., and Fink, G. R. (1981). Gene conversion between duplicated genetic elements in yeast. *Nature (London)* **292**, 306–311.

Jayaram, M., and Broach, J. R. (1983). Yeast plasmid 2μm circle promoter recombination within bacterial transposon Tn-5. *Proc. Natl. Acad. Sci. U.S.A.* **80**, 7264–7268.

Jimenez, A., and Davies, J. (1980). Expression of a transposable antibiotic resistant element in *Saccharomyces*. *Nature (London)* **287**, 869–871.

Johnson, I. S., and Burnett, J. P., Jr. (1978). Problems and potential of industrial recombinant DNA research. *In* "Genetic Engineering" (H. W. Boyer and Nicosia, S. eds.), pp. 217–226. Elsevier/North-Holland, Amsterdam.

Kado, C., and Liu, S. T. (1981). Rapid procedure for detection and isolation of large and small plasmids. *J. Bacteriol.* **45**, 365–373.

Keesey, J. K., and DeMoss, J. A. (1982). Cloning of the *trp-1* gene from *Neurospora crassa* by complementation of a *trpC* mutation in *E. coli*. *J. Bacteriol.* **152**, 954–958.

Kelbe, R. J., Harriss, R. V. Sharp, Z. D., and Douglas, M. G. (1983). A general method for polyethylene glycol-induced genetic transformation of bacteria and yeast. *Gene* **25**, 333–341.

Kingsman, A. J., Clarke, L., Mortimer, R. K., and Carbon, J. (1979). Replication in *Saccharomyces cerevisiae* of plasmid pBR313 carrying DNA from the yeast *trp1* region. *Gene* **7**, 141–152.

Kingsman, A. J., Gimlich, R. L., Clarke, L., Chinault, A. C., and Carbon, J. (1981). Sequence variation in dispersed repetitive sequences in *Saccharomyces cerevisiae*. *J. Mol. Biol.* **145**, 609–632.

Kinnaird, J., and Fincham, J. R. S. (1982). Cloning and sequence analysis of the structural gene for glutamate dehydrogenase of *Neurospora*. *Neurospora Newslett.* **29**, 11.

Kinnaird, J. H., and Fincham, J. R. S. (1983). An intron in the am (glutamate dehydrogenase) gene of *Neurospora crassa*. *Heredity* **50**, 209. (Abstr.)

Kinnaird, J. H., Keighren, M. A., Kinsey, J. A., Eaton, M., and Fincham, J. R. S. (1982). Cloning of the *am* (glutamate dehydrogenase) gene of *Neurospora crassa* through the use of a synthetic probe. *Gene* **20**, 387–396.

Kleckner, N. (1977). Transposable elements in prokaryotes. *Cell* **11**, 11–23.

Klein, H. L., and Petes, T. D. (1981). Intrachromosomal gene conversion in yeast. *Nature (London)* **289**, 144–148.

Kornberg, A. (1980). "DNA Replication." Freeman, San Francisco.

Krumlauf, R., and Marzluf, G. A. (1979). Characterization of the sequence complexity and organization of the *Neurospora crassa* genome. Genome organization of *Neurospora crassa. Biochemistry* **18**, 3705–3713.

Kuck, U., Stahl, U., and Esser, K. (1981). Plasmid-like DNA is part of mitochondrial DNA in *Podospora anserina. Curr. Genet.* **3**, 151–156.

Kushner, S. R. (1978). An improved method for transformation of *Escherichia coli* with ColEl derived plasmids. *In* "Genetic Engineering" (H. W. Boyer and S. Nicosia, eds.), pp. 17–23. North-Holland, Amsterdam.

Kushner, S. R., Hautala, J. A., Jacobson, J. W., Giles, N. H., and Vapnek, D. (1977). Expression of the structural gene for catabolite dehydroquinase of *Neurospora crassa* in *E. coli* K12. *Brookhaven Symp. Biol.* **29**, 297–308.

Langford, C., Nellen, W., Niessing, J., and Gallwitz, D. (1983). Yeast is unable to excise foreign intervening sequences from hybrid gene transcript. *Proc. Natl. Acad. Sci. U.S.A.* **80**, 1496–1500.

Lasson, R., and Lacroute, F. (1979). Interference of nonsense mutations with eukaryotic messenger RNA stability. *Proc. Natl. Acad. Sci. U.S.A.* **76**, 5134–5137.

Lauer, G. D., Roberts, T. M., and Klotz, L. C. (1977). Determination of the nuclear DNA content of *Saccharomyces cerevisiae* and implications for the organization of DNA. *J. Mol. Biol.* **114**, 507

Lauson, G., and Jund, R. (1982). Expression of a cloned yeast gene (URA1) is controlled by a bacterial promoter in *E. coli* and by yeast promoter in *S. cerevisiae. Gene* **15**, 127–137.

Letts, V. A. Klig, L. S., Baelee, M., Carman, G. M., and Henry, S. (1983). Isolation of the yeast structural gene for the membrane-associated enzyme phosphotidylserine synthase. *Proc. Natl. Acad. Sci. U.S.A.* **80**, 7279–7283.

Lowry, C., Weiss, J. L., Walthall, D. A., and Zilomer, R. S. (1983). Modulator sequence mediate regulation of CyC1 and a neighboring gene. *Proc. Natl. Acad. Sci. U.S.A.* **80**, 151–155.

Loyter, A., Scangos, G. A., and Ruddle, F. H. (1982). Mechanisms of DNA uptake by mammalian cells: Fate of exogenously added DNA monitored by the use of fluorescent dye. *Proc. Natl. Acad. Sci. U.S.A.* **79**, 422–426.

MacHattie, L. A., and Jackowski, J. B. (1977). Physical structure and deletion effects of the chloramphenicol resistance element Tn9 in phage lambda. *In* "DNA Insertion Elements, Plasmids and Episomes" (A. I. Bukhari, J. A. Shapiro, and S. L. Adhya, eds.), pp. 219–228. Cold Spring Harbor Laboratory, Cold Spring Harbor, New York.

Majors, J. (1975). Specific binding of CAP factor to lac promoter DNA. *Nature (London)* **256**, 672–673.

Mandel, M., and Higa, A. (1970). Calcium-dependent bacteriophage DNA infection. *J. Mol. Biol.* **53**, 159–162.

Mann, C., and Davis, R. W. (1983). Instability of dicenteric plasmids in yeast. *Proc. Natl. Acad. Sci. U.S.A.* **80**, 228–232.

Mark, R., Casadaban, M. J., and Botstein, D. (1981). Yeast genes fused to β-galactosidase in *Escherichia coli* can be expressed normally in yeast. *Proc. Natl. Acad. Sci. U.S.A.* **78**, 2460–2464.

Marmur, J. (1961). A procedure for the isolation of deoxyribonucleic acid from microorganisms. *J. Mol. Biol.* **3**, 208–218.

Martini, G., Grimaldi, G., and Guardiola, J. (1978). Extrachromosomal DNA in phytopathogenic fungi. *In* "Genetic Engineering" (H. W. Boyer and S. Nicosia, eds.), pp. 197–200. Elsevier/North-Holland, Amsterdam.

Maxam, A. M., and Gilbert, W. (1977). A new method for sequencing DNA. *Proc. Natl. Acad. Sci. U.S.A.* **74**, 560–564.

McBride, O. W., and Ozer, H. L. (1973). Transfer of genetic information by purified metaphase chromosomes. *Proc. Natl. Acad. Sci. U.S.A.* **70**, 1258–1262.

McBride, O. W., and Peterson, J. C. (1978). Chromosome mediated gene transfer in mammalian cells. *Annu. Rev. Genet.* **14**, 321–345.

McCarthy, B. J. (1969). The evolution of base sequence in nucleic acids. *In* "Handbook of Molecular Cytology" (A. Lima-de-Faria, ed), p. 3–20. North Holland, Amsterdam.

McClintock, B. (1941). The stability of broken ends of chromosomes in *Zea mays*. *Genetics* **26**, 234–282.

McClintock, B. (1942). The fusion of broken ends of chromosomes following nuclear fusion. *Proc. Natl. Acad. Sci. U.S.A.* **28**, 458–463.

McClintock, B. (1951). Chromosome organization and gene expression. *Cold Spring Harbor Symp. Quant. Biol.* **16**, 13–47.

McNeil, J. B., and Friesen, J. D. (1981). Expression of the herpes simplex virus thymidine kinase gene in *Saccharomyces cerevisiae*. *Mol. Gen. Genet.* **184**, 386–393.

Meagher, R. B., Tail, R. C., Betlach, M., and Boyer, H. W. (1977). Protein expression in *E. coli* minicells by recombinant plasmid. *Cell* **10**, 521–536.

Mercereau-Puijalon, D., Lacreute, F., and Kourilsky, P. (1980). Synthesis of chicken ovalbumin-like protein in the yeast *Saccharomyces cerevisiae*. *Gene* **11**, 163–167.

Messing, J., Gronenborn, B., Muller-Hill, B., and Hofschneider, P. H. (1977). Filamentous coli phage M13 as a cloning vehicle: Insertion of a *Hind*III fragment of the *lac* regulatory region in M13 replication form *in vitro*. *Proc. Natl. Acad. Sci. U.S.A.* **74**, 3642–3646.

Metzenberg, R. L., and Baisch, T. J. (1981). An easy method for preparing *Neurospora* DNA. *Neurospora Newslett.* **28**, 19–21.

Metzenberg, R. L., Stevens, J. N., Selker, E. U., and Morzycka-Wroblewska, E. (1984). A method for finding the genetic map position of cloned DNA fragments. *Neurospora Newslett.* **31**, 35–39.

Miller, W. L. (1979). Use of recombinant DNA technology for the production of polypeptides. *Adv. Exp. Med. Biol.* **118**, 153–174.

Mishra, N. C. (1967). Genetic studies in *Neurospora* and *Eudorina*. Ph.D. Thesis, McMaster University, Hamilton, Canada.

Mishra, N. C. (1976). Episome-like behavior of donor DNA in transformed strains of *Neurospora*. *Nature (London)* **264**, 251–253.

Mishra, N. C. (1977a). Genetics of certain conditional mutants which originated during transformation in *Neurospora*. *Genet. Res.* **29**, 9–19.

Mishra, N. C. (1977b). Gene transfer experiments in *Neurospora*. *Brookhaven Symp. Biol.* **29**, 161–165.

Mishra, N. C. (1979a). Gene transfer in *Neurospora crassa*. *In* "Proceedings of the Fourth

European Meeting on Bacterial Transformation and Transfection" (S. W. Glover and L. O. Butler, eds.), pp. 259–265. Costwold Press, York, England.

Mishra, N. C. (1979b). DNA-mediated genetic changes in *Neurospora crassa*. *J. Gen. Microbiol.* **113**, 255–259.

Mishra, N. C. (1981). HeLa cells lack *recA* gene. *FEBS Lett.* **137**, 175–177.

Mishra, N. C. (1983). DNA polymerase mutants of *Neurospora*. In "Proceedings of the 15th International Genetics Congress, Part II" (M. S. Swaminalham, ed.), 500 pp. Oxford and IBH Publishing Co., New Delhi.

Mishra, N. C. (1984). DNA repair defective mutants of *Neurospora*. In "Proceedings of the Symposium on Applied and Basic Mutagenesis, XV International Genetics Congress, India" (A. B. Prasad, ed.). Oxford and IBH Publishing Co., New Delhi.

Mishra, N. C., and Forsthoefel, A. M. (1983). Biochemical genetics of *Neurospora* nuclease. *Genet. Res.* **41**, 287–297.

Mishra, N. C., and Tatum, E. L. (1970). Phosphoglucomutase mutants of *Neurospora sitophila* and their relation to morphology. *Proc. Natl. Acad. Sci. U.S.A.* **60**, 638–645.

Mishra, N. C., and Tatum, E. L. (1973). Non-Mendelian inheritance of DNA-induced inositol independence in *Neurospora*. *Proc. Natl. Acad. Sci. U.S.A.* **70**, 3875–3879.

Mishra, N. C., and Threlkeld, S. F. H. (1967). Variation in expression of *rg* mutants of *Neurospora*. *Genetics* **55**, 113–121.

Mishra, N. C., Sazbo, G., and Tatum, E. L. (1973). Nucleic acids induced genetic changes in *Neurospora*. In "The Role of RNA in Reproduction and Development" (M. C. Niu and S. J. Segal, eds.), pp. 259–268. North-Holland, Amsterdam.

Miyanohara, A., Tho-e, A., Nozaki, C., Hausada, F., Ohtomo, N., and Matsubara, K. (1983). Expression of hepatitis B surface antigen in yeast. *Proc. Natl. Acad. Sci. U.S.A.* **80**, 1–5.

Montgomery, D. L., and Hall, B. D. (1978). Identification and isolation of the yeast cytochrome *c* gene. *Cell* **14**, 673–680.

Montgomery, D. L., Leung, W., Smith, M., Shatlit, P., Faye, G., and Hall, B. D. (1980). Isolation and sequence of the gene for iso-cytochrome *c* in *Saccharomyces cerevisiae*. *Proc. Natl. Acad. Sci. U.S.A.* **77**, 541–545.

Morrow, J. F. (1979). Recombinant DNA techniques. In "Methods in Enzymology" (R. Wu, ed.), Vol. 68, pp. 3–24. Academic Press, New York.

Mortimer, R. K., and Hawthorne, D. C. (1966). Genetic mapping in *Saccharomyces cerevisiae*. *Genetics* **53**, 165–173.

Mulligan, R. C., and Berg, P. (1981). Selection for animal cells that express the *Escherichia coli* gene coding for xanthine-guanine phosphoribosyltransferase. *Proc. Natl. Acad. Sci. U.S.A.* **78**, 2072–2076.

Murray, A. W., and Szostak, J. W. (1983). Construction of artificial chromosomes in yeast. *Nature (London)* **305**, 189–193.

Murray, N. E., and Murray, K. (1974). Manipulation of restriction targets in phage λ to form receptor chromosomes for DNA fragments. *Nature (London)* **251**, 476–481.

Nasmyth, K. A., and Reed, S. I. (1980). Isolation of genes by complementation in yeast: Molecular cloning of a cell-cycle gene. *Proc. Natl. Acad. Sci. U.S.A.* **77**, 2119–2123.

Neve, R. L., West, R. W., and Rodriquez, R. L. (1978). Eukaryotic DNA fragments which act as promoters for a plasmid gene. *Nature (London)* **277**, 324–325.

Ng, R., and Abelson, J. (1980). Isolation and sequence of the gene for actin in *Saccharomyces cerevisiae*. *Proc. Natl. Acad. Sci. U.S.A.* **77**, 2912–3916.

Niagro, F. D. (1981). Biochemical, genetic, morphometric characterization of an ethidium bromide induced respiratory deficient mutant of *Neurospora crassa*. Ph.D. Thesis, University of South Carolina, Columbia.

Novick, P., Field, C., and Schekman, R. (1980). Identification of 23 complementation groups required for post-translational events in the yeast secretory pathway. *Cell* **21**, 205–215.

Olsen, J., Doy, C. H., Colsen, J., Pateman, J. A., and Norris, U. (1982). Molecular biology of expression and cloning of genes in *Aspergillus nidulans. Neurospora Newslett.* **29**, 11.

Orr, W. C., and Timberlake, W. E. (1982). Clustering of spore specific genes in *Aspergillus nidulans. Proc. Natl. Acad. Sci. U.S.A.* **79**, 5976–5980.

Orr-Weaver, T. L., Szostak, J. W., and Rothstein, R. J. (1981). Yeast transformation: A model system for the study of recombination. *Proc. Natl. Acad. Sci. U.S.A.* **78**, 6354–6358.

Osley, M. A., and Hereford, L. (1982). Identification of a sequence responsible for periodic syntheis of yeast histone 2A mRNA. *Proc. Natl. Acad. Sci. U.S.A.* **79**, 7689–7693.

Patel, V. B., Schweizer, M., Dykstra, C. C., Kushner, S. R., and Giles, N. H. (1981). Genetic organization and transcriptional regulation in the Qa gene cluster of *Neurospora crassa. Proc. Natl. Acad. Sci. U.S.A.* **78**, 5783–5787.

Pederby, J. F. (1979). Fungal protoplasts, isolation, reversion and fusion. *Annu. Rev. Microbiol.* **33**, 21–39.

Penden, K. W. C., Pipas, J. M., and Pearson-White, N. D. (1980). Isolation of mutants of animal virus in bacteria. *Science* **209**, 1392–1396.

Perkins, D. D., and Barry, E. G. (1977). The cytogenetics of *Neurospora. Adv. Genet.* **19**, 133–285.

Perkins, D. D., Radford, A., Newmeyer, D., and Bjorkman, M. (1982). Chromosomal loci of *Neurospora crassa. Microbiol. Rev.* **46**, 426–570.

Peterhirth, K., Edwards, C. A., and Firtel, R. A. (1982). A DNA-mediated transformation system for *Dictyostelium discordium. Proc. Natl. Acad. Sci. U.S.A.* **79**, 7356–7360.

Petes, T. D. (1980). Molecular genetics of yeast. *Annu. Rev. Biochem.* **49**, 845–876.

Petes, T. D., Broach, J. R., Wensink, P. C., Hereford, L. M., Fink, G. R., and Botstein, D. (1978). Isolation and analysis of recombinant DNA molecules containing yeast DNA. *Gene* **4**, 37–49.

Pontecorvo, G. (1958). "Trends in Genetic Analysis." Columbia Univ. Press, New York.

Radford, A., Pope, S., Sazci, A., Fraser, M. J., and Parish, J. H. (1981). Liposome mediated genetic transformation of *Neurospora crassa. Mol. Gen. Genet.* **184**, 567–569.

Rambosek, J. A., and Kinsey, J. A. (1984). An unstable mutant gene of the *am* locus of *Neurospora* results from a small duplication. *Gene* **27**, 101–107.

Ratzkin, B., and Carbon, J. (1977). Functional expression of cloned yeast DNA in *Escherichia coli. Proc. Natl. Acad. Sci. U.S.A.* **74**, 487–491.

Roeder, G. S., Farabough, P. J., Chaleff, D. T., and Fink, G. R. (1980). The origins of gene instability in yeast. *Science* **209**, 1375–1380.

Rogers, D. T., Lemire, J. M., and Bostian, K. M. (1982). Acid phosphatase polypeptides in *Saccharomyces cerevisiae* are encoded by a differentially regulated multigene family. *Proc. Natl. Acad. Sci. U.S.A.* **79**, 2157–2161.

Roggenkamp, R., Kustermann-Kuhn, B., and Hollenberg, C. P. (1981). Expression and processing of bacterial β-lactamase in yeast, *S. cerevisiae. Proc. Natl. Acad. Sci. U.S.A.* **78**, 4466–4470.

Rose, M., Casadaban, M. J., and Botstein, D. (1981). Yeast genes fused to β-galactosidase in *E. coli* can be expressed normally in yeast. *Proc. Natl. Acad. Sci. U.S.A.* **78,** 2460–2464.

Rubin, G. M., and Spradling, A. C. (1982). Genetic transformation of *Drosophila* with transposable elements vector. *Science* **218,** 348–353.

Ruddle, F. H. (1981). A new era in mammalian gene mapping: Somatic cell genetics and recombinant DNA methodologies. *Nature (London)* **294,** 115–120.

Sakaguchi, J., and Yamamoto, M. (1982). Cloned ura-1 locus of *Schizosaccharomyces pombe* propagates autonomously in this yeast assuming a polymeric form. *Proc. Natl. Acad. Sci. U.S.A.* **79,** 7819–7823.

Sancar, A., Hack, A. M., and Rupp, W. D. (1979). Simple method for identification of plasmid-coded proteins *J. Bacteriol.* **137,** 692–693.

Sanger, F., Nicklen, S., and Coulson, A. R. (1977). DNA sequencing with chain terminating inhibitors. *Proc. Natl. Acad. Sci. U.S.A.* **74,** 5463–5467.

Scangos, C., and Ruddle, F. (1981). Mechanisms and applications of DNA-mediated gene transfer in mammalian cells—A review. *Gene* **14,** 1–10.

Schablik, M., and Sazbo, G. (1981). DNA uptake-stimulating factor in the culture medium of *Neurospora crassa.* *FEBS Lett.* **10,** 395–397.

Schablik, M., Szabolcs, M., Kiss, A., Aradi, J., Zsindely, A., and Szabo, G. (1977). Conditions of transformation by DNA of *Neurospora crassa.* *Acta Biol. Acad. Sci. Hung.* **28,** 273–279.

Schablik, M., Zsindely, A., Aradi, J., Fekete, Sz. and Szabo, G. (1978). Differences between transformed and spontaneous revertant strains of *Neurospora crassa.* *Neurospora Newslett.* **25,** 22–23.

Schablik, M., Delange, A. M. Shums, A. A., and Mishra, N. C. (1983). Construction of a genomic library of *Neurospora.* *FEBS Lett.* **16,** 321–325.

Schaffner, W. (1980). Direct transfer of cloned genes from bacteria to mammalian cells. *Proc. Natl. Acad. Sci. U.S.A.* **77,** 2163–2167.

Schechtman, M. G., and Yanofsky, C. (1983). Structure of the trifunctional *trp-1* gene from *Neurospora crassa* and its aberrant expression in *Escherichia coli.* *J. Mol. Appl. Genet.* **2,** 83–99.

Schell, M. A., and Wilson, D. B. (1979). Cloning and expression of the yeast galactokinase gene in an *Escherichia coli* plasmid. *Gene* **5,** 291–303.

Scherer, S., and Davis, R. (1979). Replacement of chromosome segments with altered DNA sequence constructed *in vitro.* *Proc. Natl. Acad. Sci. U.S.A.* **76,** 4951–4955.

Scherer, S., and Davis, R. (1980). Recombination of dispersed repeated DNA sequences in yeast. *Science* **209,** 1380–1384.

Schleif, R. F., and Wensink, P. C. (1981). "Practical Methods in Molecular Biology." Springer Verlag, Berlin and New York.

Schwartz, D. C., and Cantor, C. R. (1984). Separation of yeast chromosome-sized DNAs by pulsed field gradient gel electrophoresis. *Cell* **37,** 67–75.

Schweitzer, E., MacKecline, C., and Halverson, H. O. (1969). The redundancy of ribosomal and transfer RNA genes in *Saccharomyces cerevisiae.* *J. Mol. Biol.* **40,** 261–277.

Schweizer, M., Case, M. E., Dykstra, C. C., Giles, D. H. and Kushner, S. R. (1981a). Cloning the quinic acid (*Qa*) gene cluster from *Neurospora crassa*: Identification of recombinant plasmids containing both qa-2⁺ and qa-3⁺. *Gene* **14,** 23–32.

Schweizer, M., Case, M. E., Dykstra, C. C., Giles, N. H. and Kushner, S. R. (1981b).

Identification and characterization of recombinant plasmids carrying the complete *Qa* gene cluster from *Neurospora crassa* including the *qa-1*[+] regulatory gene. *Proc. Natl. Acad. Sci. U.S.A.* **78,** 5086–5090.

Selker, E. U., Free, S. J., Metzenberg, R. L., and Yanofsky, C. (1981). Associated pseudogene related to 5S RNA gene in *Neurospora crassa*. *Nature (London)* **294,** 576–578.

Shapiro, J. A., Adhya, S. L., and Bukhari, A. I. (eds). (1977). "DNA Insertion Elements, Plasmids and Episomes. Cold Spring Harbor Laboratory, Cold Spring Harbor, New York.

Shiba, T., and Saigo, K. (1983). Retrovirus-like particles containing RNA homologous to the transposible element copia in *Drosophila melanogaster*. *Nature (London)* **302,** 119–124.

Shoemaker, C., Goff, S., Gilboa, E., Paskina, M., Mitra, S. W., and Baltimore, D. (1980). Structure of a cloned circular Moloney murine leukemia virus DNA molecule containing an inverted segment: Implication for retrovirus integration. *Proc. Natl. Acad. Sci. U.S.A.* **77,** 3932–3936.

Shortle, D., and Nathans, D. (1978). Local mutagenesis: A method for generating viral mutants with base substitution in preselected regions. *Proc. Natl. Acad. Sci. U.S.A.* **75,** 2170–2175.

Shortle, D., Haber, J. B., and Botstein, D. (1982). Lethal disruption of the yeast actin gene by integrative DNA transformation. *Science* **217,** 371–373.

Silver, L. M., and Artzt, K. (1981). Recombination suppression of mouse t haplophytes due to chromatin mismatching. *Nature (London)* **290,** 68–70.

Simon, M., Zieg, J., Silverman, M., Mandel, G., and Doolittle, R. (1980). Phase variation: Evolution of a controlling element. *Science* **209,** 1370–1374.

Sinsheimer, R. L. (1977). Recombinant DNA. *Annu. Rev. Biochem.* **46,** 415–438.

Smarrelli, J., and Garrett, R. H. (1982). Isolation of *Neurospora* nitrate reductase structural gene: Evidence for its expression in *E. coli*. *Neurospora Newslett.* **29,** 11.

Smith, M., Leung, D. W., Gillam, S., and Astell, C. (1979). Sequence of the gene for iso-1-cytochrome c in *Saccharomyces cerevisiae*. *Cell* **16,** 753–761.

Soliday, L. L., Flurkey, W. H., Okita, T. W., and Kolatukudy, P. E. (1984). Cloning and structure determination of cDNA for cutinase, an enzyme involved in fungal penetration of plants. *Proc. Natl. Acad. Sci. U.S.A.* **81,** 3939–3943.

Southern, E. M. (1975). Detection of specific seqeuences among DNA fragments separated by gel electrophoresis. *J. Mol. Biol.* **98,** 503–517.

Specht, C. A., DiRusso, C. C., Novotny, C. P., and Ullrich, R. C. (1982). A method for extracting high molecular weight deoxyribonucleic acid from fungi. *Anal. Biochem.* **119,** 158–163.

Srb, A. M., and Horowitz, N. H. (1944). The ornithine cycle of *Neurospora* and its genetic control. *J. Biol. Chem.* **154,** 129–139.

St. John, P. T., and Davis, W. (1979). Isolation of galactose-inducible DNA sequences from *Saccharomyces cerevisiae* by differential plaque filter hybridization. *Cell* **16,** 443–452.

Stahl, F. W. (1981). "Genetic Recombination—Thinking about It in Phage and Fungi." Freeman, San Francisco.

Stahl, U., Kuck, U., Tudzynski, P., and Esser, K. (1980). Characterization and cloning of plasmid-like DNA of the ascomycete *Podospora anserina*. *Mol. Gen. Genet.* **178,** 639–646.

Stahl, U., Tudzynski, P., Kuck, U., and Esser, K. (1981). Replication and expression of a bacterial–mitochondrial hybrid plasmid in the fungus *Podospora anserina*. *Mol. Gen. Genet.* **178**, 639–646.

Stinchcomb, D. T., Struhl, K., and Davis, R. W. (1979). Isolation and characterization of a yeast chromosomal replicator. *Nature (London)* **282**, 39–43.

Stinchcomb, D. T., Thomas, M., Kelly, J., Selker, E., and Davis, R. W. (1980). Eukaryotic DNA segments capable of autonomous replication in yeast. *Proc. Natl. Acad. Sci. U.S.A.* **77**, 4559–4563.

Stohl, L. L., and Lambowitz, A. M. (1983a). Construction of a shuttle vector for the filamentous fungus *Neurospora crassa*. *Proc. Natl. Acad. Sci. U.S.A.* **80**, 1958–1062.

Stohl, L. L., and Lambowitz, A. M. (1983b). A colony filter hybridization procedure for the filamentous fungus *Neurospora crassa*. *Anal. Biochem.* **134**, 82–85.

Stohl, L. L., Collins, R. A., Cole, M. D., and Lambowitz, A. M. (1982). Characterization of two new plasmid DNAs found in mitochondria of wild-type *Neurospora* intermedia strains. *Nucleic Acids Res.* **10**, 1439–1458.

Strathern, J. N., Newton, C. S., Herskowitz, I., and Hicks, J. B. (1979). Isolation of circular derivatives of yeast chromosome. III: Implications for the mechanism of mating type interconversion. *Cell* **18**, 309–319.

Struhl, K. (1981). Deletion mapping of an eukaryotic promoter. *Proc. Natl. Acad. Sci. U.S.A.* **78**, 4461–4465.

Struhl, K. (1982). The yeast His-3 promoter contains at least two distinct elements. *Proc. Natl. Acad. Sci. U.S.A.* **79**, 7385–7389.

Struhl, K., and Davis, R. W. (1977). Production of a functional eukaryotic enzyme in *Escherichia coli*: Cloning and expression of the yeast structural gene for imidazole glycerol phosphate dehydratase (*his3*). *Proc. Natl. Acad. Sci. U.S.A.* **74**, 5255–5259.

Struhl, K., Cameron, J. R., and Davis, R. W. (1976). Functional genetic expression of eukaryotic DNA in *Escherichia coli*. *Proc. Natl. Acad. Sci. U.S.A.* **73**, 1471–1475.

Struhl, K., Stinchcomb, D., Scherer, S., and Davis, R. (1979). High frequency transformation of yeast: Autonomous replication of hybrid DNA molecules. *Proc. Natl. Acad. Sci. U.S.A.* **76**, 1035–1039.

Sugino, A., Kojo, H., Greenberg, B. D., Brown, P. O., and Kim, K. C. (1981). *In vitro* replication of a yeast 2μm plasmid DNA. *In* "The Initiation of DNA Replication—ICN-UCLA Symposium" (C. F. Fox, ed.), Vol. 22, pp. 529–553. Academic Press, New York.

Szabo, G., and Schablik, M. (1982). Behaviour of DNA-induced inositol independent transformants of *Neurospora crassa* in sexual crosses. *Theor. Appl. Genet.* **61**, 171–175.

Szabo, G., Schablik, M., Fekete, Z., and Zsindley, A. (1978). A comparative study of DNA-induced transformants and spontaneous revertants of inisotolless *Neurospora crassa*. *Acta Biol. Acad. Sci. Hung.* **29**, 375–384.

Szostak, J. W., and Blackburn, E. H. (1982). Cloning yeast telomeres on linear plasmid vectors. *Cell* **29**, 245–255.

Szostak, J. W., and Wu, R. (1979). Insertion of a genetic marker into the ribosomal DNA of yeast. *Plasmid* **2**, 536–554.

Szostak, J. W., Stiles, J. I., Tye, B. K., Chiu, P. Sherman, F., and Wu, R. (1979). Hybridization with synthetic oligonucleotides. *In* "Methods in Enzymology" (R. Wu, ed.), Vol. 68, pp. 419–427. Academic Press, New York.

Szybalska, E. H., and Szybalski, W. (1962). Genetics of human cell lines. IV. DNA-mediated heritable transformation of a biochemical trait. *Proc. Natl. Acad. Sci. U.S.A.* **48**, 2026–2034.

Szybalski, W. (1963). DNA-mediated genetic transformation of human cell lines. Proc. 12th Ann. Session of the Natl. Poultry Breeder Roundtable. Kansas City, Mo. p. 90–109.

Thomas, G. (1982). Control and cloning of acetate inducible genes of *Neurospora crassa*. *Neurospora Newslett.* **29**, 11.

Thomas, P. S. (1980). Hybridization of denatured RNA and small DNA fragments transferred to nitrocellulose. *Proc. Natl. Acad. Sci. U.S.A.* **77**, 5201–5205.

Thomas, D. Y., and James, A. P. (1980a). Genetic analysis of *Saccharomyces cerevisiae* transformed by a plasmid containing a suppressor transfer ribonucleic acid gene. *J. Bacteriol.* **137**, 1179–1186.

Thomas, D. Y., and James, A. P. (1980b). Transformation of *Saccharomyces cerevisiae* with plasmids containing fragments of yeast 2 μ DNA and a suppressor tRNA gene. *Curr. Genet.* **2**, 9–16.

Tilburn, J., Scazzocchio, C., Taylor, G. G., Zabicky-Zissman, J. H., Lockington, R. A., and Davies, R. W. (1983). Transformation by integration in *Aspergillus nidulans*. *Gene* **26**, 205–221.

Timberlake, W. E. (1980a). Developmental gene regulation of *Aspergillus nidulans*. *Dev. Biol.* **78**, 497–510.

Timberlake, W. E. (1980b). Low repetitive DNA content in *Aspergillus nidulans*. *Science* **202**, 973–974.

Timberlake, W. E., and Bernard, E. C. (1981). Organization of a gene cluster expressed specifically in the asexual spores of *A. nidulans*. *Cell* **26**, 29–37.

Tschumper, G., and Carbon, J. (1980). Sequence of a yeast DNA fragment containing a chromosomal replicator and the TRP1 gene. *Gene* **10**, 157–166.

Tschumper, G., and Carbon, J. (1981). Sequencing and subcloning analysis of autonomously replicating sequences from yeast chromosomal DNA. *In* "ICN-UCLA Symposia on Molecular and Cellular Biology" (C. F. Fox, ed.), Vol. 22, pp. 489–500. Academic Press, New York.

Tudzynski, P., and Esser, K. (1979). Chromosomal and extrachromosomal control of senescence in the ascomycete *P. anserina*. *Mol. Gen. Genet.* **73**, 71–84.

Tudzynski, P., Stahl, U., and Esser, K. (1980). Transformation to senescence with plasmid-like DNA in the ascomycete *Podospora anserina*. *Curr. Genet.* **2**, 181–184.

Turner, B. C. (1977). Resistance to spore killer genes in *Neurospora* strains from nature. *Genetics* **86**, 565–566.

Ullrich, A., Shine, J., Chirgwin, J., Pictet, R., Tischer, F., Rutter, W. J., and Goodman, H. (1977). Rat insulin genes: Construction of plasmids containing the coding sequences. *Science* **196**, 1313–1319.

Valenzuela, P., Medina, A., Rutter, W. J., Ammerer, G. and Hall, B. D. (1982). Synthesis and assembly of hepatitis B virus surface antigen particles in yeast. *Nature (London)* **248**, 347–350.

Vapnek, D., Hautala, J. A., Jacobson, J. W., Giles, N. H., and Kushner, S. R. (1977). Expression in E. coli K-12 of the structural gene for catabolic dehydroquinase of *Neurospora crassa*. *Proc. Natl. Acad. Sci. U.S.A.* **74**, 3508–3512.

Viebrock, A., Perz, A., and Sebald, W. (1983). Molecular cloning of middle-abundant mRNAs from *Neurospora crassa*. *Methods Enzymol.* **97**, 254–260.

Vosberg, H. P. (1977). Molecular cloning of DNA—an introduction to techniques and problems. *Hum Genet.* **40**, 1–72.

Vournakis, J. N., and Elander, R. P. (1983). Genetic manipulation of antibiotic-producing microorganisms. *Science* **219**, 703–708.

Walz, A., Ratzkin, B., and Carbon, J. (1978). Control of expression of a cloned yeast gene (*trp-5*) by a bacterial insertion element (IS-2). *Proc. Natl. Acad. Sci. U.S.A.* **75**, 6172–6176.

Watson, J. D. (1972). Origin of concatameric T7 DNA. *Nature New Biol. (London)* **239**, 197–201.

Watson, J. D., Tooze, J., and Kurtz, D. T. (1983). ·Recombinant DNA—A short course. Freeman, New York.

Weissmann, C., Weber, H., Tanguchi, T., Mueller, W., and Meyer, F. (1978). Site-directed mutagenesis as a tool in genetics: Application to RNA and DNA genomes. *In* "Genetic Engineering" (H. W. Boyer and S. Ricosia, eds.), pp. 65–77. Elsevier/North-Holland, Amsterdam.

Wiley, W. R. (1979). Isolation of spheroplasts and membrane vesicles from yeast and filamentous fungi. *Methods Enzymol.* **31**, 609–626.

Williamson, V. M., Bennetzen, J., Young, E. T., Nasmyth, K., and Hall, B. D. (1980). Isolation of the structural gene for alcohol dehydrogenase by genetic complementation in yeast. *Nature (London)* **283**, 214–216.

Wootton, J. C., Fraser, M. J., and Baron, A. J. (1980). Efficient transformation of germinating *Neurospora* conidia using total nuclear fragments. *Neurospora Newslett.* **27**, 33.

Woudt, L. P., Pastnik, A., Veenstra, A. E. K., Jansen, A. E. M., Mager, W. H., and Planta, R. J. (1983). The genes coding for histones H_3 and H_4 in *Neurospora crassa* are unique and contain intervening sequences. *Nucleic Acids Res.* **11**, 5347–5360.

Yang, H. L., and Zubay, G. (1978). Expression of the cel gene in ColEl and certain hybrid plasmids derived from EcoRl treated ColEL. *In* "Microbiology" (D. Schessinger, ed.), pp. 154–155. American Society for Microbiology, Washington, D.C.

Yelton, M. M., Hamer, J. E., DeSouza, E. R., Mullaney, E. J., and Timberlake, W. E. (1983). Developmental regulation of the *Aspergillus nidulans trpC* gene. *Proc. Natl. Acad. Sci. U.S.A.* **80**, 7576–7580.

Yelton, M. M., Hamer, J. E., and Timberlake, W. E. (1984). Transformation of *Aspergillus nidulans* by using a *trpC* plasmid. *Proc. Natl. Acad. Sci. U.S.A.* **81**, 1470–1474.

Young, R. A., and Davis, R. W. (1983). Efficient isolation of genes by using antibody probes. *Proc. Natl. Acad. Sci. U.S.A.* **80**, 1194–1198.

Yoshikawa, H. (1966). Mutations resulting from the transformation of *Bacillus subtilis*. *Genetics* **54**, 1201–1213.

Zakian, V. A. (1981). Origin of replication from *Xenopus laevis* mitochondrial DNA promotes high frequency transformation of yeast. *Proc. Natl. Acad. Sci. U.S.A.* **78**, 3128–3132.

Zakian, V. A. (1983). Control of chromosome behavior in yeast. *Nature (London)* **305**, 275.

Zamir, A., Marina, C. V., Fink, G. R., and Szalay, A. A. (1981). Stable chromosomal integration of the entire nitrogen fixation gene cluster from *Klebsiella pneumoniae* in yeast. *Proc. Natl. Acad. Sci. U.S.A.* **78**, 3496–3500.

Zimmerman, C. R., Orr, W. C., Leclere, R. F., Barnard, E. C., and Timberlake, W. E. (1980). Molecular cloning and selection of genes regulated in *Aspergillus* development. *Cell* **21**, 709–715.

Zinder, N. D., and Boeke, J. D. (1982). The filamentous phage (FG) as vectors for recombinant DNA. *Gene* **19**, 1–10.

Zsindeley, A., Szabolcs, M., Dava, M., Schablik, M., Aradi, J., and Szabo, G. (1979). Demonstration of myo-inositol-1-phosphate synthase and its assumed defective variant in various *Neurospora crassa* strains by immunological methods. *Acta Biol. Acad. Sci. Hung.* **30**, 141–149.

Y CHROMOSOME FUNCTION AND SPERMATOGENESIS IN *Drosophila hydei*

Wolfgang Hennig

Department of Genetics, Katholieke Universiteit
Toernooiveld, Nijmegen, The Netherlands

I. Introduction

The differentiation between germ line and soma is one of the fundamental differentiation processes in eukaryotic organisms. Cells in the germ line are responsible for maintaining the continuity of the genetic material through the generations. In both sexes the development of the gametes is a highly specialized process, but our knowledge of the genetic, molecular, and morphogenetic parameters of these developmental pathways is still surprisingly sparse. In general much more is known about egg development, probably because it appears more important in the context of the early embryonic development. Spermatogenesis, on the other hand, is a morphogenetic process that is primarily designed to produce a structure that permits transfer of the paternal genomic information into the egg. It is generally accepted

179

ADVANCES IN GENETICS, Vol. 23

that no other major functions in embryonic development are performed by the male gametes. The difficulties in studying spermatogenesis, as in investigating any other developmental pathway in eukaryotes, arise from the necessity to approach several levels of investigation (genetical, cytological, ultrastructural, and molecular parameters) simultaneously to achieve more than a purely descriptive picture. In most systems studied so far, an integration of the different levels of research has not been achieved yet. Obviously, each organism studied will offer advantages and disadvantages for experimentation. An exceptional situation seems to exist for *Drosophila,* which moves more and more into the focus of molecular and developmental studies. The possibility of carrying out genetic manipulation, the relatively small size of the genome, and the extensive classic knowledge of this organism make it the favored subject of modern molecular biology.

The investigation of spermatogenesis in *Drosophila* has a long tradition. Its origin reaches back to early days of *Drosophila* genetics, when it was recognized that the Y chromosome plays an important role in this developmental process (Bridges, 1916; Stern, 1927, 1929). A particular advantage of spermatogenesis in *Drosophila* is the autonomy of germ cell development in testes (Stern and Hadorn, 1938), which is not dependent on regulatory signals from other tissues as in mammals. Two major reviews deal with the spermatogenesis of *Drosophila melanogaster* (Cooper, 1950; Lindsley and Tokuyasu, 1980). They assess the state of knowledge of the cytogenetics of the Y chromosome and the more general genetic and morphogenetic aspects of spermatogenesis. Molecular aspects of spermatogenesis were, however, not considered because of the lack of information.

The work of our laboratory is dedicated to the study of spermatogenesis in *Drosophila hydei.* It is our intention to integrate the knowledge obtained at the levels of genetics, cytology, and ultrastructural research with molecular data. All data available for *D. hydei* were restricted to certain more specialized topics that can be particularly well studied in *D. hydei* (for reviews, see Hess and Meyer, 1968; Hess, 1976, 1980). We started to investigate systematically the ultrastructural and cytological events in the male germ line of *D. hydei* and extended this approach to the fine structure of the genes involved in spermatogenesis from the genetic and molecular points of view. The purpose of this review is to summarize the knowledge on spermatogenesis in *D. hydei* and develop some working concepts for further studies. It will also become apparent that the study of spermatogenesis in *D. hydei* provides a particularly favorable opportunity to explore molecular events at the chromosomal level by using cytological parameters. Unfortunately, the limitations of space permit us to draw only a few comparisons to *D. melanogaster*. Spermatogenesis in both species may, however, in general not be that much different as prior studies have implied.

II. Development of the Male Gamete

A. GENERAL DESCRIPTION

The primordial germ cells of *D. hydei* are located in the apex of the testes, which are long, tubular, highly coiled organs, approximately 25 mm in length. By mitotic divisions groups of spermatogonia arise from the stem cells and become surrounded by two cyst cells. All cells derived by mitotic divisions from one primary spermatogonium remain thus clustered within one cyst and develop in synchrony (see Fig. 1b). The cells within one cyst are connected by intercellular bridges (Meyer, 1963). Such intercellular connections are found through most stages of spermatogenesis. At an eight-cell stage the cells enter the meiotic prophase and are designated as primary spermatocytes. During the spermatocyte stage not only the amount of cytoplasm increases, but also the size of the nuclei enlarges. This enlargement of the nuclei is accompanied by the attachment of the nucleolus to the nuclear membrane and the development of conspicuous nuclear structures. The latter can be seen in phase contrast microscopy (Figs. 1b and 2a). These structures were identified as lampbrush loops, formed by the Y chromosome (Meyer, 1963). The five autosome pairs and the X chromosome remain almost invisible and can only be recognized after specific staining (Yamasaki, 1977) or by immunofluorescence techniques (Fig. 3b). Chromosome condensation occurs only late during the first meiotic prophase. Different stages of the meiotic prophase cannot be distinguished in *Drosophila* spermatogenesis. It is hence not justified to designate the lampbrush phase of the Y chromosome as diplotene, as sometimes suggested. Further drastic changes can be seen in the cytoplasm: the mitochondria increase in number and size (see also Fig. 12a and b). Before the first meiotic division they assemble around the nuclei, while the nuclear structures disappear (Figs. 1c and 5h). The eight-cell primary spermatocyte stage is followed by the two meiotic divisions. The aster of each spindle pole is formed around the two centrioles [Figs. 1c and 4a (insert)], which are located at each pole

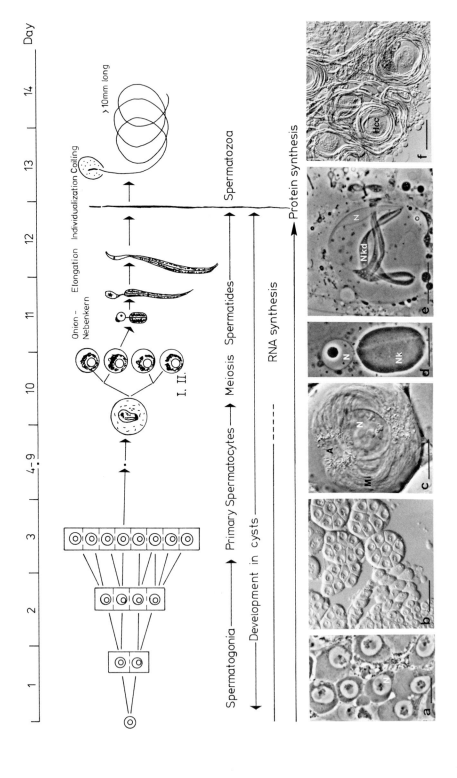

of the nucleus. After the second meiotic division the mitochondria fuse (Fig. 4b) to form an onion-like *nebenkern* (Fig. 1d), which is closely associated with the nucleus of the spermatid. During all these events the cysts formed during early spermatogonial development persist. As a consequence of the meiotic divisions, each cyst (of originally eight spermatocytes) contains 32 haploid spermatids (Meyer, 1963). Each spermatid develops into a spermatozoon. There is some variability in the number of spermatids per cyst (Fig. 1b), which may be due to asynchronous development in the premeiotic phase (Hanna *et al.*, 1982; Liebrich *et al.*, 1982).

The morphogenetic processes that finally produce a mature spermatozoon are initiated in the early spermatid. Our recent investigations (Grond *et al.*, 1984b) have shown that this differentiation process in *D. hydei* closely resembles that described for *D. melanogaster* (Tates, 1971; for review, see Lindsley and Tokuyasu, 1981). This is in contrast to prior reports from other laboratories (Liebrich, 1981a,b) where *in vitro*-grown testes (Fowler, 1973) were studied. Our study was carried out with *in vivo* material dissected from pupae or from adult flies.

Unlike the situation in *D. melanogaster,* only primary spermatocytes are found in the (elliptoid) larval testes of *D. hydei* until puparium formation. Four days after puparium formation the adults hatch. At this time the testes are transformed into tubes. They now contain in addition to all earlier stages of spermatogenesis immature spermatozoa, which have hardly passed the step of individualization. Only after an additional 5 days has fertilization been achieved at 23°C breeding temperature.

The postmeiotic development of the spermatozoon can be subdivided

---

FIG. 1. Spermatogenesis of *Drosophila hydei*. The upper scale shows the time course (Hennig, 1967). The various stages are schematically indicated below this scale, and the time of RNA and protein synthesis is indicated. Photographs: (a) Spermatogonia. Note the nucleolus and the partially condensed chromatin associated with the nucleolus. (b) Cysts with primary spermatocytes. Note the variable number of spermatocytes per cyst. (c) First meiotic division. The mitochondrial layer around the nuclear membrane is clearly visible. Note the residues of the lampbrush loops within the nucleus. (d) Young spermatid. Onion-nebenkern stage. The nucleus contains one large refractive body with smaller spheres associated with it. (e) Elongating spermatid. The nucleus is elongated and no refractive material is present. (f) Coiled spermatozoa after individualization. They are still associated in bundles. The heads are inserted in the head cyst cell. N, Nucleus; Nk, nebenkern; Nkd, nebenkern derivatives; Hcc, head cyst cell; Mi, mitochondria; A, aster. Bars represent 10 μm in (a), (c)–(e), and 50 μm in (b) and (f). Photographs by W. Kühtreiber, C. J. Grond, W. Hennig.

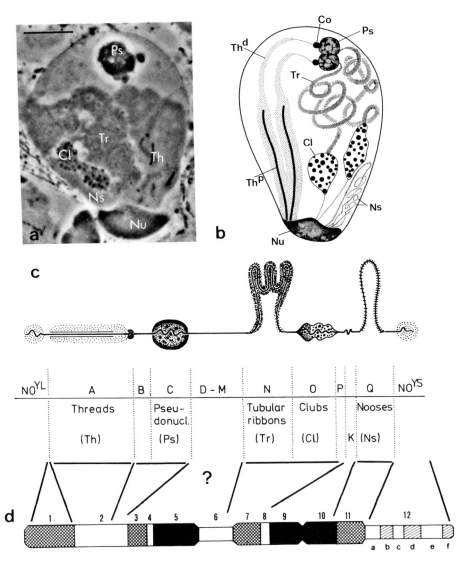

Fig. 2. Primary spermatocyte nucleus with the Y-chromosomal lampbrush loops. (a) *D. hydei,* wild-type in phase contrast. (b) Diagram (Hennig, 1967). (c) Chromatid with the loops and the genetic map according to Hackstein *et al.* (1982). Loci *A–Q,* and the nucleolus organizer regions (NO^YL and NO^YS) in their relationship to the lampbrush loops are indicated. (d) Metaphase Y chromosome with fluorescent and dark regions (1–12) after staining with Hoechst 33258 (Bonaccorsi *et al.,* 1981). The positions of the genetic loci and the lampbrush loops are indicated by black lines (S. Bonaccorsi, J. H. P. Hackstein, and W. Hennig, unpublished data). Nu, Nucleolus; Th, threads (d, distal; p, proximal); Ps, pseudonucleolus; Co, cones; Tr, tubular ribbons; Cl, clubs; Ns, nooses; K, kinetochore. (a)–(c) From Hackstein *et al.* (1982) and (d) from Bonaccorsi *et al.* (1981) with permission.

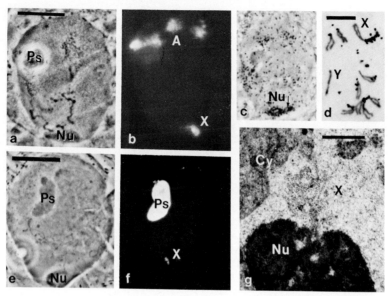

FIG. 3. The X chromosome in primary spermatocytes. (a), (b) Spermatocyte nucleus reacted with histone H2A antiserum. The antiserum was kindly supplied by M. Frasch, Tübingen. The autosomes (A) and the X chromosome (X) are fluorescent; other nuclear structures do not display significant amounts of fluorescence. The H2A concentration must therefore be very low in other chromosome regions. (c) *In situ* hybridization of a spermatocyte nucleus with cRNA complementary to total genomic DNA (Hennig *et al.*, 1974a). The most intense hybridization is in the nucleolus (Nu). (d) Neuroblast metaphase plate after *in situ* hybridization with the same cRNA used in (c). The X heterochromatin is preferentially labeled, as well as the kinetochore regions of some autosomes (Hennig, 1973). (c) and (d) show that the X heterochromatin is in the nucleolus of the spermatocyte nucleus. (e), (f) Spermatocyte nucleus after reaction with the Ps-protein antiserum (Hulsebos *et al.*, 1984). Fluorescence is in the pseudonucleolus (Ps) and in a region close to the nucleolus (Nu), which according to (a)–(d) and (g) must be the X chromosome. (g) Electron micrograph of the nucleolus region of a primary spermatocyte. A structure extends from the nucleolus that represents the X chromosome (Grond *et al.*, 1984a). Cy, Cytoplasm. Bars represent 10 μm in (a), (b), (e), and (f), 5 μm in (d), and 2 μm in (g). (c) and (d) From Hennig *et al.* (1974a) with permission.

into four major developmental steps: (1) the preelongation period, (2) the elongation of the spermatid, (3) the individualization, and (4) the coiling of the spermatozoa. To understand these processes, we focused on the differentiation of three different structural components of the spermatozoon: (1) the head (nucleus) (with the chromosomal material), (2) the axoneme, which forms the central part of the sperm tail, and (3)

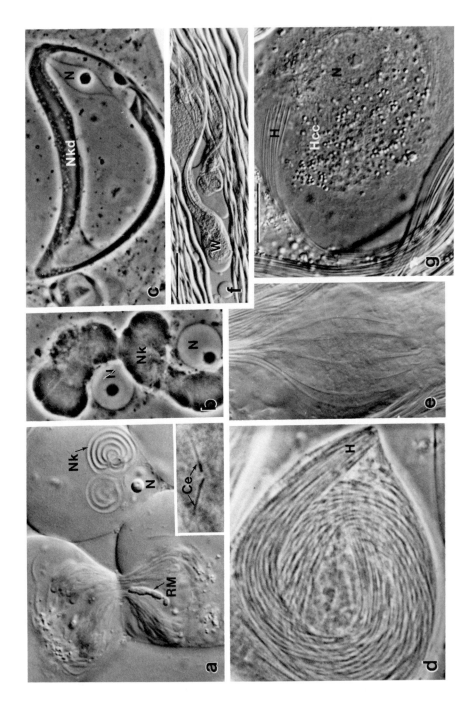

the nebenkern derivatives, which form lateral elements of the sperm tail. All these structures undergo characteristic developmental transformations during the postmeiotic development that can be seen even on the light microscopic level (Figs. 1d–f and 4a–g).

The time course of spermiogenesis in *D. hydei* has been studied by Hennig (1967, 1968) and Leoncini (1977). Their data are summarized in Fig. 1. The initial mitotic activity of the spermatogonia is completed within 2 days, while the subsequent growing phase (spermatocyte stage) requires almost 7 days, nearly 50% of the entire time necessary for spermatogenesis. The major postmeiotic differentiation processes are completed within approximately 2 days, but the spermatozoa need another 5 days to become available for fertilization.

The general metabolism during spermatogenesis has been studied by Hennig (1967). RNA synthesis is restricted to premeiotic stages while protein synthesis proceeds far into postmeiotic development (Fig. 1). The rates of RNA and protein synthesis differ at various stages. The activity of RNA synthesis in the nucleus and nucleolus is at its maximum in the early and middle primary spermatocyte stages (stages I and II, see Fig. 5), and it decreases drastically during late spermatocyte stages (III and IV). Protein synthesis is highest during the middle spermatocyte stage (II) and during early elongation of the spermatids. The occurrence of postmeiotic protein synthesis implies the presence of stable messenger RNA fractions since RNA synthesis has ceased before meiosis.

A surprising phenomenon is the size of the mature spermatozoa of *D.*

FIG. 4. Postmeiotic stages of spermatogenesis of *D. hydei*. (a) Second meiotic division and young spermatid with nebenkern (Nk) and nucleus (N) with refractive bodies. In the meiotic spindle, refractive material of unknown character is visible (RM). Differential interference contrast according to Nomarski. The inset displays two centrioles (Ce) associated with one of the asters before the first meiotic division. Phase contrast. (b) Young spermatids with nucleus (N) and the fusing mitochondria (Nk) forming the nebenkern. This stage is prior to the stage shown in Fig. 1d. Phase contrast. (c) Elongating spermatid. The elongating nucleus (N) still contains refractive bodies (cf. Fig. 1e). Nkd, nebenkern derivatives. Phase contrast. (d) Elongating spermatid in an advanced stage. The nucleus (H) is compact and the nebenkern derivatives are tightly associated with the axoneme. Phase contrast. (e) Cystic bulge (the investment cones are slightly visible). Differential interference contrast. (f) Waste bag (W) at the end of the sperm tails. Differential interference contrast. (g) Head cyst cell (Hcc) with sperm heads (H) and nucleus (N). Magnified from Fig. 1f (lower right, head cyst cell). Differential interference contrast. Bar represents 10 μm. Photographs by W. Kühtreiber, C. J. Grond, and W. Hennig.

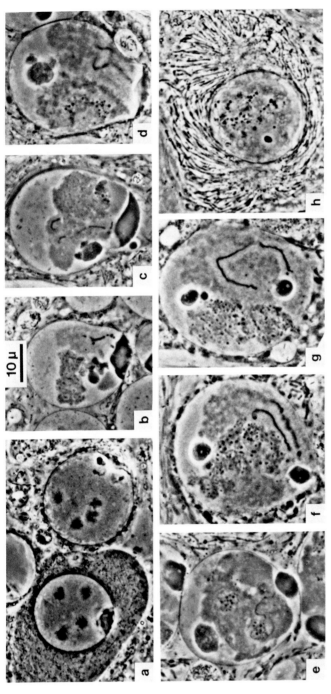

FIG. 5. Primary spermatocyte development. (a) Young spermatocyte nuclei. Only a small nucleus, attached to the nuclear membrane, and refractive material representing the four large autosome pairs is visible. In the right nucleus, the Y-chromosomal loops begin to develop from the nucleolus. (b), (c) Developing Y-chromosomal loops. Stage I. primary spermatocyte nuclei. (b) Stage II. primary spermatocyte nuclei. (b) Stage II. At this stage RNA synthesis is at its highest level. (e), (f) Stage III. The nucleolus begins to detach from the nuclear membrane. The loops are now fully developed (see also Fig. 2a). Stage II. At this stage RNA synthesis is at its highest level. (e), (f) Stage III. The nucleolus begins to detach from the nuclear membrane. The loops are still fully developed, but the activity in RNA synthesis decreases drastically. The autosomes are located in the region distal to the nucleolus (cf. Fig. 3b). (g) Stage IV. The nucleolus has detached from the nuclear membrane, and the loops begin to degrade. RNA synthesis has ceased. (h) First meiotic division. Note the residues of the loops (cf. Fig. 1d). The refractive materials surrounding the nuclear membrane are mitochondria. Phase contrast from Hennig (1967).

hydei. In the literature a length of 6.6 mm has been reported for wild-type spermatozoa. A direct influence of the Y chromosome on the size of the spermatozoa was deduced from data indicating that XYY males have spermatozoa 12 mm in length and that partial duplications of the Y chromosome lead to an intermediate length of the spermatozoa (Hess and Meyer, 1963a, 1968; Hess, 1966). Our recent investigations prove these data incorrect and imply that spermatozoa of XY males are considerably longer than reported before. We isolated spermatozoa from wild-type testes by use of a micromanipulator directly under microscopic control and measured their length under a scanning microscope. It turned out to be almost impossible to recover intact spermatozoa because of their extreme length and their positioning in bundles within the testis. The longest spermatozoa we measured were not completely extracted from the testis, but their length exceeds 10 mm (Grond *et al.*, 1984b). This also implies that the increase in length reported for (partial) Y duplications must be reestablished.

B. PREMEIOTIC DEVELOPMENT: THE PRIMARY SPERMATOCYTE STAGE

1. General Description

From the general description of spermatogenesis it is evident that the primary spermatocyte stage is a prominent period. Not only does it extend over almost half of the entire time course of spermatogenesis, but also its cytology suggests an important role in the development. In early spermatocytes the Y chromosome begins to form five lampbrush loops, which remain present during all the spermatocyte development (Fig. 5) (Meyer, 1963). Lampbrush loops have conventionally been interpreted as active genetic loci (for review, see Callan, 1982) comparable to the puffs of polytene chromosomes (Beermann, 1952). Based on the consideration that the Y chromosome carries genetic information important for spermatogenesis in *D. melanogaster* (Bridges, 1916; Stern, 1927, 1929; Brosseau, 1960; cf. also Gatti and Pimpinelli, 1983) as well as in *D. hydei* (Meyer *et al.*, 1961; see also Hackstein *et al.*, 1982) in addition to ribosomal DNA (Meyer and Hennig, 1974; Hennig *et al.*, 1975; Schäfer and Kunz, 1975) the importance of the spermatocyte stage becomes evident. Since the postmeiotic genome is inactive and 50% of the spermatids lose their Y chromosome due to their segregation during meiosis, the Y chromosome is expected to become active before meiosis. (It could of course be argued that it is not essential that the genetic information of the Y chromsome is made available before meiosis since the postmeiotic development occurs within cysts with many intercellular connections between the individual spermatids. This would permit exchange of molecules not available in each cell.)

From cytological, ultrastructural, and molecular investigations of the five Y-chromosomal lampbrush loops the following picture can now be composed. During the early spermatocyte stage (I) (Fig. 5a and b) the lampbrush loops emerge from the region of the nucleolus, which is attached to the nuclear membrane. The individual loops with their characteristic morphology cannot yet be identified. However, *in situ* hybridization experiments with loop-specific DNA sequences (see below) allow visualization of their active transcription during this developmental stage (Fig. 9c). It is not unexpected that the loops develop from a region associated with the nucleolus since the nucleolus is supposedly formed by all three nucleolus organizer regions present in the genome: the two at the ends of the Y and the one in the X heterochromatin. The X chromosome in primary spermatocytes is always attached to the nucleolus (Fig. 3) (see also Section II,A).

The loops develop rapidly into their final distinct morphology and length (cf. Hennig *et al.*, 1974a) as seen in spermatocytes of late stage I and stages II and III (Fig. 5d–f) (see also Grond *et al.*, 1984a). The unfolding of the loops leads to a characteristic and unique structure of each of the loops but also to a characteristic location within the nucleus. This general pattern of the loop location within the nucleus is diagrammatically shown in Fig. 2b. Each loop has been designated with a name according to its morphology in phase contrast (Meyer, 1963; Hess, 1965). The *nooses, clubs,* and *threads* are always in the neighborhood of the nucleolus; the *tubular ribbons* and the *pseudonucleolus* are distal compared to the nucleolus. From the cytogenetic analysis of the sequence of the loop pairs within the Y chromosome (Hess, 1965), their relative locations in the Y are known (see Fig. 2). Hennig (1967) has observed that all the loop pairs are directly connected with the next following loop pair by thin connecting threads, which can be more easily recognized in spermatocyte nuclei of males partially deficient for parts of the Y (Fig. 6). This led to a model of the Y chromosome that is based on chromatid separation all along its length (except for the kinetochore and the nucleolus organizer regions) and on the postulate of a continuous extension of the chromatids throughout the loops without returning to an axial chromomere. The likelihood of such a structural state has been discussed in detail by Hennig (1967). Essential arguments are the positional arrangement of

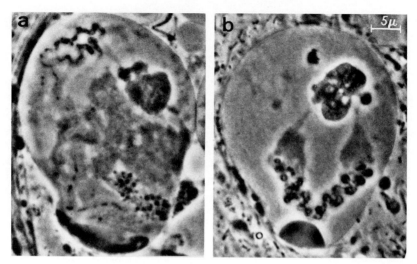

FIG. 6. Connections between threads and pseudonucleolus. (a) Mutant Y*tu-d* (Hess, 1964), (b) Mutant Y*tu-p* (Hess, 1964); only the pseudonucleolus and the modified threads (cf. Fig. 15a) are present in this spermatocyte nucleus. Phase contrast, from Hennig (1967).

the various loops in wild-type and mutant spermatocyte nuclei and the general phenomenon that meiotic anaphase chromosomes are usually only attached in their kinetochores. This separation of the chromatids in *D. hydei* could take place during the prophase. Additional arguments for this model are mentioned in Section II,B,5 of this review. The model of Hess (1980) accepts direct connections between some of the loops but assumes the existence of chromomeres and partially condensed regions of the Y chromosome. However, cytological evidence strongly conflicts with the (see also Section II,B,5) existence of axial chromomeres. There is also no experimental evidence for the location of major parts of the Y chromosome in the nucleolus, as postulated by Hess (1980).

The occurrence of lampbrush looplike structures in primary spermatocyte nuclei appears to be a more general phenomenon in *Drosophila* spermatogenesis (Hess and Meyer, 1963a; Hess, 1967a). However, only a few species display a loop morphology that would permit a detailed cytological description. *D. hydei* and two sibling species, *D. neohydei* and *D. eohydei*, are exceptional in this respect (cf. I. Hennig, 1978). Since the presence of Y-chromosomal lampbrush loops in *D.*

hydei represents the only substantial reason to study spermatogenesis in this species, their nature will be discussed in detail in this review.

2. *The Structure of the Lampbrush Y Chromosome*

The process of loop development has been followed for each loop individually (Grond *et al.*, 1984a). The typical cytological and ultra-structural elements of each loop appear soon after the first steps of unfolding. Developmental changes in the loop structure are restricted to an increase in volume of some of their components such as the Cl-grana and the Ps-channels (Fig. 7). No new elements appear during the subsequent spermatocyte development. Contrary to the results of earlier studies (discussed by Hess, 1980, p. 19), the ultrastructure of the various loops has, however, been found to display defined loop-specific differences.

a. Nooses. The loop pair nooses (and probably also the tubular ribbons) resemble the "classic" type of lampbrush loops. Their DNA is copied into transcripts of enormous size that in Miller spreading experiments have been visualized (Grond *et al.*, 1983). The growing transcripts display a high degree of secondary structure with knobby and branched regions (Fig. 8). They are densely packed along the DNP axis and indicate that the entire loop forms a single transcription unit of at least 260-kb DNA. The structure seen in the Miller spreads closely resembles the *in vivo* structure, as can be concluded from the ultrastructural analysis of spermatocytes (Grond *et al.*, 1983). We assume therefore that no structural components of the loop pair nooses have been lost during spreading.

The analysis of this loop could be extended to the molecular level. With the aid of recombinant DNA clones derived from this loop it has been demonstrated that the loop is composed of repetitive DNA sequences (Fig. 9). The shortest subrepeat, which is found within larger superimposed repeats, is ~70 nucleotides. DNA sequencing data indicate a high degree of homology between various subrepeats (R. Bremers and P. Vogt, unpublished data).

The transcripts homologous to the cloned DNA sequences are highly heterogeneous in size and range between 1000 and 100,000 nucleotides, without displaying a preference for a discrete size class. The transcripts must be processed from the giant primary transcripts of at least 260,000 nucleotides, as they are seen in Miller spreads. DNA sequencing data reveal no indications for the presence of long protein-coding regions in the DNA sequences. At least parts of the transcripts are therefore not translated, a situation which is common for other

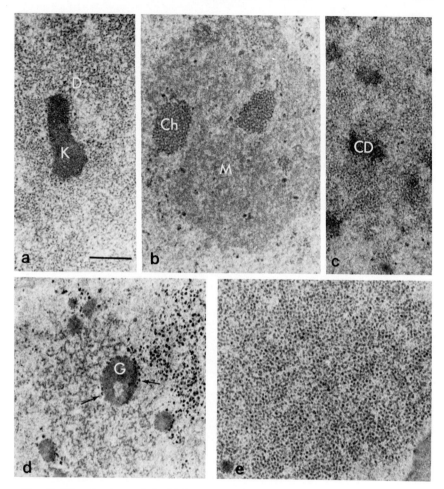

Fig. 7. Ultrastructure of the Y-chromosomal lampbrush loops of *D. hydei*. (a) the threads, with Th-compact-part (K) and the Th-diffuse-part (D) with 35- to 40-nm particles. (b) Pseudonucleolus, with the Ps-channels (Ch) embedded within a fibrillar matrix (M) with 25- to 30-nm particles. (c) Tubular ribbons, with Tr compact dots (CD) embedded in diffuse tubular material. (d) Clubs, with Cl-grana (G) surrounded by diffuse fibrillar material; 40- to 45-nm particles are attached to the Cl-grana (arrow). (e) Nooses, composed of Ns granulae and connecting fibers. Each loop can be identified on basis of its particular structure. EM micrographs: C. J. Grond. Bar represents 0.1 μm.

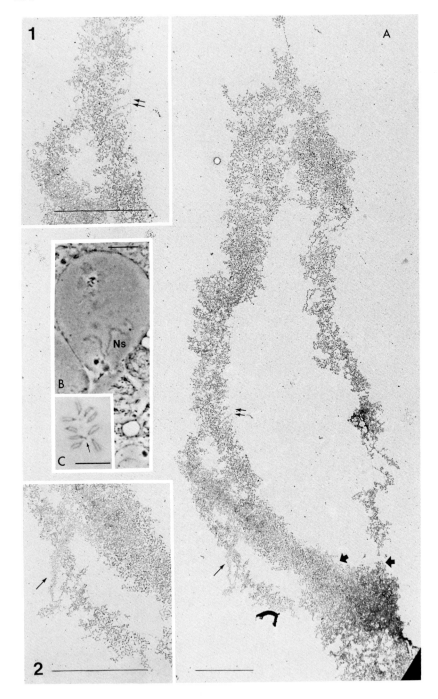

eukaryotic genes. Although the molecular structure of the nooses reminds one at first glance of the structure of the lampbrush loop coding for histones in *Notophthalmus* (Diaz *et al.*, 1981; Stephenson *et al.*, 1981), their properties are not directly comparable. In *Notophthalmus* the histone gene clusters are separated by long runs of a fairly regular, tandemly repeated DNA sequence. In the nooses, the repeated DNA sequences are much more divergent, and we have no evidence for the presence of gene clusters between these DNA repeats. It might therefore be questioned whether this loop pair contains translatable sequences at all. An alternative biological function is considered in Section II,B,4.

b. *Tubular Ribbons.* The tubular ribbons have so far been analyzed on an ultrastructural level (Meyer, 1963; Grond *et al.*, 1984a) and by recombinant DNA clones (Lifschytz *et al.*, 1983). The molecular data indicate a composition of repeated DNA sequences and their transcription (Lifschytz *et al.*, 1983). It cannot be determined from these data whether the loops have a repeat structure comparable to the structure of the Nooses, since no restriction analysis or mapping data are available. The ultrastructural data argue for a conventional lampbrush structure (Grond *et al.*, 1984a).

c. *Threads, Pseudonucleolus, and Clubs.* A different picture emerges for the residual three loop pairs: threads, pseudonucleolus, and clubs (Grond *et al.*, 1984a). It is based on ultrastructural arguments. Each of these loop pairs contains several distinctly different components (Fig. 7). Small particles of a diameter between 35 and 45 nm, most likely representing RNP complexes, occur in all three loop pairs. Cytochemical experiments indicate that these are the only constituents of the loops accommodating significant amounts of RNA. Other, more voluminous structures of a rather uniform composition, such as the Th-compact-part and the Cl-grana and the Ps-channels (Fig. 7), must mainly or exclusively be composed of protein. These loop compo-

---

FIG. 8. Lampbrush loop nooses, visualized with the Miller spreading technique (Grond *et al.*, 1983). The length of the loop between the two thick arrows is 47.5 μm. The DNA content is estimated to be 260 kb. There is only one initiation point within the loop (right side) and the transcripts grow over all the length of the loop. Insets 1 and 2 display some sections of the loop in higher magnification. In inset 1 the DNA axis is visible, and in inset 2 a ribosomal RNA gene is seen that is displaced from the nucleolus. (B) T(X;Y)58/0 spermatocyte nucleus. Only the nooses (Ns) are seen because of the deletion of the rest of the Y chromosome. Such nuclei were spread to obtain the picture displayed in (a). (C) Neuroblast metaphase of a T(X;Y)58/0 male, displaying the small Y translocation fragment (arrow). Bar represents 10 μm. From Grond *et al.* (1983).

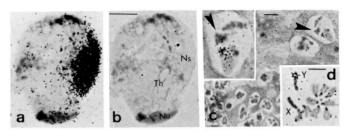

FIG. 9. *In situ* hybridization with a Y-chromosomal DNA sequence cloned in pBR322 (clone PY9) (Vogt *et al.*, 1982). This sequence is derived from the nooses as *in situ* hybridization to the transcripts in spermatocyte nuclei indicate (a and b). In young spermatocytes with developing loops (stage I, Fig. 5a and b), the first transcripts are found by this method (c). The inset displays a magnified spermatocyte nucleus as indicated with the arrowhead. In younger spermatocyte and spermatogonial nuclei (lower part of the picture) no *in situ* hybridization to transcripts is obtained, which indicates that the nooses are inactive in these developmental stages. (d) *In situ* hybridization to metaphase chromosomes from neuroblasts. The cloned DNA sequences hybridize heavily to the short arm of the Y chromosome, which carries the loop-forming site for the nooses (Fig. 2c). No reaction is obtained with polytene chromosomes. Nu, Nucleolus; Th, threads; Ns, nooses. Phase contrast after Giemsa staining. Bar represents 10 μm in (a)– (c) and 5μm in (d).

nents increase in volume during the spermatocyte development (cf. W. Hennig, 1977, 1978; Grond *et al.*, 1984a), which argues for a progressive accumulation of protein within these loops.

Evidence on transcription patterns of the threads and the pseudonucleoli comes from Miller spreading experiments (de Loos *et al.*, 1984) and allows further conclusions on the structure of these genetic loci. Chromatin spreading displays a transcriptional pattern highly characteristic for each of both loops and different from the pattern found for the nooses (Fig. 10). Both loops carry transcripts of giant sizes but at much lower density than the nooses. They can be easily distinguished from their secondary structure, a finding that differs fundamentally from earlier observations on the transcript structure of lampbrush loops in other organisms (Scheer *et al.*, 1976; Angelier and Lacroix, 1975). The complex structure of Y-chromosomal transcripts has already been demonstrated by Glätzer (1975) and Glätzer and Meyer (1981). These authors suggested some correlations between certain types of transcripts and distinct loops, but generally refused to accept a loop specificity of transcript structures. However, from our studies it is evident that the secondary and tertiary structure of the transcripts is highly characteristic for each of the loops studied so far. The threads carry transcripts at an average distance of 2 μm, composed of a long linear portion and a highly folded part close to the DNA axis (i.e., in the final parts of the transcription unit). The transcripts in the pseudonucleolus are in an average distance of 3 μm. They have a highly branched secondary structure over all their length. The length exceeds the length of the transcripts of the nooses (260 kb) considerably and is probably more than 1000 kb. The threads as well as the pseudonucleolus seem to contain only one main transcription unit, although in the pseudonucleolus secondary initiation points for transcription are often used.

The transcription patterns of the threads and the pseudonucleolus are not easily recognized in the cytological and ultrastructural picture of these loops. The major portion of material in the loops appears unrelated to the transcript structure seen in the Miller spreads. Since a preferentially proteinaceous nature was derived for the electron-dense

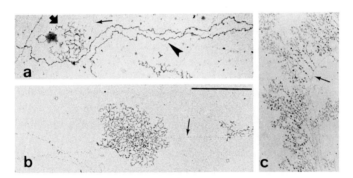

FIG. 10. Transcripts from (a) threads, (b) pseudonucleolus, and (c) nooses. Each loop displays a highly characteristic secondary structure in the transcripts and in the distances between the transcripts. The transcripts of the threads are composed of a bushlike portion (thick arrow) and long distal RNA molecule, which is thicker than the RNA in the bushlike region and therefore probably associated with relatively large amounts of protein (arrowhead). The pseudonucleolus contains only bushlike transcripts, which are much bigger than the bushlike portion in the thread transcripts (a). Often smaller transcripts are found between the great bushlike transcripts indicating secondary initiation sites within the loop (right side of b). Such secondary initiation points have not been found in other loops. The distance between transcripts in the threads and pseudonucleolus is large compared with the transcript density in the nooses (c). The nooses' transcripts also display a high degree of secondary structure but they are not as large as the transcripts of the other two loops. They initiate more frequently than do the other loops. The arrows in (a)–(c) indicate the DNA axis. Bar represents 1 μm. Electron micrographs by R. Dijkhof, F. de Loos, and I. Siegmund. (c) From Grond *et al.* (1983).

components of these loops (Grond *et al.*, 1984a), the structure of the threads may be described schematically as shown in Fig. 11.

In evaluating these various features we arrive at the conclusion that the thread, pseudonucleolus, and club loop pairs are different from the type of the nooses (and probably the tubular ribbons), and therefore different from classic lampbrush loops. It must be recalled that also in amphibian oocytes, which contain the classic lampbrush chromosomes, exceptional loop pairs can be distinguished that are cytologically comparable to the threads, clubs, or pseudonucleolus (see Meyer, 1963; Callan, 1982). We may therefore assume that a small number of loci have special properties in chromosomal metabolism during transcription. Whether they are of any relevance with respect to the post-transcriptional processing, packaging, or other modifications of RNA (cf. Risau *et al.*, 1983) or have an entirely different significance remains an open question. It can, however, not be questioned that these loci are actively transcribed, but contrary to the situation in most of the other active loci they accumulate proteins. From our data it is unlikely that the proteins are coded by these genetic loci (see Section III). Also, Callan (1982) considers the possibility that some of the

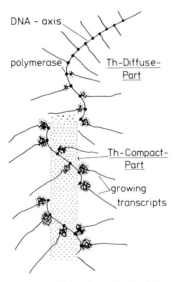

FIG. 11. Diagram of the structure of the threads. The Th-compact-part is proposed to be due to proteins assembled by the transcripts due to their secondary structure (compare to Fig. 10a). For details, see text.

lampbrush loops represent loci required for the accumulation of products of other genetic loci. Such an accumulation of particular proteins, probably not coded in the particular loops, has indeed been observed for one of the Y-chromosomal loops, the pseudonucleolus (Hulsebos *et al.*, 1984, Kremer and Hennig, 1984), but seems also to be true for the threads and clubs (Grond *et al.*, 1984a). Although these findings coincide with Callan's idea, it is difficult to agree with his interpretation that the accumulation of proteins is related to the stabilization and packaging of messenger RNAs from other chromosomal loci. In view of the relatively small proportion of RNA found in these loops it seems more likely that the proteins themselves are important, possibly as chromosomal proteins during meiosis and the postmeiotic development or as proteins related to some other nuclear functions. This question will be discussed in more detail in Section II,B,4.

d. *Cones.* An unclear situation exists with respect to the character of the cones. These structures are associated with the pseudonucleolus (Fig. 2), but may also exist separately in appropriate deletion strains such as T(X;Y)74, which displays only the thread loop pair. In genetic terms, the Y chromosome fragment of this strain carries loci *A* and *B*. The ultrastructural properties of the cones as well as their reaction with Ps-serum (Section II,B,2–3) suggest, however, that they may be parts of the pseudonucleolus. Their real nature will probably only be clarified in connection with the analysis of the molecular fine structure of loci *A–C*.

3. Nature and Origin of Loop Constituents

We obtained some evidence for the origin of the proteins accumulated in some of the Y-chromosomal lampbrush loops. From microelectrophoretic studies it had earlier been determined that loop proteins are major constituents of the spermatocyte nucleus (Fig. 13) (W. Hennig, unpublished; see Hennig, 1977). At least some of the major loop proteins turned out to be loop specific as the studies of nuclei from species hybrids indicated. It was also concluded that such proteins are not coded for by the Y chromosome.

Hulsebos *et al.* (1984) described the identification of a loop-specific chromosomal protein (Ps-protein) in primary spermatocytes that is heavily enriched in the pseudonucleolus (Fig. 3e and f). This Ps-protein ($M_r = 80,000$) is tissue specific and present also in other, distantly related *Drosophila* species such as *D. melanogaster, D. virilis,* and *D. mulleri.* At least in *D. melanogaster* it is not exclusively associated with a Y-chromosomal site since it is also found in spermatocyte nuclei

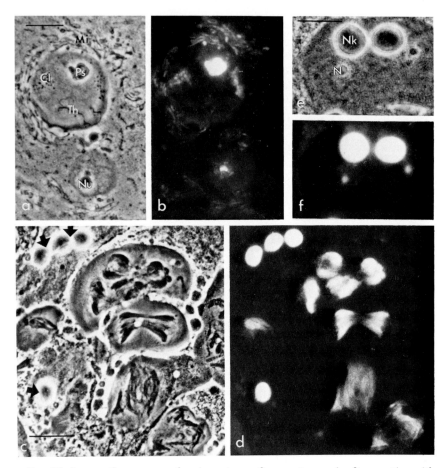

FIG. 12. Immunofluorescence of various stages of spermatogenesis after reaction with histone H1 antiserum. The antiserum was kindly supplied by Dr. H. Saumweber, Tübingen. (a), (b) Spermatocyte nuclei (stages I and IV). Note the strong fluorescence with the pseudonucleolus (Ps) and the Cl-grana (Cl, Clubs) in the stage IV spermatocyte nucleus. Mitochondria (Mi), which also display strong fluorescence (cf. Fig. 5h), are assembled around this nucleus. In the spermatogonial nucleus a distinct region close to the nucleolus (Nu) is fluorescent. (c), (d) Meiotic spindles (cf. Fig. 4a) and young spermatids (onion-nebenkern stage) strongly reacting with the H1 antiserum. The spindles are fluorescent because of the large numbers of mitochondria assembled around the nuclear membrane (Figs. 5h and 1c). In young spermatids only the onion-like nebenkern (Nk) reacts initially, but soon thereafter one of the refractive bodies inside the nucleus (N) is also fluorescent (e and f). Phase contrast and fluorescence photographs by J. M. J. Kremer and W. Hennig. Bars represent 10 μm.

of XO males in a distribution indicating an association with distinct parts of the chromatin. Our studies imply that this protein is not coded by the Y chromosome but by an autosome or the X.

By immunofluorescence staining with histone H1 antiserum (obtained against H1 of *D. melanogaster*) a strong reaction is obtained with the pseudonucleolus and a less strong reaction with the Cl-grana (J. M. J. Kremer and W. Hennig, 1984). It represents thus another example of a loop-specific protein. In state III and IV spermatocytes and in the meiotic spindle (Fig. 12c and d) a strong fluorescence is, in addition, found in mitochondria (Fig. 12a and b). After meiosis the mitochondrial fusion product, the nebenkern, reacts strongly with H1 antiserum (Fig. 12e,f). This reaction is further found in the nebenkern derivatives, which are major constituents of the sperm tail. It is so far not certain whether this reaction with histone H1 antiserum indicates another protein carrying the same antigenic determinant or whether the pseudonucleolus and the mitochondria and their derivatives accumulate somatic histone H1 or a (testis-specific ?) variant. The later assumption is supported by our observations (Hulsebos *et al.*, 1983) that a protein of M_r 35,000 (*sph35*) is found as a constituent of spermatozoa: histone H1 in *Drosophila* has a molecular weight of ~35,000 (Oliver and Chalkley, 1972). Such a protein component is also found as a major protein fraction in spermatocyte nuclei (Fig. 13). It seems likely that sph35 represents a testis-specific histone H1 variant, which is first accumulated in the pseudonucleoli, later in development transferred to the mitochondria, and last maintained in the mitochondrial derivatives in the sperm tails. This raises the question of a possible function of histone H1 in the sperm tail. It should be recalled that during fertilization of the egg the spermatozoon in its entirety is incorporated into the egg in *D. melanogaster* (Huettner, 1930; see Sonnenblick, in Demerec, 1950; Hildreth and Lucchesi, 1963). Observations in our group (Grond *et al.*, 1984b) confirm that also in *D. hydei* a complete spermatozoon is found in the egg after fertilization. The H1 might be required to release the chromatin in the male nucleus from the high degree of condensation achieved during the spermatid development (cf. Bode *et al.*, 1977). An alternative possibility is that it is needed during the early development, where considerable amounts of histones are required during the fast cleavages, which in *D. hydei* occur only slightly more slowly than in *D. melanogaster* (unpublished data of D. Engelen and W. Hennig).

The chromosomal origin for the potential histone H1 fraction found in the testis is not yet established. In *D. melanogaster* the major his-

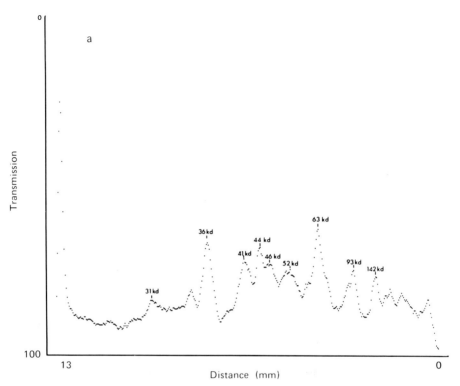

FIG. 13. Microdensitometric scans of protein patterns from (a) *D. hydei* and (b) *D. neohydei* wild-type spermatocyte nuclei after electrophoresis in SDS–polyacrylamide gels (7.5%). Spermatocyte nuclei were isolated and dissolved in 1% SDS–1% β-mercaptoethanol and separated in 5 μl capillaries on 7.5% polyacrylamide gels according to

tone genes occur clustered in chromosome 2 (39E) (Pardue and Birnstiel, 1972). In *D. hydei* a histone cluster is in chromosome 3 (50A) (see Boender, 1983). However, the sensitivity of the of the experiments carried out to localize histone genes is insufficient to detect minor numbers of histone genes in other positions, for example in the Y chromosome. In particular, separated copies of an H1 gene would not be detected in the earlier experiments. It remains thus to be established whether an H1 gene is located on the Y chromosome.

Other evidence for an autosomal origin of loop-associated proteins comes from genetic data (Hackstein *et al.*, 1985). An autosomal recessive male-sterile mutation [*ms(2)1*] has been recovered that in a homozygous state interacts with the loop pair threads. The threads

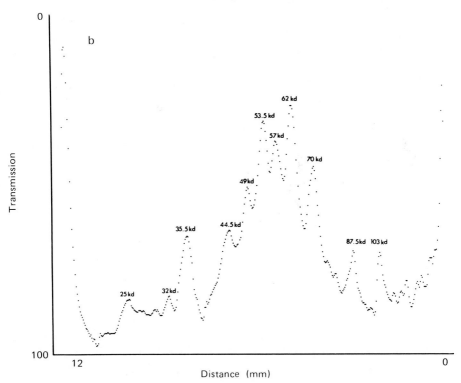

Lugtenberg. The gels were stained with Coomassie brilliant blue R250 and scanned with the aid of an Zeiss Axiomat scanning microscope. A relatively small number of prominent protein bands are seen, including a M_r 35,000 fraction. It is assumed that the majority of the protein bands are derived from the lampbrush loops.

display no Th-compact-part and become almost invisible. If threads, mutated to a "tube-like" morphology, as in the case of *tu-p* (see Fig. 15), (Hess, 1964) or *tu-g* (Hackstein *et al.*, 1982), are combined with the autosomes homozygous for the mutation *ms(2)1*, their compact regions are drastically reduced and almost disappear. Whether the male sterility (due to an interruption of spermatogenesis) is caused primarily by the autosomal locus or secondarily by its effects on thread morphology is so far unclear. In any case, modified thread morphology usually results in the sterility of the male (see Section IV).

Different arguments for an autosomal origin of some of the major loop proteins have already earlier been made from cytogenetic observations (Hennig, 1977, 1978). Between *D. hydei* and a closely related species, *D. neohydei* (see Section V), fertile hybrids can be obtained (Hess and Meyer, 1963b; I. Hennig, 1972, 1978) that permit backcrossing to either parent species. By using autosomal markers the genetic constitution of backcross hybrids can be controlled. In this way hybrid males carrying only the *neohydei* Y chromsome in an autosomal background of *hydei* chromosomes were constructed (Hennig, 1977). The cytology of the lampbrush loops of the *neohydei* Y chromosome is drastically changed compared with the *neohydei* wild-type (Fig. 14) and the structure of the *neohydei* loops in such hybrids approaches the structure of *hydei* loops. This indicates an autosomal origin of at least some of the loop constituents.

Entirely contrasting conclusions regarding loop-associated proteins have been drawn by Kloetzel *et al.* (1981) and Hess (1980). Kloetzel *et al.* studies spermatocyte nuclei from larval testes after labeling proteins in an *in vitro* culture. They found no evidence for differences in the protein content of different loops. The methodological approach in these experiments is, however, insufficient and the results are probably misinterpreted (cf. also Hulsebos *et al.*, 1984).

Further considerations on the origin of loop constituents are based on the occurrence of mutations leading to a modified loop morphology (Hess, 1964; Leoncini, 1977; Hackstein *et al.*, 1982; cf. Callan, 1982, p. 424; see also discussion in Hess, 1980, p. 19). It was argued by Hess that the loop morphology is solely determined by the genetic constitution of the locus itself since the expression of a mutated loop in a combination with a wild-type loop in YY spermatocyte nuclei is determined in a cis fashion. In other words, both the mutated Y chromosome and the wild-type Y chromosome maintain their original structure without apparent trans effects on the other chromosome; i.e., they behave as codominants. These observations of Hess agree with our results (Leoncini, 1977; Hackstein *et al.*, 1982). The Mendelian inheritance of loop morphology has also been demonstrated in *Triturus* lampbrush chromosomes by Callan (1982). All Y-chromosomal mutants with structurally modified loops recovered so far behave in this way. However, contrary to the observations of Hess (1980), almost all of the loop mutants with modified morphology obtained by us are sterile. The autonomy of the loop structure is however not so absolute as assumed by Hess (W. Hennig, unpublished data; cf. also Callan, 1982, p. 423). If duplications of mutated loops of *D. hydei*, such as XY*tu-pYtu-p* are constructed, a clear-cut trans effect is observed (Fig. 15). For both chromosomes the *tu-p* character of the threads is less prominent than in a XY*tu-p* constitution (Fig. 15). Such a picture would be

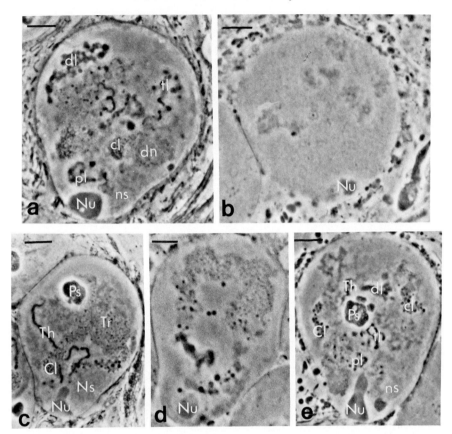

FIG. 14. Spermatocyte nuclei of *D. hydei* (c), *D. neohydei* (a), and various hybrids (d and e). In (d) a spermatocyte nucleus from a hybrid male with *D. hydei* autosomes 2–4 (homozygous) and a *D. neohydei* Y chromosome is shown. Note that the loop morphology has been changed compared with the wild-type *D. neohydei* (a)—it seems more like the structures seen in the *D. hydei* nucleus (c). In (e) a spermatocyte nucleus of a X(*hydei*) Y(*hydei*) Y(*neohydei*) male is shown. The Y loops of both species can be identified, but they are less voluminous than in the respective wild-type spermatocytes. In (b) a XO spermatocyte nucleus of *D. neohydei* is shown. No structures besides the small nucleolus and some autosomal material are seen. Nu, Nucleolus; Th, threads, Ps, pseudonucleolus; Tr, tubular ribbons; Cl, clubs; Ns, nooses; pl, proximal loop; dl, distal loop; tl, threadlike loop; cl, clublike loop; dn, diffuse loop; ns, nooses (small letters indicate loops of the *D. neohydei* Y chromosome). Phase contrast, from I. Hennig, 1978, 1982; W. Hennig, 1978. Bar represents 5 μm.

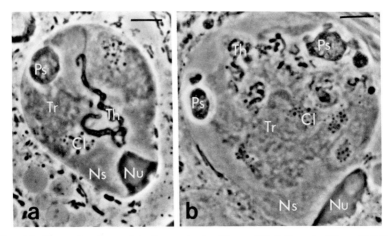

Fig. 15. Spermatocyte nuclei of a mutant with modified threads (Hess, 1964). (a) XY*tu-p* nucleus, (b) XY*tu-p*Y*tu-p* nucleus. Note that the volume of the Th-compact-parts is much smaller in (b) than in (a). This is interpreted as an exhaustion of loop protein in the nucleus. Loop identifications as in Fig. 2. The nucleus in (a) is early stage II and therefore smaller than the nucleus in (b) (stage II/III). However, the XYY nuclei are somewhat larger than wild-type nuclei. Bar represents 5 μm. (a) From Hennig (1967); (b) W. Hennig, original photograph.

expected if the amount of loop-associated proteins in the cell is limited. The *tu-p* mutation is most likely the result of an increased accumulation of protein in the th-compact-part (see Fig. 11). If the synthesis of autosomally coded loop protein is not controlled by the number of Y-chromosomal loops, then an exhaustion of such a protein should be achieved. This should lead to a decrease in the amount of this protein in each loop compared with a XY situation.

It has already been argued by Hennig (1967) that the specific structure of the loops is not in the first instance directed by the loop-associated proteins but rather by the conformation of the transcripts. Such an interpretation would be more consistent with the data available so far. It would explain why the loop morphology is autonomously dependent on the genetic locus itself. Also, the occurrence of mutations of the loop morphology could be explained by assuming a changed primary or secondary structure of the transcripts or a changed length of the transcripts (see Fig. 10 and de Loos *et al.*, 1984).

Our observations on the interference of autosomal male-sterile mutations with the morphology of distinct loops (Section II,B,3) is not compatible with such an interpretation. We therefore prefer to assume that the specific interaction between proteins and the transcripts determines the specific loop morphology. Our interpretation is that autosomally coded, loop-specific chromosomal proteins are required during transcription of the Y-chromosomal loci and contribute to the maintenance of the specific loop morphology. In cases of loop duplications with one loop of a wild-type morphology and another loop of modified morphology, the autosomal proteins would associate with the wild-type or mutated transcripts and thus lead to the differing cytological appearance. This interpretation is the more supported as in certain mutations the Ps-protein, usually associated with the pseudonucleolus, is found in the adjacent loop pair threads, which also display a modified loop morphology (Hackstein *et al.*, 1984).

4. Conclusions on the Structure and Function of the Lampbrush Loops

The discussion of the structure and function of the Y-chromosomal lampbrush loops reveals that their study provides essential insights into the structure and function of lampbrush chromosomes or, more generally, euchromatic genes, that cannot easily be obtained in other systems. This will become even more evident from the subsequent discussion. Here, some general conclusions on the Y-chromosomal lampbrush loops are summarized. Probably all loops consist of single giant transcription units, each of a specific and characteristic pattern of transcription. The length of the transcription units ranges between several hundred and at least 1000 kb. While some of the loops (nooses and probably tubular ribbons) are of a conventional lampbrush loop type, others have different properties. They accumulate molecules, preferentially proteins, of autosomal (or X-chromosomal) origin. Each loop seems to contain at least some loop-specific proteins. The morphology of each loop is primarily determined by the genetic constitution of the loop itself but results from an interaction with proteins of other chromosomal origins.

One of the central question of our work concerns the biological significance of lampbrush loop formation in primary spermatocytes. At present, only speculations can be made in this respect. It has in the past been suggested that the loops are required for the stabilization of messengers that will be used in the postmeiotic phase and are being packaged into proteins. However, based on our observations a different possibility may be envisaged. The highly specific secondary structure of the transcripts in the various loops might be required to provide specific binding sites for distinct proteins, as for example for histone

H1 or a H1-like protein in the pseudonucleolus. These proteins may be required in the postmeiotic condensation of the chromatin, or may, without or together with associated RNA, be transferred into other cellular compartments, for example the mitochondria. RNA might in this context not be of relevance for processes other than binding a particular protein for some time during development. This could explain why at least some of the RNA fractions coded by loops (Lifschytz *et al.*, 1983) appear to be restricted to the nucleus. Also the length heterogeneity of RNA homologous to loop DNA sequences (Vogt *et al.*, 1982; Lifschytz *et al.*, 1983) could be explained on this basis, since the repetitive nature of the nucleotide sequences may not require highly specified processing sites to fulfill the function of protein binding. Such a working model requires further evaluation but seems related to the idea of Callan (1982), who proposed a protein-binding function for some of the lampbrush loops in amphibian oocytes.

5. *The Autosomes and the X Chromosome during the Spermatocyte Stage*

The autosomes cannot be easily visualized in spermatogonia and spermatocytes. However, the use of antibodies against histone core proteins permits a demonstration of the behavior of the autosomes during the premeiotic stages (Fig. 3a and b). In spermatogonia the nuclei are entirely filled with decondensed chromosomal material reacting with core histone antibodies. Transcript *in situ* hybridization with Y-specific cloned DNA sequences provides no evidence for an activation of the Y chromsome in these cells (see Fig. 9). It is therefore reasonable to assume an inactivity of the Y chromosome in spermatogonia (except probably its nucleolus organizer regions). Transcriptionally active chromosomal material in spermatogonial nuclei represents, therefore, the X chromosome and the autosomes.

A different picture emerges in primary spermatocytes. The Y chromosomal loops display no significant interaction with core histone antibodies while the autosomes strongly react. In early spermatocytes (stage I), autosomes and the X chromosome are found close to the nucleolus; in all later stages the autosomes reside in the distal region of the nucleus opposite to the nucleolus and outside the area accommodating the Y-chromosomal loops. Since usually five distinct regions reacting with the core histones can be distinguished in this region of the spermatocyte nucleus, we conclude that the homologous autosomes are paired. A sixth region of fluorescence is consistently found in the neighborhood of the nucleolus (Fig. 3a and b). Since the X chromosome carries a nucleolus organizer region (NOR), which is supposedly active in spermatocytes, it is likely that this fluorescent structure represents the X chromosome. In agreement with this interpretation, in *in situ* hybridization experiments genomic cRNA reacts in spermatocytes exclusively with a region in the nucleolus (Fig. 3c). The same cRNA reacts in metaphase chromosomes preferentially with the heterochromatic arm of the X chromosome (Fig. 3d) (Hennig *et al.*, 1974a; Hennig, 1973). The heterochromatic arm carries the nucleolus organizer region and should thus be present in the nucleolus if the X-chromosomal NOR is active. Also, *in situ* hybridization with rRNA is only found in the nucleolus and the ultrastructural analysis displays chromosomal material associated with the nucleolus (Fig. 3g) that has been interpreted as the X chromosome (Grond *et al.*, 1984a).

The behavior of the autosomes is rather unusual for the meiotic prophase. They are decondensed throughout spermatocyte development. Only at the time of the condensation of the Y chromosome, demonstrated by the breakdown of the Y chromosomal lampbrush loops, they contract. This decondensed condition is in agreement with earlier observations on the high transcriptional activity of the autosomes in spermatocytes (Hennig, 1967). It might be that the entire genome is activated in the spermatocyte stage, as has been postulated for the lampbrush stage in amphibian oocytes (cf. Davidson, 1976).

It is remarkable that the Y-chromosomal loops do not react with core histone antibodies. This is additional evidence in support of a total decondensation of the Y chromosome (see Section II,B). In a decondensed Y chromosome the core histones should be strongly diluted, particularly since actively transcribed chromosome regions are poor in nucleosomes (cf. Grond *et al.*, 1983; Franke *et al.*, 1976). The histones should thus represent a negligible fraction of the total loop proteins. Consistent with this, the histones are not prominent in the protein pattern of wild-type spermatocyte nuclei (Fig. 13).

Contrary immunological evidence has been published by Runggjer-Brändle *et al.* (1981). These authors claim reactions of histone antisera with the Y-chromosomal loops in spermatocytes of *D. hydei*. They observed patterns for the different core histone fractions (H2B, H3, H4) not consistent with the pattern of H2A found by us (Fig. 3a and b). They do not see any reaction with the autosomes of H3 or H4 antisera, while in the H2B pattern the autosomes were not recognized. It must first be pointed out that the cytology of the spermatocytes studied by these authors is entirely inadequate for such studies and must seriously interfere with the identification of the fluorescent regions.

Moreover, as judged from their published figures, the level of fluorescence must be very low. This is reminiscent of nonspecific reactions observed by us with preimmunesera and other antisera not specifically reacting with Y-chromosomal proteins or with antisera applied without much dilution (unpublished data of M. H. J. Ruiters *et al.*). Since the nuclei in the figures published by Rungger-Brändle *et al.* are isolated from the cytoplasm, no judgment of the background fluorescence is possible. We interpret their results as artifacts such as are often connected with cytological immunofluorescence studies (cf. Kurth *et al.*, 1983; Silver and Elgin, 1976) if inadequate cytological conditions or fixation procedures are chosen. The reactions with histone antisera also raise doubts of the validity of the reaction obtained against polymerase B antiserum. The distribution of polymerase B found by Rungger-Brändle *et al.* is in disagreement with the distribution of actively transcribed regions of the loops as established by ultrastructural studies (Grond *et al.*, 1984a) and with the density of polymerases expected from Miller spreading studies (Fig. 10) (de Loos *et al.*, 1984).

C. MEIOSIS

The meiotic divisions are difficult to study because they pass very rapidly (Fig. 1). As a consequence only a few meiotic cells can be found in testes. Some description of meiotic stages has been given by Meyer (1963). A more detailed picture has only recently been obtained by Grond *et al.* (1984c). Only some remarkable points are summarized here.

Meiotic chromosomes are difficult to visualize. Their morphology is usually rather vague and does not allow much differentiation of chromosomal details. Nevertheless, the X heterochromatin and the Y chromosome appear more condensed than the rest of the genome, as expected for heterochromatic chromosome regions (N. Yamasaki, personal communication; Grond *et al.*, 1984c). Remarkably, the X chromosomal nucleolus organizer region is often less condensed than other chromosomal regions, which argues for its activity in primary spermatocytes (see Section II,B,5).

The chromosomes in the secondary spermatocytes could so far not be analyzed. It remains questionable whether any decondensation occurs. The second meiotic division follows very quickly. Immediately thereafter the chromosomes decondense in the spermatid nucleus.

To our suprise, we have not been able to obtain any immunological reaction of meiotic chromosomes with core histone or H1 antibodies (Kremer and Hennig, 1985). This cannot be explained on the basis of a substitution of the histones by other basic proteins, as occurs often during spermiogenesis, since in spermatid nuclei a reaction with core histones and histone H1 is obtained. We therefore assume either that the core histones (as the chromosomal H1, see Section II,C) are masked by chemical modification or that the chromosomes are so tightly packed that even after acetic acid pretreatment the histones do not become available for the immune reaction.

In contrast, the mitochondrial sheet that is formed around the dividing nuclei (Figs. 1c, 4a, 5h, and 12b) strongly reacts with H1 antiserum. Within the mitochondria the reaction is found in all meiotic stages without interruption.

In the meiotic spindle refractive components can often be found that seem not to contain nucleic acids (Fig. 4a). They appear to be of proteinaceous nature and might be derived from the Y-chromosomal lampbrush loops. A direct relationship has so far not been demonstrated. They could not be correlated with constituents of the pseudonucleolus with either the Ps-protein-recognizing antiserum nor with the histone H1 antiserum. Our data exclude the possibility that these components represent chromosomes. The fate of these structures in the postmeiotic development is unclear. For reasons which will be discussed later (Section II,D) it is unlikely that they are homologous to the intranuclear bodies found in spermatid nuclei.

D. POSTMEIOTIC DEVELOPMENT

1. Young Spermatids

After the meiotic divisions, 32 spermatids form a cyst and continue to develop synchroneously. The mitochondrial layer around the meiotic nuclei condenses quickly to form the onionlike nebenkern (Figs. 1c, 4b). Each nebenkern becomes associated with one nucleus and both components are the basic structures giving rise to one spermatozoon. Each of the four spermid nuclei released from the meiotic divisions becomes associated with one of the four centrioles of the primary spermatocyte. From here, the axoneme grows and dissects the nebenkern into two portions of unequal size (Fig. 1c). Both portions will subsequently develop into the nebenkern derivatives associated with the axoneme (Figs. 1d and 4c).

The nucleus of the spermatids undergoes distinct morphological changes. In young spermatids it is round and contains one or a few refractive bodies (Figs. 1d and 4b and c). The number of these bodies increases initially. Then the nuclei extend into a sicklelike shape

(Figs. 4c and 1e). Late in this extension period the nuclear bodies disappear (Fig. 1e). Cytologically, the nuclei then appear homogeneous in their content. Nuclear bodies occur in spermatid nuclei of many other organisms, and have often been designated as nucleoli. It is, however, clear that they have no correlation to the synthesis of ribosomal RNA, as will become evident from the following description.

The membrane of spermatid nuclei is, as in *D. melanogaster,* initially composed of two different regions, one without pores, the other with many pores (the "fenestrated" side). During early spermatid development the pores spread successively all over the nuclear membrane, a process that is probably related to the process of chromatin condensation (Grond *et al.,* 1984b). At the ultrastructural level not only the bodies recognized by light microscopy can be seen within the nuclei but also some regional condensation of the otherwise homogeneous chromatin. During the elongation of the nuclei (Fig. 1e), the chromatin condensation proceeds rapidly, starting at the nuclear sides close to the fenestrated region. At a certain level of condensation the nuclei contract very rapidly, a process that is caused or accompanied by a tight packaging of the chromatin into parallel fibrils. Simultaneously, parts of the nuclear membrane, which now become excessive, are discarded. No further transformations of the chromatin can be recognized in the nuclei during subsequent development. Their fungal length in the mature spermatozoon is 70–75 μm (Grond *et al.,* 1984b).

The process of chromatin condensation in the spermatid nuclei is an important event during the postmeiotic development. It is generally accepted that during spermatogenesis chromosomal proteins are replaced by more basic proteins, probably for protective reasons. For *D. melanogaster* this has been inferred from cytochemical studies (Hauschteck-Jungen and Hartl, 1982). It seems attractive to postulate some Y-chromosomal function in this context. We therefore studied the composition of postmeiotic nuclei by histochemical and immunological means (Kremer and Hennig, 1985). A reaction with core histone antisera in the postmeiotic development is only found in young spermatid nuclei prior to late elongation stages of the nuclei. In the beginning only a very low, diffuse fluorescence is seen. In the elongated nuclei an enhanced fluorescence is found along one side of the nuclear membrane, probably identical with the part of the membrane where chromatin condensation initiates and in a small body close to the insertion site of the centriole. This region is enriched in chromosomal material (Grond *et al.,* 1984b). In all subsequent stages no reaction can be obtained.

Comparable results were obtained with sulfoflavine, a dye specifically interacting with the amino groups of basic proteins (Leemann and Ruch, 1972). The staining patterns obtained in postmeiotic stages resemble in principle those obtained with anti-core histone sera. No reaction is found in more advanced stages of spermatogenesis of *D. hydei*, in contrast to the data of Hauschteck-Jungen and Hartl (1982) for *D. melanogaster,* which were confirmed by us. It is possible that the chromosomal protein constitution and replacement differs in both species. However, we prefer another interpretation, which is based on differential accessibility for the dye or the antisera. As with core histone antiserum we do not obtain any reaction of the sulfoflavine with meiotic chromosomes. During the meiotic divisions the histones are therefore obviously not available for histochemical or immunological reactions, either due to modifications or due to a particularly high degree of packaging. The same arguments could apply to the failure to achieve histochemical or immunological reactions with nuclei during advanced stages of spermiogenesis. From the ultrastructural studies it is obvious that late spermatid nuclei and all subsequent stages contain extremely highly condensed chromatin. It seems in fact not very plausible to expect protein substitution in such a condensed state of the chromatin. Preliminary biochemical studies of the histone content of testes of *D. hydei* give no evidence for the presence of major amounts of protamine-like or other strongly basic proteins in the nuclei. Although this cannot yet be excluded we prefer to assume that either no such substitution takes place at all or that testis-specific histones may be involved in the replacement process that do not differ much in their electrophoretic properties from the somatic histones. Alternatively, or in addition, modifications of the existing histones may contribute to the changed pattern of condensation as found in the meiotic and postmeiotic chromatin.

Of interest in this context is the behavior of histone H1 in the postmeiotic nuclei. In the early spermatid, immunofluorescent staining occurs over all the spermatid nuclei, indicating the presence of H1. Then increasing amounts of H1 are found in the thickest of the refractive bodies inside the spermatid nucleus, while the fluorescence of the nucleoplasm decreases (Fig. 12e and f). In elongation spermatid nuclei the reaction with H1 antiserum has entirely disappeared. This distribution of H1 during the early postmeiotic development suggests that H1 is initially present in the postmeiotic chromatin but is gradually eliminated before the chromatin becomes compacted. It remains an open question whether a replacement by other basic proteins occurs. The nuclear bodies may be considered as accumulations of histones (and other chromosomal proteins) removed from the condensing chromatin. They can therefore not directly related to the refractive material found in the meiotic spindles, which has been interpreted as Y-chromosomal loop material (see Section II,C).

The other main components of the spermatozoon, the axoneme and the nebenkern, undergo fundamental structural changes in the early spermatid. The nebenkern divides into two unequal portions, which become laterally associated with the growing axoneme and elongate in parallel with the axoneme (Fig. 4d). The material inside the nebenkern derivatives is gradually transformed into a paracrystalline structure (Meyer, 1964; Grond *et al.*, 1984b). The axoneme grows out of the centriole, which in the young spermatid becomes associated with the nuclear membrane. Initially, the axoneme contains the characteristic 9 + 2 microtubule structure. Later in development an outer circle of accessory microtubules develops from the A subunit of the peripheral doublets. Additional material is associated with them. The development of these components occurs in a strictly defined sequence (Grond *et al.*, 1984b).

2. Advanced Spermatid Development

After the fundamental structure of a spermatozoon is developed, secondary processes are required to achieve a functional state of the spermatozoon. One of the most obvious processes is the detachment of membranes and other waste accumulated during the spermatid development with the aid of a "cystic bulge" into a "waste bag" (Tokuyasu *et al.*, 1972). This cystic bulge (Fig. 4f) moves along the spermatid bundles (Fig. 4e) (see Lindsley and Tokuyasu, 1981). By this process the elongated spermatids, which are still within a common cyst, become individualized. For *D. melanogaster* it has been assumed that the cystic bulge is actively moving along the spermatid bundle. Our observations in *D. hydei* suggest however, that it, at least in this species, may be the inactive part. The spermatid bundle may, with the aid of the coiling process, actively move toward the proximal end of the testis tubes and in this way pass the cystic bulge, which is in a fixed position inside the testis tube (Grond *et al.*, 1984b). At this time the heads of the spermatozoa are inserted in the head cyst cell, which has a so far unknown function (Figs. 4g and 1f). From this point, the coiling process is initiated that transforms the initially stretched spermatozoa into a circular bundle after they have become individualized by passing through the cystic bulge (Fig. 4e).

The exact description of the postmeiotic spermiomorphogenetic processes are essential for an analysis of the developmental effects induced by genetic defects. In the past it has been postulated that inactivity of genes involved in the postmeiotic development lead to rather unspecific, general defects that can hardly be associated with the absence or misformation of distinct structural components of the spermatozoon (Hess, 1967b; Meyer, 1968; Hess and Meyer, 1968). Doubts about such a view have recently been raised by Hardy *et al.* (1981) for *D. melanogaster.* Our studies seem to support such a view also for *D. hydei.* In genetically defective males usually a wide range of postmeiotic stages can be found without any obvious block of the development at a distinct stage, but ultrastructural studies show that in certain mutants very specific defects in particular morphological components occur. For example, Leoncini (1977) described a temperature-sensitive mutation within the Y chromosome that leads to defective inner doublets in the axoneme. We therefore suspect that more detailed investigation of postmeiotic development in mutants will reveal rather specific defects in the morphogenetic process. Such defects do not necessarily result in a block of the development at a defined stage.

III. Further Molecular Aspects of Spermatogenesis

Our insight into the molecular processes during spermiomorphogenesis is still fragmentary. The first biochemical studies showed that testes contain RNA species complementary to repetitive Y-chromosomal DNA sequences and that such RNA species are still found in the postmeiotic stages (Hennig, 1968; Hennig *et al.,* 1974a). This was consistent with the observation that after meiosis a considerable level of protein synthesis occurs (Hennig, 1967). However, it had not been possible to identify distinct RNA fractions originating from the Y chromosome that could be used to analyze the nature of the genetic information associated with the Y chromosome. The reasons for this difficulty have only recently become evident: Y-chromosomal genes code for giant transcripts that cannot be isolated by any conventional method in an intact form (see Section II,D,2). Such a large size of the transcripts was, in fact, already suspected by Hennig *et al.* (1974a). In RNA preparations from total testes, the transcripts of Y-chromosomal origin occur in heterogeneous sizes rather than in defined molecular weight fractions (Vogt *et al.,* 1982), while (parts of) the primary transcripts may be restricted to the spermatocyte nucleus (Lifschytz *et al.,* 1983).

Even the recovery of transcribed Y-chromosomal DNA sequences has so far not enabled us to understand their function. Searching for translation products has been unsuccessful to this point, but the presence of protein-coding sequences can also not be excluded. A speculative alternative function of loop-forming genes has been discussed in Section II,B,4.

The study of proteins involved into the postmeiotic development has been more informative. The comparison of protein patterns in testes of wild-type and mutant males revealed at least three major protein fractions as constituents of the spermatozoon. They accumulate during the postmeiotic development (Hennig *et al.*, 1974a; Hulsebos *et al.*, 1983). These three proteins (sph35, sph55, and sph155) are structural components of the sperm tails, as immunological studies demonstrated. The sph155 protein resides probably in the nebenkern derivatives (J. Hendrikx *et al.*, unpublished). The sph55 fractions represents tubulins, which must obviously be located in the axonemal microtubuli. The sph 35 fraction is possible the histone H1, which by immunological means has been identified in the mitochondria and their derivatives (see Section II,B,3).

In the first study of the sph155 protein, it was concluded that it is coded for by a Y-chromosomal site (Hennig *et al.*, 1974a). This conclusion was based on the protein patterns found in species hybrids. However, our more detailed investigations showed that all three testis proteins including the sph155 protein are probably of autosomal origin (W. Hennig, unpublished data; Hulsebos *et al.*, 1983). For the tubulin genes it could be established by *in situ* hybridization that they are located in chromosomes 2 and 4 (T. J. M. Hulsebos, W. Hennig, and D. Hacker, unpublished).

The Y chromosome must, however, play an important role in the expression of these proteins. If specific Y chromosomal loci (*O* and *P*) (see Fig. 2) are deleted or mutated, none of the three sph proteins are found in significant amounts (Hulsebos *et al.*, 1983). Preliminary experiments showed that the mRNA, at least of the tubulins, is still present in considerable amounts (T. J. M. Hulsebos and R. C. Brand, unpublished data). This suggests that the Y-chromosomal loci act on a posttranscriptional level in the expression of these genes.

It has already been pointed out that at least some of the chromosomal proteins associated with distinct Y-chromosomal lampbrush loops are of autosomal origin. Such proteins contribute, obviously, to the specific morphology of the loops. They may be accumulated where they are required inside the nucleus (see Section II,D) or may be of regulatory importance in the posttranscriptional metabolism of the RNA at these active loci (see also Risau *et al.*, 1983). The identification and isolation of such genes as well as the isolation of autosomal genes coding for structural components of the spermatozoon open a wide field for further studies. The special properties of the Y-chromosomal lampbrush loops in *D. hydei* will obviously be of fundamental advantage for investigations of chromosomal proteins.

IV. Genetics of the Y Chromosome

The earlier sections of this review imply a central role of the Y chromosome in spermiogenesis. Such a function of the Y chromsome was recognized in the early days of *Drosophila* genetics (Bridges, 1916). It contains male fertility genes essential for spermatogenesis, since XO males do not produce fertile spermatozoa in *D. melanogaster* (Stern, 1927). In *D. hydei* spermiogenesis is interrupted early during spermatid elongation in the absence of the Y chromosome (Hennig *et al.*, 1974b). However, the exceptional genetic properties of the Y chromosome, which does not take part in crossing-over, has no phenotypic marker genes, and carries relatively few genes compared with its DNA content, has in the past interfered with the exploration of its genetic and functional properties (Brosseau, 1960; Cooper, 1950; Kennison, 1981; Gatti and Pimpinelli, 1983). The Y chromosome of *D. hydei* offers in this respect some advantages due to the presence of lampbrush loops formed by some of the fertility genes. These loops can be used as phenotypic markers at the cytological level in genetic experiments and allow some fine mapping of the fertility genes (Hackstein *et al.*, 1982). The initial studies of the Y chromosome were restricted to the loop-forming loci (Hess, 1965, 1967b, 1968) and solely concerned the location of these genes by cytogenetic means. With the aid of translocation chromosomes the linear sequence of the loops and their approximate positions within the Y chromosome were established (Hess, 1965, 1967b). It could also be demonstrated that all loop-forming loci must be present and active to allow normal spermatogenesis (Hess, 1967c; Meyer, 1968) and that duplications of one or more loops cannot functionally compensate for the absence of others (Hess, 1968).

Consequently, studies of the genetic content of the Y chromosome of *D. hydei* were started in our group by Leoncini (1977). He isolated conditional male-sterile mutations in the Y chromosome. They in part turned out not to be associated with loop-forming sites. In continuation of this study, an extended investigation of the genetics of the Y chromsome of *D. hydei* including several hundred newly recovered mutations, which were tested in approximately 4000 crosses, was carried out by Hackstein *et al.* (1982). In this analysis we not only made use of translocation fragments but included X-ray- and EMS-induced male-sterile mutations. The availability of a strain suitable for producing large numbers of XO males (Beck, 1976) was of additional advantage for the cytological part of the experiments. This study led to the genetic map shown in Fig. 2. Several points are obvious from this map. First, there are genetic loci related to lampbrush loop-forming sites and others not forming such loops. Second, each loop is associated with only one complementation group. Third, the number of fertility genes does not exceed 16 and is thus small compared with the size of this chromosome, which contains 9.5% of the genomic DNA (Zacharias *et al.*, 1982). No dominant mutations and no mutations not associated with male fertility were found.

These results are generally consistent with the results obtained for *D. melanogaster* (see Gatti and Pimpinelli, 1983). Also, the distribution of the genetic loci in the Y chromosome of *D. hydei* (Fig. 2d) (Bonaccorsi *et al.*, 1985) resembles closely that of *D. melanogaster* (Gatti and Pimpinelli, 1983). All loci are contained in relatively restricted chromosome regions, which in *D. hydei* are identified as weakly fluorescent after Hoechst 33258 staining (Bonaccorsi *et al.*, 1981). Some blocks of DNA in the Y chromosome must be devoid of genes. It seems, on other hand, that the chromosome regions carrying fertility genes include very long DNA stretches activated during spermatogenesis. This is particularly evident for region 2 (Bonaccorsi *et al.*, 1981) carrying the loci *A, B,* and *C* (associated with the loops threads and pseudonucleolus, see Fig. 2c) of *D. hydei*. From Miller spreading experiments we found that loci *A* and *C* together contain at least 1500 kb DNA. The entire chromosome region 2 accommodating genes *A–C* cannot contain more than 3000 kb of DNA since it represents less than 10% of the Y chromosome. The transcribed portion of the DNA must therefore include a major part of the chromosomal DNA in this chromosome region. These conclusions are consistent with conclusions drawn for *D. melanogaster* from different arguments (Gatti and Pimpinelli, 1983).

A principal question in the early studies of *D. hydei* was whether the expression of lampbrush loops is directly correlated with the activity of a fertility gene or whether the location of fertility genes in loop-forming chromosome regions is a coincidence. The complementation studies of Hess (1967b, 1968) indicated that a full set of active loops is required to complete spermatogenesis, thus suggesting a direct relationship between fertility genes and loop function. This conclusion was substantiated by Hackstein *et al.* (1982) who used EMS-induced male-sterile mutations with a modified loop morphology or the complete absence of loops in spermatocyte nuclei. They showed that in these mutants single complementation groups were affected, which must therefore be directly related to the function of the particular loop. Loops are therefore the morphological expression of the active fertility genes.

Loci *A–C* (which include the threads and pseudonucleolus loop pairs) are of particular interest. The mutations recovered from this chromosome region often include two or even three of the loci. Moreover, double mutations of loci *A* and *C* with an active locus *B* in between have been obtained. Most of these mutations are connected with a modified loop morphology or result in an entire inactivation. The cytogenetic data suggest a strong functional relationship between these loci. With the aid of an antiserum that recognizes a pseudonucleolus-specific protein (the Ps-protein, see Section II,B,3), it could be demonstrated that in two double mutants, in which loci *A* and *C* are affected and produce morphologically modified loops, the Ps-protein is also found in the threads. This suggests strongly that in these mutants transcripts from the pseudonucleolus are found associated with threads, possibly by a defective termination in the pseudonucleolus-forming locus *C* (Hackstein *et al.*, 1984).

Leoncini (1977) has pointed out that mutagenesis of the Y chromosome yields three types of male-sterile mutations: (1) mutations without visible cytological effects on the spermatocyte nuclei, (2) mutations that lead to the disappearance of one or several Y-chromosomal lampbrush loops, and (3) mutants with modified loop morphology. Our recent immunological studies (Hulsebos *et al.*, 1984) allow a more refined distinction of these mutant types. We could demonstrate that even in mutants in which single loops cannot be seen in phase contrast some gene activity is retained. The use of such mutations will be of relevance in studying regulatory events since they allow discrimination between entirely inactive loci and loci with a defective function.

The genetic studies of the Y chromosome of *D. hydei* revealed another kind of information that cannot be easily obtained with other genetic systems. Hess (1968) reported occasional inactivation of lampbrush loops in X–Y translocations that occurred some time after the initial translocation step. In our experiments (Hackstein *et al.*, 1982) we noticed that such inactivations were frequent in certain strains and certain crosses. Similarly, I. Hennig (1982) found that inactivation of translocated Y chromosome fragments is the rule in hybrid translocation chromosomes constructed from a *hydei* X chromosome and a *neohydei* Y chromosome. Recent observations in our laboratory showed that inactivations are also the rule within the pure species *D. neohydei* and in a related species, *D. eohydei* (J. H. P. Hackstein *et al.*, unpublished data). It seems that the inactivation is not a sudden event but that adjacent loci gradually become inactive in subsequent generations and that this process may proceed over a period of 20 generations or more (I. Hennig, 1982; J. H. P. Hackstein, unpublished data). We assume that the inactivation is a gradual but regular process in X–Y translocations and that it is at least in part dependent on the particular structure of the translocation chromosomes.

Other inactivation processes have been observed in another genetic situation (Hackstein *et al.*, 1984). Van Breugel (1970) described the isolation of a Y chromosome of *D. hydei* that is associated with a w^mCo (w^m, white mottled; Co, Confluence) marker derived from the X chromosome w^{m1}. This marker displayed frequent exchanges with the X heterochromatin (van Breugel, 1971). The exchanges involve a region of about seven polytene chromosome bands derived from the w region of the X chromosome (van Breugel, 1971). Beck (1978) used the Y w^mCo chromosome for inducing deletions and recovering free Y chromosome fragments (see Hackstein *et al.*, 1982). From cytogenetic experiments we determined that the X–Y exchanges of the w^mCo region are due to a transposition process in which an insertion occurs within or close to the nucleolus organizer regions (Hackstein *et al.*, 1984). The transposition event is accompanied by the transfer of 28 S IVS sequences normally not present in the Y chromosomal NORs (W. Hennig, P. Vogt, and J. H. P. Hackstein, unpublished data) and usually found only in the X-chromosomal ribosomal RNA genes of *D. hydei* (Renkawitz *et al.*, 1980, 1981). Since IVS sequences in *D. melanogaster* (Long and Dawid, 1980) and in *D. hydei* and *D. neohydei* behave similarly to transposable genetic elements it is not unlikely that IVS sequences are directly involved in the exchanges of the w^mCo segment between X and Y chromosomes (Hennig *et al.*, 1982; W. Hennig, and P. Vogt, unpublished data).

An important observation in the context of these exchanges of the w^mCo region is that they are frequently accompanied by mutations in the Y chromosome (Hackstein *et al.*, 1984). Either modified Y-chromosomal lampbrush loops, inactive loci, or nonfunctional loci without cytologically affected loops can be recovered after an insertion into, or an excision from, the Y chromosome of the w^mCo fragment. The molecular processes underlying the exchanges of the w^mCo fragments between X and Y chromosomes are unknown.

In reviewing the genetics of spermatogenesis we should include the consideration of fertility genes other than on the Y chromosome. The problems of this topic has been discussed at length by Lindsley and Tokuyasu (1980). The information on non-Y-chromosomal fertility genes in *D. hydei* is limited. Some X-linked male-sterile mutants have been recovered and described in Lifschytz (1974, 1975). In our laboratory we isolated X-linked and autosomal male sterile mutants (H. Beck, A. Funk, J. H. P. Hackstein, W. Hennig, G. Jacob, O. Leoncini, and G. Rhode, unpublished data), which can be recovered with high frequency after EMS treatment or X-irradiation of the parents. This suggests that, as in *D. melanogaster*, fertility is influenced by many genetic loci. Many of these effects are, however, pleiotropic, as has been pointed out by Lindsley and Tokuyasu.

An important aspect of X-chromosomal and autosomal male-sterile mutations is that some of them interfere with the activity of Y-chromosomal genes, which may result in the modification of the morphology of one or several loops or in their complete inactivity (Lifschytz, 1974, 1975; Hackstein *et al.*, 1984). Such effects would not be recognized in *D. melanogaster,* although regulatory mutants affecting Y-chromosome activation should also be expected in this species. Mutations of this type will be of great relevance for studying the regulation of Y-chromosomal genes.

V. Use of Species Hybrids in the Study of the Y Chromosome

Some of the species of the *repleta* group are so closely related to *D. hydei* that in crosses between them and *D. hydei* viable hybrids can be obtained (Hess and Meyer, 1963b; I. Hennig, 1972, 1978, 1982; Schäfer, 1978, 1979). Hybrids from crosses with *D. neohydei,* the most closely related species, are fertile in both sexes and allow backcrossing to either parent species. Offspring of such crosses decrease in their fertility in so far as more of the autosomes are homozygous for one species in the presence of a Y chromosome of the other species (I. Hennig, 1972; W. Hennig, 1977; Schäfer, 1978). Such hybrids are nevertheless of interest since they allow various kinds of mapping experiments making use of species-specific properties of each of both species (cf. W. Hennig *et al.*, 1970, 1974a, 1982, 1983; Hulsebos *et al.*, 1983).

The existence of fertile hybrids of *D. hydei* and *D. neohydei* is surprising in view of the differences in their genome sizes. The genome of *D. neohydei* is 20% smaller than that of *D. hydei* (Hennig *et al.*, 1974a; Zacharias *et al.*, 1982). Since the banding patterns of the polytene chromosomes of both species are almost identical (Berendes 1963; I. Hennig, 1978), the major difference in the genome size must be attributed to heterochromatic chromosome regions. The main differences in the karyotype occur in the sex chromosomes, in particular between the X chromosomes. The X chromosome of *D. neohydei* misses the heterochromatic arm almost entirely, which in *D. hydei* is as long as the euchromatic arm. But differences are also found in the kinetochore, associated and in the intercalary heterochromatin of the autosomes (cf. Hennig *et al.*, 1970, 1972). The DNA composition of both species varies considerably with respect to their repetitive components (Hennig *et al.*, 1970). Recently also, more delicate differences have been detected by molecular studies. It has, for example, been demonstrated that three autosomes of *D. neohydei* carry ribosomal RNA genes in the regions of their telomeres (Hennig *et al.*, 1983), while in *D.*

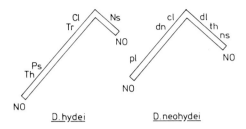

F𝐼G. 16. Comparison of the Y chromosomes of *D. hydei* and *D. neohydei*. The approximate positions of the loop-forming sites are indicated. The data for *D. neohydei* are from I. Hennig (1982) and B. Link (unpublished) (see Hennig, 1977). The positions of the nucleolus organizer regions (NO) are from Hennig *et al.* (1975, 1982).

hydei these sequences are not present. A cytological difference in the telomere regions of autosomes 2, 3, and 4 had been recognized before (I. Hennig, 1978), which demonstrates the potential of cytology to recognize minor genomic differences. Also, the location of repetitive elements, probably mobile genetic elements, such as the 28S-IVS varies between both species.

The spermatocyte nuclei of *D. neohydei* display as prominent structures as they are seen in *D. hydei* (Fig. 14). Their Y chromosomal origin could, similar as for *D. hydei* (Hess and Meyer, 1963b), be derived from the cytology of XO and XYY spermatocytes (I. Hennig, 1978; Hennig *et al.*, 1974) (Fig. 14). While the spermatocyte nuclei of XO males are cytologically almost "empty" and only some autosomal and X chromosome structures are seen besides the nucleolus, the XYY nuclei show all loops in duplicate.

Cytologically, the lampbrush loops in *D. hydei* spermatocyte nuclei differ almost entirely in their loop morphology from those of *D. neohydei* (Fig. 14). Nevertheless, the number of loops is probably identical in both species (I. Hennig, 1982; see also W. Hennig, 1977) (Fig. 16). This suggested some homology between the loops of *D. hydei* and *D. neohydei* despite of the differences in loop morphology (I. Hennig, 1982). Experimental evidence for this has been accumulated in various ways. First, histochemical staining experiments demonstrate comparable staining reactions, especially of the pseudonucleolus of *D. hydei* and the proximal loop of *D. neohydei* (Yamasaki, 1977, 1981). Such a homology had been postulated on the basis of genetic experiments (I. Hennig, 1978, 1982). After translocations between a *D. hydei* X chromosome and a *D. neohydei* Y chromosome were induced, the recovered translocations were used to complement Y-chromosomal deficiencies in *D. hydei*. From successful complementation in flies with such a genetic constitution it was inferred that at least two of the genetic loci in the Y chromosomes of *D. hydei* and *D. neohydei* are sufficiently homologous in their function to allow complementation between both species. Unfortunately, at the time of these experiments the availability of adequate Y-chromosomal mutations in *D. hydei* was restricted. The data are therefore not as complete as data that could be obtained in experiments with presently available mutations (Hackstein and Hennig, 1982).

Additional confirming evidence for the homology of the pseudonucleolus with the proximal loop has recently been obtained by immunological means (Hulsebos *et al.*, 1984). An antiserum, specifically reacting with a loop-specific protein of the pseudonucleolus (the Ps-protein), reacts also with the proximal loop in *D. neohydei* and an identical staining behavior is obtained with Mo^{2+}–hematoxilin (J. H. P. Hackstein, unpublished).

DNA sequences homologous to the nooses' DNA sequences (Vogt *et al.*, 1982) have been found in the Y chromosome of *D. neohydei* and *D. eohydei*. This agrees with the observation that in *D. neohydei* a loop exists with a morphology comparable to that of the nooses of *D. hydei* (I. Hennig, 1982). Moreover, some of the hybrid translocation chromosomes complement *ms(Y)1*, a mutant of the nooses gene in *D. hydei*. The function of the nooses in both species must hence be homologous.

Both species, *D. hydei* and *D. neohydei*, have also been used to identify species-specific proteins in testes (Hennig *et al.*, 1974a; Hulsebos *et al.*, 1983). Since species-specific proteins can be used in species hybrids of different genetic constitutions (Hennig, 1977) to assign the genetic origin of the proteins to certain chromosomes, this is of great value in explicating the poorly developed genetics of *D. hydei* and its sibling species.

VI. Y-Chromosomal DNA Sequences

A final consideration of this review is dedicated to the properties of Y-chromosomal DNA sequences. For a long time the isolation of Y-chromosomal sequences has been almost impossible, although many efforts have been made (Hennig, 1972a,b; Renkawitz, 1978a,b, 1979). Only recombinant DNA techniques allowed the recovery of Y-chromosomal DNA sequences (Lifschytz, 1979; Vogt *et al.*, 1982; Vogt and Hennig, 1983; Hennig *et al.*, 1983). The identification of such sequences and their location within defined segments of the Y chromosome, however, is still an experimental problem (for discussion, see Vogt and Hennig, 1983; Hennig *et al.*, 1983). This is due to a (probably rather general) feature of Y-chromosomal DNA sequences. Y-chromosomal sequences are usually members of families of repeated sequences that are spread over different sites in the genome. According to the restriction analyses of cloned sequences the various members of these sequence families are rather conserved in their nucleotide sequences, which adds additional problems to identification of their chromosomal origin (see Hennig *et al.*, 1983). These experimental problems can now be overcome by applying the microcloning technique (Scalenghe *et al.*, 1981; see Hennig *et al.*, 1983), which allows the isolation of DNA sequences from metaphase Y chromosomes (W. Hennig, unpublished).

The finding that RNA species in testes are homologous to repeated Y-chromosomal DNA sequences (Hennig, 1968; Hennig *et al.*, 1974a) and the consideration that Y-chromosomal genes should represent unique DNA sequences (Williamson, 1972) induced the idea that Y-chromosomal DNA sequences are amplified in testes (Williamson, 1976). It has in the meantime been demonstrated that loop-associated, transcribed DNA sequences belong to the repetitive class of DNA (Vogt *et al.*, 1982; Lifschytz *et al.*, 1983). Amplification could not be confirmed for these DNA sequences (E. Lifschytz, personal communication; P. Huijser, unpublished data).

Whether the dispersed distribution of Y chromosome-related DNA sequences is of any functional or evolutionary relevance remains to be seen in further studies. An important question in subsequent studies of the Y-chromosomal DNA sequences and their related copies in other parts of the genome will concern the significance of the observation that DNA sequences that are found in the Y as well as in other chromosomes are homologous to RNA fractions. It remains to be elucidated which sequences are actually transcribed and whether the active loci are different in different tissues. The investigation of prophase Y chromosomes in neuroblasts (Bonaccorsi *et al.*, 1981) revealed an intriguing property of the Y. Contrary to the expectation from the classic view about heterochromatin, it appears not entirely condensed in the interphase nuclei of somatic tissues. The prophase Y chromosome clearly shows the signs of differential condensation, as do the other chromosomes. More remarkably, the chromosome regions carrying the fertility genes are less condensed than the residual parts of the chromosome. A similar conclusion about a decondensed state of the Y chromosome in mitotic interphase cells must be drawn from *in situ* hybridization studies with Y chromosome-specific or Y chromosome-associated

DNA sequences (Vogt and Hennig, 1983; Hennig *et al.*, 1983). In metaphase chromosomes these sequences are located in a small and well-defined region of the Y, but in interphase nuclei they are found spread over a large area of the nucleus. Since this observation has now been made for several cloned DNA sequences from different Y-chromosomal positions, it appears to be a more general phenomenon. It remains an open question whether this is of any functional significance, but it certainly deserves some attention in further studies.

The occurrence of Y chromosome-associated DNA sequences in other genomic positions is interesting in another context. It has so far generally been accepted that "genes essential for survival" (Hess, 1980) are not carried by the Y chromosome. To maintain such a conclusion it will first have to be established that no loci of an allellic character reside elsewhere in the genome. This again raises the old question of the presence of polygenes in heterochromatin. It might in this context be of interest that certain Y-chromosomal duplications lead to sterility (see, for example, Hackstein *et al.*, 1982; Grell, 1969) and that XYYY constitutions (in contrast to T(X;Y)YY constitutions) are usually not viable in *D. hydei* (W. Hennig, unpublished data) and are sterile in *D. melanogaster* (Williamson and Meidinger, 1979).

The Y-chromosomal DNA sequences of *D. hydei* are to some extent conserved in related species (see Hennig *et al.*, 1983; unpublished data of P. Vogt and P. Huijser). However, their restriction patterns are not identical with those of *D. hydei*. It seems, nevertheless, that also in other species these sequences form a rather homogeneous family of repeated sequences. This implies some mechanisms maintaining the sequence homogenity during the occurrence of genomic rearrangements (Boender, 1983). All these properties are reminiscent of the properties of mobile genetic elements, which are usually considered as much less conservative (see Dowsett, 1983).

From the present knowledge of Y-chromosomal DNA sequences an old speculation on the origin of the Y chromosome can perhaps be rejected. The Y chromosome has often been considered a "degenerated" X chromosome (see Wright, 1976). The sequence relationship found so far does not indicate any particular relationships with the X chromosome nor with any other chromosome.

One might, however, consider another correlation in the evolution of a Y chromosome. The existence of sex chromsomes is a widespread phenomenon in eukaryotes. Nevertheless, often their presence is not strictly correlated with the mechanisms of sex determination (see Wright, 1976). One could then ask why the maintenance or evolution of particular sex chromosomes is evolutionarily favored. Our and other authors' observations on the composition of Y chromosomal DNA sequences and their occurrence in other genomic positions may point to another function connected with sex chromosomes. Sex chromosomes are characteristically rich in heterochromatin. The assumption, initially made on the basis of rather crude experimentation, that heterochromatin is composed simply of large blocks of highly repetitive sequences has now been generally modified to a more refined picture, and a much more complex constitution of heterochromatic chromosome regions is recognized (Hennig *et al.,* 1970; Peacock *et al.,* 1978; and other authors). Blocks of satellite DNA in heterochromatin are interspersed with moderately repetitive DNA sequences. The moderately repetitive DNA fraction in *Drosophila* is now assumed to be composed preferentially of movable genetic elements. If significant amounts of Y-chromosomal repetitive DNA sequences are also transposable sequences, one might draw connections to known functions of transposable elements. For one of these elements, the *P* element, it has been demonstrated that it is involved in the phenomenon of hybrid dysgenesis (for review, see Kidwell, 1983) and other movable genetic elements may have similar properties. Similar to hybrid dysgenesis as observed in different strains of the same species, interspecies hybrids are often sterile or seriously reduced in their fertility. Since hybrids from reciprocal crosses usually differ in their fertility, this effect must primarily be related to their different sex chromosome constitutions. Nevertheless, the involvement of autosomal factors has been demonstrated. One might therefore consider that the introduction of a Y chromosome (or, more generally, any sex chromosome carrying transposable elements or interfering with their activation) into a genome not adapted to this particular sequence type could lead to reduced fertility as a consequence of mechanisms comparable to those responsible for hybrid dysgenesis. The evolution of sex chromosomes may in this way facilitate sympatric isolation within a population and hence contribute to an efficient speciation process.

VII. Conclusion

The description of spermatogenesis reveals that there are many similarities between *D. melanogaster* and *D. hydei.* The entire postmeiotic morphogenetic process is comparable in both species, and the genetic system involved appear to be rather similar, as far as can be determined from our present knowledge. This raises the question whether the study of *D. hydei* is of any advantage or even of any interest. To answer this question I have tried to concentrate in this review on features of the *hydei* system that are of definite advantage for studying gene structure, gene activity, and regulation in general, and the process of spermatogenesis in particular. These advantages are mainly based on the occurrence of Y-chromosomal lampbrush loops. Such loops can be used as cytological markers in genetic experiments. This permits not only a more precise fine mapping of the fertility genes in the Y chromosome but also the detection of genetic effects such as synthetic sterility and loop inactivation processes. Lampbrush loops offer a unique opportunity in the study of regulatory interactions in the eukaryotic genome. Non-Y chromosomal loci, interacting with the activity of the Y chromosome, can be recovered by isolating male-sterile mutations and by screening these mutations cytologically for effects on Y-chromosomal lampbrush loops in spermatocyte nuclei. This permits not only the recovery of loci involved in the general regulation of Y chromosomal gene activity but also, for example, the identification of loci responsible for the production of Y-specific chromosomal proteins. Our present knowledge on chromosomal proteins and their function in relation to chromosome structure and gene expression is minute, even though proteins are major components in gene regulation. Evidently, loop morphology can serve as a tool to dissect such molecular parameters of gene activation. Comparable studies are difficult in *D. melanogaster* since no adequate cytological manifestation of Y-chromosomal gene activity is present.

The unusual length of the spermatozoa of *D. hydei* is another advantage for biochemical studies. It requires the production of considerable amounts of their various structural components. This has already enabled us to identify several sperm constituents while similar approaches in *D. melanogaster* have failed (Ingman-Baker and Candido, 1980). The early interruption of spermiogenesis in XO males and in some other genetic constitutions (for example in the mutants in locus *Q*) is advantageous for studying morphogenesis and the role of the various genetic elements in this developmental process in *D. hydei*. Together with more extended genetic studies in *D. melanogaster* and the transfer of knowledge obtained in the *hydei* system to *D. melanogaster,* the study of *Drosophila* should lead to basic information on the morphogenesis of spermatozoa. This will be of general interest since spermatogenesis in animals exhibits many conserved features that should be based on a common genetic system.

An entirely different aspect of the study of spermatogenesis in *D. hydei* represents the extension of our knowledge on structure and function of lampbrush loops. In particular, genetic experiments as de-

scribed in this review may answer fundamental questions on lampbrush loop function, as they are raised in the reviews of Macgregor
(1980) and Callan (1982).

It is clear that the study of spermatogenesis of *Drosophila* has, due to
the invention of new molecular methodology, come into a new phase. It
is expected that in the forthcoming years essential insight into the
morphogenesis of the spermatozoon can be achieved and that some
elements of the regulation of this process can be identified and understood in their function. Therefore, spermatogenesis also represents an
outstanding system for studying developmental processes in general.
An exciting property of this developmental system is that it allows the
integration of classic biological approaches such as cytology, genetics,
and ultrastructural research with modern biochemical techniques.

ACKNOWLEDGMENTS

This review is dedicated to the discoverer of the Y chromosomal lampbrush loops,
Professor Günther F. Meyer, at the occasion of his retirement, which terminates 20 years
of pleasant and fruitful scientific communication.

I am much indebted to Drs. C. Grond, J. Hackstein, H. Jäckle, and P. Vogt for their
critical reading of the manuscript. The extensive discussions with them and Drs. S.
Bonaccorsi, R. Brand, P. Boender, T. Hulsebos, P. Huijser, O. Leoncini, and M. Ruiters
contributed to the experimental concepts presented in this review. For supplying material I am grateful to Caspar Grond, Hannie Kremer, Wiel Kühtreiber, and Frans de
Loos. I highly appreciate the excellent and engaged assistance provided over many years
by Christiane Wierichs, Ingrid Siegmund, Dorette ten Hacken, and Rosilde Dijkhof. Also
Sigrid Jaweed-Hertling, Veronika Böhme-Hangelberger, Renate Gniffke, Heike van
Zadelhof, Angelika Funk, Ilse Dahlström, Annerose Kurz, Rob Rutten, and Willi Jan-
ssen supported the work of our group with much engagement. Some of the drawings have
been made by Mr. E. Freiberg. For many years Dr. Ingrid Hennig has been of indispensi-
ble help in this work.

For parts of the work financial support was provided by the Deutsche Forschungs-
meinschaft and the EMBO.

REFERENCES

Angelier, N., and Lacroix, J. C. (1975). Complexes de transcription d'origines nucléolaire
    et chromosomique d'ovocytes de *Pleurodeles waltii* et *P. poiretti* (Amphibiens,
    Urodeles). *Chromosoma* **51**, 323–335.
Beck, H. (1976). New compound (1) chromosomes and the production of large quantities
    of X/O males in *Drosophila hydei*. *Genet. Res.* **26**, 313–317.
Beck, H. (1978). Das Y-Chromosom von *Drosophila hydei*: Lokalisation und Isolation der
    Gene mittels Y-Fragmenten. *Arch. Genet.* **51**, 109.
Beermann, W. (1952). Chromomerenkonstanz und spezifische Modifikationen der Chro-
    mosomenstruktur in der Entwicklung und Organdifferenzierung von *Chironomus
    tentans*. *Chromosoma* **5**, 139–198.
Berendes, H. D. (1963). The salivary gland chromosomes of *Drosophila hydei* Sturtevant.
    Chromosoma **14**, 195–206.

Bode, J., Willmitzer, L., and Opatz, K. (1977). On the competition between protamines and histones: Studies directed towards the understanding of spermiogenesis. *Eur. J. Biochem.* **72**, 393–403.

Boender, P. J. (1983). Genomic organization of a family of repeated sequences in *Drosophila hydei*. Ph.D. Thesis, Universiteit Nijmegen.

Bonaccorsi, S., Pimpinelli, S., and Gatti, M. (1981). Cytological dissection of sex chromosome heterochromatin of *Drosophila hydei*. *Chromosoma* **84**, 391–403.

Bonaccorsi, S., Hennig, W., and Hackstein, J. H. P. (1984). Cytogenetic dissection of the Y chromosome of *Drosophila hydei*. Submitted.

Bridges, C. B. (1916). Non-disjunction as proof of the chromosome theory of heredity. *Genetics* **1**, 1–52, 107–163.

Brosseau, G. E. (1960). Genetic analysis of the male fertility factors on the Y chromosome of *Drosphila melanogaster*. *Genetics* **44**, 257–274.

Callan, H. G. (1982). Lampbrush chromosomes. *Proc. Soc. London, B Ser.* **214**, 417–448.

Cooper, K. W. (1950). Normal spermatogenesis in *Drosophila*. *In* "The Biology of *Drosophila*" (G. Demerec, ed.), pp. 1–61. Hafner Publ. Co., New Haven, Connecticut.

Davidson, E. H. (1976). "Gene Activity in Early Development," 2nd ed. Academic Press, New York.

de Loos, F., Dijkhof, R., Grond, C. J., and Hennig, W. (1984). Lampbrush chromosome loop-specificity of transcript morphology in spermatocyte nuclei of *Drosophila hydei*. *EMBO J.* **3**(12).

Demerec, M. (1950). "The Biology of *Drosophila*." Hafner Publ. Co., New Haven, Connecticut.

Diaz, M. O., Barsacchi-Pilone, G., Mahon, K. A., and Gall, J. G. (1981). Transcripts from both strands of a satellite DNA occur on lampbrush chromosome loops of the newt *Notophthalamus*. *Cell* **24**, 649–659.

Dowsett, A. P. (1983). Closely related species of *Drosophila* can contain different libraries of middle repetitive DNA sequences. *Chromosoma* **88**, 104–108.

Fowler, G. L. (1973). *In vitro* cell differentiation in the testes of *Drosophila hydei*. *Cell Differ.* **2**, 33–41.

Franke, W. W., Scheer, U., Trendelenburg, M., Spring, H., and Zentgraf, H. (1976). Absence of nucleosomes in transcriptionally active chromatin. *Cytobios* **13**, 401–434.

Gatti, M., and Pimpinelli, S. (1983). Cytological and genetic analysis of the Y chromosome of *Drosophila melanogaster*. I. Organization of the fertility factors. *Chromosoma* **88**, 349–373.

Glätzer, K. H. (1975). Visualization of gene transcription in spermatocytes of Drosophila hydei. *Chromosoma* **53**, 371–379.

Glätzer, K. H., and Meyer, G. F. (1981). Morphological aspects of the genetic activity in primary spermatocyte nuclei of *Drosophila hydei*. *Biol. Cell.* **41**, 165–172.

Grell, R. F. (1969). Sterility, lethality and segregation rations in XYY males of *Drosophila melanogaster*. *Genetics* **61**, s23–s24.

Grond, C. J., Siegmund, I., and Hennig, W. (1983). Visualization of a lampbrush loop-forming fertility gene in *Drosophila hydei*. *Chromosoma* **88**, 50–56.

Grond, C. J., Rutten, R. G. J., and Hennig, W. (1984a). Ultrastructure of the Y chromosomal lampbrush loops in primary spermatocytes of *Drosophila hydei*. *Chromosoma* **89**, 85–95.

Grond, C. J., Kühtreiber, W., and Hennig, W. (1984b). Spermiogenesis in *Drosophila hydei*. Submitted for publication.

Hackstein, J. H. P., and Hennig, W. (1982). Mutants of Drosophila hydei. *Drosophila Inf. Serv.* **58**, 195–203.

Hackstein, J. H. P., Leoncini, O., Beck, H., Peelen, G., and Hennig, W. (1982). Genetic fine structure of the Y chromosome of *Drosophila hydei*. *Genetics* **101**, 257–277.

Hanna, P. J., Liebrich, W., and Hess, O. (1982). Evidence against a (2)n synchronous increase of spermatogonia to produce spermatocytes in *Drosophila hydei*. *Gamete Res.* **6**, 365–370.

Hardy, R. W., Tokuyasu, K. T., and Lindsley, D. L. (1981). Analysis of spermatogenesis in *Drosophila melanogaster* bearing deletions for Y-chromosome fertility genes. *Chromosoma* **83**, 593–617.

Hauschteck-Jungen, E., and Hartl, D. L. (1982). Defective histone transition during spermiogenesis in heterozygous segregation distorter males of *Drosophila melanogaster*. *Genetics* **101**, 57–69.

Hennig, I. (1972). Vergleichende Untersuchungen am Genom von *Drosophila hydei*, *Drosophila neohydei* und *Drosophila pseudoneohydei* unter besonderer Berücksichtgung des Lampenbürsten-Y-Chromosoms in der Spermatogenese. Diplomarbeit Fachbereich Biologie, Universität Tübingen.

Hennig, I. (1978). Vergleichend-zytologische und genetische Untersuchungen am Genom der Fruchtfliegen-Arten *Drosophila hydei, D. neohydei* und *D. eohydei. Entomol. Ger.* **4**, 211–223.

Hennig, I. (1982). "Hybrid" X–Y translocation chromosomes of *Drosophila hydei* and *D. neohydei. Chromosoma* **86**, 491–508.

Hennig, W. (1967). Untersuchungen zur Struktur und Funktion des Lampenbürsten-Y-Chromosoms in der Spermatogenese von *Drosophila. Chromosoma* **22**, 294–357.

Hennig, W. (1968). Ribonucleic acid synthesis of the Y-chromosome of *Drosophila hydei. J. Mol. Biol.* **38**, 227–239.

Hennig, W. (1972a). Highly repetitive DNA sequences in the genome of *Drosophila hydei*. I. Preferential localization in the X chromosome heterochromatin. *J. Mol. Biol.* **71**, 407–417.

Hennig, W. (1972b). Highly repetitive DNA sequences in the genome of *Drosophila hydei*. II. Occurrence in polytene tissues. *J. Mol. Biol.* **71**, 419–431.

Hennig, W. (1973). Molecular hybridization of DNA and RNA *in situ. Int. Rev. Cytol.* **36**, 1–44.

Hennig, W. (1977). Gene interactions in germ cell differentiation of *Adv. Enzymine Regul.* **15**, 363–371.

Hennig, W. (1978). The lampbrush Y chromosome of the fruit fly species *Drosophila hydei* (Diptera: Drosophilidae). *Entomol. Ger.* **4**, 200–210.

Hennig, W., and Leoncini, O. (1972). Heterochromatin, simple sequence DNA and differential staining patterns in *Drosophila* chromosomes. *In* "Modern Aspects of Cytogenetics: Constitutive Heterochromatin in Man" (R. A. Pfeiffer, ed.), pp. 88–99. Schattauer Verlag, Stuttgart.

Hennig, W., Hennig, I., and Stein, H. (1970). Repeated sequences in DNA of *Drosophila* and their localization in giant chromosomes. *Chromosoma* **32**, 31–63.

Hennig, W., Meyer, G. F., Hennig, I., and Leoncini, O. (1974a). Structure and function of the Y-chromosome of *Drosophila hydei. Cold Spring Harbor Symp. Quant. Biol.* **38**, 673–683.

Hennig, W., Hennig, I., and Leoncini, O. (1974b). Some observations on spermatogenesis of *Drosophila hydei. Drosophila Inf. Ser.* **51**, 127.

Hennig, W., Link, B., and Leoncini, O. (1975). The location of the nucleolus organizer regions in *Drosophila hydei. Chromosoma* **51**, 57–63.

Hennig, W., Vogt, P., Jacob, G., and Siegmund, I. (1982). Nucleolus organizer regions in *Drosophila* species of the *repleta* group. *Chromosoma* **87**, 279–292.

Hennig, W., Huijser, P., Vogt, P., Jäckle, H., and Edström, J.-E. (1983). Molecular cloning of microdissected lampbrush loop DNA sequences of *Drosophila hydei*. *EMBO J.* **2**, 1741–1736.

Hess, O. (1964). Strukturdifferenzierungen im Y-Chromosom von *Drosophila hydei* und ihre Beziehungen zu Genaktivitäten. I. Mutanten der Funktionsstrukturen. *Verh. Dtsch. Zool. Ges.* pp. 156–163.

Hess, O. (1965). Struktur-Differenzierungen im Y-Chromosom von *Drosophila hydei* und ihre Beziehungen zu Gen-Aktivitäten. III. Sequenz und Lokalisation der Schleifenbildungsorte. *Chromosoma* **16**, 222–248.

Hess, O. (1966). Structural modifications of the Y-chromosome in *Drosophila hydei* and their relation to gene activity. *Chromosomes Today* **1**, 167–173.

Hess, O. (1967a). Morphologische Variabilität der chromosomalen Funktionsstrukturen in den Spermatocytenkernen von *Drosophila*- Arten. *Chromosoma* **21**, 429–445.

Hess, O. (1967b). Complementation of genetic activity in translocated fragments of the Y chromosome in *Drosophila hydei*. *Genetics* **56**, 283–295.

Hess, O. (1968). The function of the lampbrush loops formed by the Y chromosome of *Drosophila hydei* in spermatocyte nuclei. *Mol. Gen. Genet.* **103**, 58–71.

Hess, O. (1976). The genetics of *Drosophila hydei* Sturtevant. *In* "The Genetics and Biology of *Drosophila*" (M. Ashburner and E. Novitski, eds.), Vol. 1C, pp. 1343–1363. Academic Press, London.

Hess, O. (1980). Lampbrush chromosomes. *In* "The Genetics and Biology of *Drosophila*" (M. Ashburner and T. R. F. Wright, eds.), Vol. 2D, pp. 1–38. Academic Press, London.

Hess, O., and Meyer G. F. (1963a). Artspezifische funktionelle Differenzierungen des Y-Heterochromatins bei *Drosophila*-Arten der *D. hydei*-Subgruppe. *Port. Acta Biol., Ser. A* **8**(1–2), 29–46.

Hess, O., and Meyer, G. F. (1963b). Chromosomal differentiations of the lampbrush type formed by the Y chromosome in *Drosophila hydei* and *Drosophila neohydei*. *J. Cell Biol.* **16**, 527–539.

Hess, O., and Meyer, G. F. (1968). Genetic activities of the Y chromosome in *Drosophila* during spermatogenesis. *Adv. Genet.* **14**, 171–223.

Hildreth, P. E., and Lucchesi, J. C. (1963). Fertilization in *Drosophila*. I. Evidence for the regular occurrence of monospermy. *Dev. Biol.* **6**, 262–278.

Huettner, A. F. (1930). The spermiogenesis of *Drosophila melanogaster*. *Z. Zellforsch. Mikrosk. Anat.* **11**, 615–637.

Hulsebos, T. J. M., Hackstein, J. H. P., and Hennig, W. (1983). Involvement of Y chromosomal loci in the synthesis of *Drosophila hydei* sperm proteins. *Dev. Biol.* **100**, 238–243.

Hulsebos, T. J. M., Hackstein, J. H. P., and Hennig, W. (1984). Lampbrush loop-specific protein of *Drosophila hydei*. *Proc. Natl. Acad. Sci. U.S.A.* **81**, 3404–3408.

Ingman-Baker, J., and Candido, E. P. M. (1980). Proteins of the *Drosophila melanogaster* male reproductive system: Two-dimensional gel patterns of proteins synthesized in the XO, XY, and XYY testis and paragonial gland and evidence that the Y chromosome does not code for structural sperm proteins. Biochem. Genet. **18**, 809–828.

Kennison, J. G. (1981). The genetic and cytological organization of the Y chromosome of *Drosophila melanogaster*. *Genetics* **98**, 529–548.

Kidwell, M. G. (1983). Intraspecific hybrid sterility. *In* "The Genetics and Biology of *Drosophila*" (M. Ashburner, H. L. Carson, and J. N. Thompson, Jr., eds.), Vol. 3E, pp. 125–154. Academic Press, London.

Kloetzel, P. M., Knust, E., and Schwochau, M. (1981). Analysis of nuclear proteins in primary spermatocytes of *Drosophila hydei*: The correlation of nuclear proteins with the function of the Y chromosome loops. *Chromosoma* **84**, 67–86.

Kremer, J. H. Y., and Hennig, W. (1985). In preparation.

Kurth, P. D., Reisch, J. C., and Bustin, M. (1983). Selective exposure of antigenic determinants in chromosomal proteins upon gene activation in polytene chromosomes. *Exp. Cell Res.* **143**, 257–269.

Leemann, K., and Ruch, F. (1972). Cytofluorometric determination of basic and total proteins with sulfoflavine. *J. Histochem. Cytochem.* **20**, 659–671.

Leoncini, O. (1977). Temperatursensitive Mutanten im Y Chromosom von *Drosophila hydei*. *Chromosoma* **63**, 329–357.

Liebrich, W. (1981a). *In vitro* differentiation of isolated single spermatocyte cysts of *Drosophila hydei*. *Eur. J. Cell Biol.* **24**, 335.

Liebrich, W. (1981b). *In vitro* spermatogenesis. I. Development of isolated spermatocyte cysts from wild-type *D. hydei*. *Cell Tissue Res.* **220**, 251–262.

Liebrich, W., Hanna, P. J., and Hess, O. (1982). Evidence for asynchronous mitotic cell divisions in secondary spermatogonia of *Drosophila*. *Int. J. Invertebr. Reprod.* **5**, 305–310.

Lifschytz, E. (1974). Genes controlling chromosome activity. An X-linked mutation affecting Y-lampbrush loop activity in *Drosophila hydei*. *Chromosoma* **47**, 415–427.

Lifschytz, E. (1975). Genes controlling chromosome activity: The role of genes blocking Y chromosomal loop propagation. *Chromosoma* **53**, 231–241.

Lifschytz, E. (1979). A procedure for the cloning and identification of Y-specific middle repetitive sequences in *Drosophila hydei*. *J. Mol. Biol.* **133**, 267–277.

Lifschytz, E., Hareven, D., Azriel, A., and Brodsly, H. (1983). DNA clones and RNA transcripts of four lampbrush loops from the Y chromsome of *Drosophila hydei*. *Cell* **32**, 191–199.

Lindsley, D., and Tokuyasu, K. T. (1980). Spermatogenesis. *In* "The Genetics and Biology of *Drosophila*" (M. Ashburner and T. R. F. Wright, eds.), Vol. 2D, pp. 226–294. Academic Press, London.

Long, E. O., and Dawid, I. B. (1980). Repeated genes in eukaryotes. *Annu. Rev. Biochem.* **49**, 727–764.

Macgregor, H. C. (1980). Recent developments in the study of lampbrush chromosomes. *Heredity* **44**, 3–35.

Meyer, G. F. (1963). Die Funktionsstrukturen des Y-Chromosoms in den Spermatocytenkernen von *Drosophila hydei, D. neohydei, D. repleta* und einigen anderen *Drosophila*-Arten. *Chromosoma* **14**, 207–255.

Meyer, G. F. (1964). Die parakristallinen Körper in den Spermienschwänzen von *Drosophila*. *Z. Zellforsch. Mikrosk. Anat.* **62**, 762–784.

Meyer, G. F. (1968). Spermiogenese in normalen und Y-defizienten Männchen von *Drosophila melanogaster* und *D. hydei*. *Z. Zellforsch. Mikrosk. Anat.* **84**, 141–175.

Meyer, G. F., and Hennig, W. (1974). The nucleolus in primary spermatocytes of *Drosophila hydei*. *Chromosoma* **46**, 121–144.

Meyer, G. F., Hess, O., and Beermann, W. (1961). Phasenspezifische Funktionsstrukturen in Spermatocytenkernen von *Drosophila melanogaster* und ihre Abhängigkeit vom Y- Chromosom. *Chromosoma* **12**, 676–716.

Oliver, D., and Chalkley, R. (1972). An electrophoretic analysis of *Drosophila histones*. *Exp. Cell Res.* **73**, 295–302.

Pardue, M. L., and Birnstiel, M. L. (1972). Cytological localization of repeated gene sequences. *In* "Modern Aspects of Cytogenetics: Constitutive Heterochromatin in Man" (R. A. Pfeiffer, ed.), pp. 75–85. Schattauer Verlag, Stuttgart.

Peacock, W. J., Lohe, A. R., Gerlach, W. L., Dunsmuir, P., Dennis, E. S., and Appels, R. (1978). Fine structure and evolution of DNA in heterochromatin. *Cold Spring Harbor Symp. Quant. Biol.* **42**, 1121–1135.

Renkawitz, R. (1978a). Characterization of two moderately repetitive DNA components within the β-heterochromatin of *Drosophila hydei*. *Chromosoma* **66**, 225–236.

Renkawitz, R. (1978b). Two highly repetitive DNA satellites of *Drosophila hydei* localized within the α-heterochromatin of specific chromosomes. *Chromosoma* **66**, 237–248.

Renkawitz, R. (1979). Isolation of twelve satellite DNAs from *Drosophila hydei*. *Int. J. Biol. Macromol.* **1**, 133–136.

Renkawitz-Pohl, R., Glätzer, K. H., and Kunz, W. (1980). Characterization of cloned ribosomal DNA from *Drosophila hydei*. *Nucleic Acids Res.* **8**, 4593–4611.

Renkawitz-Pohl, R., Glätzer, K. H., and Kunz, W. (1981). Ribosomal RNA genes with an intervening sequence are clustered within the X chromosomal ribosomal DNA of *Drosophila hydei*. *J. Mol. Biol.* **148**, 95–101.

Risau, W., Symmons, P., Saumweber, H., and Frasch, M. (1983). Nonpackaging and packaging proteins of hnRNA in *Drosophila melanogaster*. *Cell* **33**, 529–541.

Rungger-Brändle, E., Jamrich, M., and Bautz, E. K. F. (1981). Localization of RNA polymerase B and histones in the nucleus of primary spermatocytes of *Drosophila hydei*, studied by immunofluorescence microscopy. *Chromosoma* **82**, 399–407.

Scalenghe, F., Turco, E., Edström, J. E., Pirrotta, V., and Melli, M. (1981). Microdissection and cloning of DNA from a specific region of *Drosophila melanogaster* polytene chromosomes. *Chromosoma* **82**, 205–216.

Schäfer, U. (1978). Sterility in *Drosophila hydei* × *D. neohydei* hybrids. *Genetica* **49**, 205–214.

Schäfer, U. (1979). Viability in *Drosophila hydei* × *D. neohydei* hybrids and its regulation by genes located in the sex heterochromatin. *Biol. Zentralbl.* **98**, 153–161.

Schäfer, U., and Kunz, W. (1975). Two separated nucleolus organizers on the *Drosophila hydei* Y chromosome. *Mol. Gen. Genet.* **137**, 365–368.

Scheer, U., Franke, W. W., Trendelenburg, M., and Spring, H. (1976). Classification of loops of lampbrush chromosomes according to the arrangement of transcriptional complexes. *J. Cell Sci.* **22**, 503–519.

Silver, L. M., and Elgin, S. C. (1976). A method for determination of the *in situ* distribution of chromosomal proteins. *Proc. Natl. Acad. Sci. U.S.A.* **73**, 423–427.

Stephenson, E. C., Erba, H. P., and Gall, J. G. (1981). Histone gene clusters of the newt *Notophthalmus* are separated by long tracts of satellite DNA. *Cell* **24**, 639–647.

Stern, C. (1927). Ein genetischer und zytologischer Beweis für Vererbung im Y-Chromosom von *Drosophila melanogaster*. *Z. Indukt. Abstamm.- Vererbungsl.* **44**, 187–231.

Stern, C. (1929). Untersuchungen über Aberrationen des Y-Chromosoms von *Drosophila melanogaster*. *Z. Indukt. Abstamm.- Vererbungs* **51**, 254.

Stern, C., and Hadorn, E. (1938). The determination of sterility in *Drosophila* males without a complete Y chromosome. *Am. Nat.* **72**, 42–52.

Tates, A. D. (1971). Cytodifferentiation during spermatogenesis in *Drosophila melanogaster*. Proefschrift Universiteit Leiden.

Tokuyasu, K., Peacock, W. J., and Hardy, R. W. (1972). Dynamics of spermiogenesis in

Drosophila melanogaster. I. Individualization process. *Z. Zellforsch. Mikrosk. Anat.* **124**, 479–506.

van Breugel, F. (1970). An analysis of white-mottled mutants in *Drosophila hydei* with observations on X–Y exchanges in the male. *Genetica* **41**, 589–625.

van Breugel, F. (1971). X–Y exchanges in males of *Drosophila hydei* carrying the w-mCo duplication. *Genetica* **42**, 1–12.

Vogt, P., and Hennig, W. (1983). Y chromosomal DNA of *Drosophila hydei*. *J. Mol. Biol.* **167**, 37–56.

Vogt, P., Hennig, W., and Siegmund, I. (1982). Identification of cloned Y chromosomal DNA sequences from a lampbrush loop of *Drosophila hydei*. *Proc. Natl. Acad. Sci. U.S.A.* **79**, 5132–5136.

Williamson, J. H. (1972). Allelic complementation between mutants in the fertility factors of the Y chromosome in *Drosophila melanogaster*. *Mol. Gen. Genet.* **119**, 43–47.

Williamson, J. H. (1976). The genetics of the Y chromosome. *In* "The Genetics and Biology of *Drosophila*" (M. Ashburner and E. Novitski, eds.), Vol. 1B, pp. 667–699. Academic Press, London.

Williamson, J. H., and Meidinger, E. (1979). Y chromosome hyperploidy and male fertility in *Drosophila melanogaster*. *Can. J. Genet. Cytol.* **21**, 21–24.

Wright, M. J. D. (1976). "Animal Cytology and Evolution," 3rd ed. Cambridge Univ. Press, London and New York.

Yamasaki, N. (1977). Selective staining of Y chromosomal loops in *Drosophila hydei, D. neohydei* and *D. eohydei*. *Chromosoma* **60**, 27–37.

Yamasaki, N. (1981). Differential staining of Y chromosome lampbrush loops of *Drosophila hydei*. *Chromosoma* **83**, 679–684.

Zacharias, H., Hennig, W., and Leoncini, O. (1982). Comparative studies of *Drosophila hydei* and related species. I. Microspectrophotometry of genome sizes. *Genetica* **58**, 153–157.

NOTE ADDED IN PROOF: The data summarized in Section VI have recently been considerably extended. The distinction of a Y-specific and Y-associated class of DNA sequences has been further confirmed by studies of microcloned Y chromosomal DNA sequences. Y-specific sequences appear generally to consist of short, tandemly repeated DNA sequences of a satellite-DNA type, but of moderately repetitive nature. These sequences are transcribed and appear on RNA blots as a heterogeneous mixture of RNA molecules of high molecular weight. Y-associated DNA sequences, i.e., sequences sharing homology with autosomal and X chromosomal sites, are found complementary to discrete size class RNA species. Transcript RNA hybridization indicates that they are preferentially transcribed from their Y chromosomal sites. But also non-Y chromosomal loci must be expressed since in some instances these RNA fractions were also detected in XO testes, although at lower frequencies. All evidence indicates that the discrete class RNA sequences occur also in the cytoplasm, while the heterogeneous RNA species are, at least preferentially, restricted to the spermatocyte nuclei (R. C. Brand *et al.*, in preparation). Also, the conclusions drawn in Section II regarding the loop-specificity of chromosomal proteins have been further supported by the recovery of other antisera, reacting with loop-specific proteins (W. Hennig *et al.*, in preparation).

RECENT DEVELOPMENTS IN POPULATION GENETICS*

Michael T. Clegg and Bryan K. Epperson

Department of Botany and Plant Sciences
University of California, Riverside, California

I. Introduction

During much of the 1970s research in population genetics was dominated by a single issue, the neutrality controversy. The situation today stands in marked contrast. A number of divergent research areas, spanning a wide range of biological interests from ecology at one ex-

---

*This article is dedicated with respect and affection to our teacher, Professor R. W. Allard.

ADVANCES IN GENETICS, Vol. 23

treme to molecular biology at the other extreme, are flourishing. This diversity of research activity has been identified as tangible evidence of the health of the population genetics research program in the published summary of a recent National Institutes of Health-sponsored workshop on population genetics (Anonymous, 1983). Our goal in this review is to provide a brief account of several divergent areas of recent research activity in population genetics to illustrate the breadth of contemporary research activity in this field.

II. DNA Sequence Variation in Populations

The major goal of experimental population genetics has been the description and analysis of genetic variation (e.g., see Lewontin, 1974). Genetic variation is central to evolutionary theory because it governs the rate of population improvement under selection (i.e., Fisher's Fundamental Theorem of Natural Selection). In the past few years it has become possible to examine the fine structure of genetic differences at the DNA sequence level. This new approach to the study of genetic variation has advantages over all previous methods in permitting the detection of genetic differences at the level of primary DNA sequences, unconfounded by gene expression. Applied to the study of individual genes, this approach allows the construction of fine physical maps of sequence variants directly from the primary data. Consequently, genetic variants can be associated with various functional regions of a gene and statistical associations (linkage disequilibria) between different polymorphic sites can be investigated. On a broader scale DNA sequence variation can be used to investigate the evolution of gene families, as well as the evolution of entire genomes, such as those of eukaryotic cell organelles.

In the following we will confine our attention to the analysis of DNA sequence variation within populations as distinct from evolutionary comparisons of homologous genes among different taxa. For examples of evolutionary comparisons among diverse taxa for pseudogenes, plant actin genes and the 5.8 S ribosomal RNAs see Li (1983), Shah et al. (1983), and Olsen and Sogin (1982).

A. SINGLE COPY SEQUENCES

In the study of DNA sequence polymorphisms among single copy genes, three different issues have been predominant. First, attention has focused on the problem of describing the different types of mutational events that occur and how they are distributed among introns, exons, and flanking sequences. Second, the pattern of statistical association among adjacent molecular polymorphisms has been exploited to make inferences about the origin of major polymorphisms. And third, DNA sequence polymorphisms have been employed as genetic markers to follow the transmission of the other genes of major phenotypic effect (e.g. genes, responsible for genetic diseases in man).

1. Surveys of DNA Sequence Polymorphism

The first large-scale survey of DNA sequence polymorphism involved the β-globin gene family in a sample of 60 Europeans (Jefferys, 1979). Cloned globin genes were used as probes to detect restriction endonuclease fragment pattern variation. Polymorphic HindIII restriction sites were located in the 3' end of the large intervening sequence of both the γ^G and γ^A genes. Because these sites occur at the same relative location in the two duplicate genes they may have a common origin. Three possibilities exist. First, the HindIII polymorphism may predate the γ-globin gene duplication. Second, two independent mutations may have occurred in the two duplicate genes. Third, the polymorphic site may have arisen in one gene copy and been subsequently transferred to the second copy via unequal crossing over or gene conversion. This latter process, whereby two genes appear to evolve in concert because of the lateral transfer of mutant sites, has been termed "concerted evolution" by Arnheim et al. (1980).

Subsequent, more extensive surveys (samples of 400 genomes) of restriction site polymorphism in the β-globin gene family have uncovered 12 polymorphic restriction sites (Kazazian et al., 1983). Approximately half of the polymorphic sites have been verified to be the result of single nucleotide substitutions by DNA sequencing. The estimated nucleotide sequence diversity per nucleotide site is approximately 0.002 for these data.

Nucleotide substitutions are not the only events responsible for restriction site variation. A variety of studies have shown high levels of polymorphism in noncoding DNA sequences due to insertion/deletion events. For example, the 5' flanking region of the human insulin gene is highly polymorphic for insertions ranging from 1.5 to 3.4 kilobase pairs (kb) (Rotwein et al., 1981). These insertional variants occur with significantly higher frequency in the genome of individuals with type 2 diabetes and may play a role in insulin gene expression. In addition, Wyman and White (1980) reported up to eight allelic variants in a

Utah population, arising from insertion/deletion variants associated with an uncharacterized single copy region of the human genome.

Reports of high levels of insertion/deletion polymorphism are not restricted to man. The *Adh1* (alcohol dehydrogenase 1) gene of maize appears to have insertional polymorphisms 3' to the transcription unit (Johns *et al.*, 1983). Similarly, insertion/deletion variants associated with the *Adh* locus of *Drosophila melanogaster* are highly polymorphic (Langley *et al.*, 1982). While the mechanisms generating insertion/deletion polymorphisms are unknown, transposable elements are an obvious candidate (see below).

The ultimate approach to the study of genetic variation is the complete DNA sequencing of a gene and associated flanking regions from a number of independent samples. Kreitman (1983) has sequenced and compared 11 independent clones of a 2.7-kb region containing the *Drosophila melanogaster Adh* gene. These data allow a complete description of all genetic differences among sampled genes. Interestingly, the only amino acid substitution is the lysine–threonine substitution responsible for the major electrophoretic alleles (*f, s*). A total of 13 silent substitutions (= 6.7% of potential silent changes) occurred in coding (translated) regions and the distribution of silent changes among the three coding exons was statistically homogeneous. The percentage of substitutions in introns was significantly less (2.4%) than the 6.7% observed for silent changes in exons. This difference can be accounted for by a low fraction of substitutions (1.7%) in intron 1, which may be conserved because it codes for larval promoter sequences. It is more remarkable that 767 base pairs compared for the 3' nontranscribed region show only 0.6% nucleotide differences. The reasons for this high degree of conservation are unknown but may indicate important functional constraints.

The information content of the *Adh* sequence data is much greater than the simple statistics presented above suggest. The joint distribution of polymorphic sites is highly correlated, reflecting the historical origin of each sequence. Kreitman (1983) infers from these data that *Adh-f* is derived from *Adh-s*. In addition, the distribution of polymorphic sites strongly suggests that two independent intragenic recombination events have occurred among the 11 sequences compared. Thus intragenic recombination may be an important process in the generation of allelic novelties.

2. *Use of Linkage Disequilibria to Make Historical Inferences*

The idea that historical information can be extracted from the pattern of linkage disequilibrium among polymorphic sites has been exploited in the study of globin mutants in man. Kan and Dozy (1978a) first reported a case of strong linkage disequilibrium between the β^s allele (responsible for sickle-cell anemia) and a polymorphic *Hpa*I site approximately 5 kb from the 3′ end of the β-globin locus. This study was based upon a San Francisco population of black sickle-cell disease patients. Later studies of other United States black populations revealed a lower association than first noted in the San Francisco population (e.g., Panny *et al.*, 1981). Kan and Dozy (1980) then conducted a geographic survey of the *Hpa*I association with the β^s, β^c, and β^A alleles. They found a high association in populations of West African origin, but little association in samples from India and Saudi Arabia. While recombination between β^s and the *Hpa*I site cannot be excluded as an explanation for the different patterns of linkage disequilibrium, the recombination event would have to have occurred prior to the increase in β^s frequency caused by malarial selection (Asmussen and Clegg, 1982). The possibility that the β^s mutant arose separately on fragments with and without the polymorphic *Hpa*I site may be equally likely.

Even stronger evidence for a nonunique origin of the β^E gene, which causes reduced β-globin message production and behaves like a mild β-thalassemia gene, comes from the work of Antonarakis *et al.* (1982). These workers have examined the joint distribution of nine β-globin restriction site polymorphisms and the β^E polymorphism in individuals of Asian origin. They show that β^E is associated with two distinctly different restriction site frameworks (haplotypes). A double recombination event must be invoked to account for the β^E association, which is considerably less likely than a second β^E mutation event.

As data on restriction site polymorphisms continue to accumulate, it will become possible to identify the geographic origin and process of spread of major mutant genes. Because evolution is an historical science, these kinds of retrospective approaches promise to revolutionize the study of evolutionary change. An obvious extension of these ideas is to use molecular markers for the prediction of the transmission of other genes of major phenotypic effect.

3. Use of Molecular Markers

The notion of using linked genes to predict the transmission of another gene is not new. However, practical applications have been limited, because relatively few markers have been available in man or in most other organisms. Molecular techniques promise to radically alter this situation. Botstein *et al.* (1980) outlined a program to construct a high-density linkage map of the human genome using naturally occurring DNA sequence polymorphisms. The estimates of DNA sequence diversity cited above make it likely that polymorphic restriction sites can be associated with any region of 20 or more kilobase pairs.

An important practical application arises in genetic counseling when a marker gene can be used to predict the transmission of a gene responsible for a genetic disease. A great advantage of restriction site polymorphisms is the fact that they can be assayed early in development. For instance, Huntington's disease, which has a late onset (30–50 years of age), has recently been associated with polymorphic *Hind*III sites tightly linked to the disease locus (Gusella *et al.*, 1983). This means that the prenatal diagnosis of the transmission of the dominant Huntington's gene may now be possible. Similarly, molecular markers associated with phenylalanine hydroxylase allow the prenatal diagnosis of phenylketonuria (Woo *et al.*, 1983) and molecular polymorphisms associated with the insulin gene are also available (Rotwein *et al.*, 1981). Successful prenatal diagnosis of globin gene pathologies has already been accomplished using associated restriction site polymorphisms (Kan and Dozy, 1978b; Kazazian *et al.*, 1980).

Because the use of linked markers in genetic counseling rests on naturally occurring genetic polymorphisms, several population genetic problems arise. The theoretical issues of primary concern are, (1) what fraction of the matings involving affected parents will have useful marker associations? And (2) how is this fraction affected by linkage, allele multiplicity, and linkage disequilibrium between the marker and target loci? The fraction of informative matings for diallelic loci can range from one, when there is absolute association between the marker and target loci, to very small values when the marker gene is rare and there is no association (Asmussen and Clegg, 1982; Clegg and Asmussen, 1983; Chakravarti, 1983). When an arbitrary number (n) of alleles occurs at the marker locus, the importance of linkage disequilibrium diminishes and the proportion of families in which a diagnosis can be made is roughly approximated by $1 - (1/n)$ (Asmussen and Clegg, 1984). The effect of recombination between the marker loci and target loci has been investigated by Nei (1979) and Asmussen (1984), who give expressions for the accuracy of diagnosis when linkage is incomplete.

Beyond the genetic counseling application, linkage maps based upon DNA sequence polymorphisms are likely to have important applications in quantitative genetics as well. It may be feasible to dissect the genetic architecture of complex phenotypes using marker genes as outlined by Mather and Jinks (1971). The availability of abundant markers in any organism could make many complex traits of agronomic importance susceptible to simple Mendelian analysis.

B. MULTIGENE FAMILIES

The evolutionary dynamics of multigene families pose several interesting mathematical problems. These problems arise because the phenomenon of concerted evolution (discussed above) necessitates the incorporation of gene conversion and unequal recombination into mathematical models. Ohta (1980) has pioneered the study of this class of problems. More recent developments can be found in the work of Ohta (1984), Nagylaki (1984), and Nagylaki and Petes (1982). The basic methods of analysis have rested on the use of diffusion approximations and on the use of gene identity coefficients [see Ohta (1980) for a detailed review of the theoretical foundations of this important area of recent research activity].

A multigene family can be defined as a group of genes with the properties of "multiplicity, close linkage, sequence homology and related or overlapping phenotypic functions" (Ohta, 1980). This very broad definition encompasses a wide spectrum of repeated gene families ranging from the globin genes at one extreme to highly reiterated ribosomal RNA genes at the other extreme (reviewed by Long and Dawid, 1980). (In the preceding section we considered the β-globin gene family under the heading of single copy sequences because restriction site polymorphisms could be mapped to individual loci.) In the following, we briefly discuss two important consequences of the concerted evolution of multigene families.

Because genetic exchange among repeat units takes place, mutant sites arising in one copy can spread to other copies. As a result, copies at different genetic loci will be more similar in nucleotide sequence than if they were accumulating mutations independently. The equilibrium distribution of gene identity among genes within and between loci is governed by the mutation, gene conversion, and crossover rates, as well as the population size and number of repeating units. Equilibrium expressions are given by Nagylaki (1984).

The joint distribution of mutant sites within a gene is also affected by gene conversion and unequal crossover rates. This follows because several mutant sites can be transmitted together among repeating units, in contrast to their independent accumulation under a pure mutation process. Linkage disequilibria among mutant sites within repeating units can be generated by this process. A. H. D. Brown and

Clegg (1983) analyzed the joint distribution of mutant sites within repeating units for a highly repeated sequence associated with knob heterochromatin in maize. They found that the joint distribution of mutant sites within copies was markedly nonrandom, indicating substantial linkage disequilibria among polymorphic nucleotide sites.

The second important consequence of concerted evolution is the (possibly rapid) change in the number of repeating units in a multigene family. In some cases distinct phenotypic effects may be associated with changes in copy number. A particularly dramatic example of a phenotype–copy number correspondence is associated with the ribosomal RNA (rRNA) gene family of *Drosophila melanogaster*. The *bobbed* locus of *D. melanogaster* corresponds to the rRNA tandon on the X chromosome (Ritossa, 1976). The mutant *bb* allele is associated with a partial deficiency of the rRNA gene family. The recessive *bb/bb* phenotype is characterized by a reduced abdominal bristle size.

Frankham *et al.* (1980) established replicate selection lines of *D. melanogaster* that were initially homozygous for the X chromosome. Selection for high and low abdominal bristle number was carried out over a period of nearly 100 generations. A selection response was observed and much of the response in low selected lines was shown to be due to X chromosomal factors allelic to the *bobbed* locus. Frankham *et al.* (1980) interpret these results as evidence for the *de novo* origin of genetic variation caused by the reduction in rRNA gene copy number due to unequal crossing over. They calculated a lower bound estimate for the frequency of unequal crossover events in the rRNA tandon as 2.96×10^{-4} changes per gamete per generation. Coen and Dover (1983) have recently shown that the low selected lines have a reduced rDNA copy number due to unequal exchange between X and Y chromosome rDNA arrays. Even more dramatic changes in rRNA copy number have been reported to occur within a single generation in flax (Cullis, 1976). These changes are also associated with marked phenotypic effects. Thus concerted evolution may result in the *de novo* origin of selectable phenotypic variation. This possibility has important implications for quantitative genetic theories (discussed below). It also argues for a reevaluation of the traditional view that selection response is dependent on preexisting genetic variation [see Nei (1983) for further arguments in favor of such a reevaluation].

C. ORGANELLE GENOME VARIATION

The area of molecular population genetics that has experienced the most rapid growth is the analysis of DNA sequence variation in organelle genomes. In this section we briefly discuss the animal and plant mitochondrial genomes and the plant chloroplast genome. We then consider two new problem areas presented by organelle gene transmission. The first problem area concerns the development of population genetic models to describe the transmission of organelle genes. The second area involves the use of maternal transmission and restriction site maps as a means of inferring the pattern of genetic relationships among different maternal lineages within and between populations.

The animal mitochondrial genome (mtDNA) is a closed circular molecule that ranges in size from 16.5 kb in man (Anderson *et al.*, 1981) to 15.7–19.5 kb in various *Drosophila* species (Fauran and Wolstenholme, 1976). The great majority of work on the population genetics and molecular evolution of mtDNA has been with mammalian species. While the structural organization of mammalian mtDNA is highly conserved, rates of nucleotide substitution are high, averaging about four times the substitution rate for single-copy nuclear sequences in primates (reviewed by W. M. Brown, 1983). Moreover, comparisons of homologous mtDNA sequences from seven different human lineages show a 24-fold excess of transition over transversion substitutions (Aquadro and Greenberg, 1983). Comparisons of complete nucleotide sequence data for the mitochondrial gene that encodes cytochrome oxidase subunit II between *Rattus norvegicus* and *R. rattus* reveal that 94% of nucleotide substitutions are synonymous substitutions (G. G. Brown and Simpson, 1982). These results suggest that the high rate of mtDNA evolution can, in part, be accounted for by a high rate of synonymous substitution. W. M. Brown (1983) provides a lucid discussion of the various factors that may account for the high rate of nucleotide substitution associated with the mammalian mtDNA. Finally, short addition–deletion events appear to be relatively frequent in the small intergenic regions of human mtDNA (Cann and Wilson, 1983).

A high rate of nucleotide substitution has, as a consequence, a high level of mtDNA polymorphism within populations, because many mutants will be drifting towards fixation at any point in time. Indeed, Avise *et al.* (1983) estimate a maximum level of sequence divergence among individuals of *Peromyscus leucopus* of 4%. This means that many genetically distinct mtDNA types exist in populations. Despite this, individual animals appear to be homogeneous for mtDNA type (*homoplasmic*) [see Avise and Lansman (1983) for a critical discussion of these results]. This observation presents a theoretical problem: what are the sampling processes involved in organelle transmission that lead to individual homogeneity, despite high levels of variation between individuals? We will consider theoretical investigations of this problem following a brief account of plant mtDNA and chloroplast (cpDNA) variation.

The plant mtDNA is much more complex and variable in its structural organization than the animal mitochondrial genome. For instance, the size of this genome ranges from 250 to 2500 kb among plant species [reviewed by Levings (1983) and Sederhoff (1984)]. Palmer and Shields (1984) have recently shown that the mtDNA of *Brassica campestris* exists as three circular chromosomes of 218, 135, and 83 kb. The 218-kb chromosome is postulated to be a master chromosome that contains all the genetic information of the mitochondrial genome. The smaller 135- and 83-kb chromosomes are believed to arise from the 218-kb chromosome via a recombinational mechanism. The suggestion of intermolecular recombination is consistent with studies of mtDNA variation in populations of maize. Sederhoff *et al.* (1981) have surveyed mtDNA restriction fragment variation in maize and teosinte and conclude that a substantial proportion of the variation observed is due to sequence rearrangements, rather than nucleotide substitution.

Even more intriguing is evidence for the incorporation of some plant chloroplast DNA sequences into the plant mtDNA. Stern and Lonsdale (1982) have found an apparent pseudogene for the maize chloroplast 16 S rRNA in the maize mtDNA. These findings have led to the speculation that the large size variations in the plant mtDNA may arise because of the ability of this genome to take up foreign DNA (Fox, 1984), perhaps through recombinational mechanisms. The evolutionary and population genetic implications of these phenomena have yet to be explored.

The plant chloroplast genome (cpDNA) stands in sharp contrast to plant mtDNA in being conservative in structural organization. Most studies of the evolution of this molecule have concentrated on interspecific or wider taxonomic comparisons. A major conclusion from these investigations is that cpDNA is conservative in evolution and rates of nucleotide substitution are more than an order of magnitude below that observed for mammalian mtDNA (reviewed by Curtis and Clegg, 1984).

The plant chloroplast genome consists of a single closed circular molecular averaging about 150 kb in size. Because of its larger size, digestion of cpDNA with a given restriction endonuclease will detect approximately 10 times as many sites as in animal mtDNA (assuming comparable G + C content). However, cpDNA polymorphism is much less frequent. The largest survey of intraspecific variation for cpDNA published to date yields a maximum level of sequence divergence of $\sim$ 0.3% in barley (Clegg *et al.*, 1984), as compared to the estimate of 4% cited above for *P. leucopus*. The low rates estimated from restriction fragment comparisons have been verified by the comparison of complete DNA sequences among three barley lineages (Zurawski and Clegg, 1984). The reasons for a more conservative rate of cpDNA evolution are unknown; however, it seems likely that a lower mutation rate is implicated.

With this background, we now briefly consider theoretical problems related to organelle gene transmission. The features of organelle transmission that depart from nuclear transmission are (1) cytoplasmic transmission and frequently uniparental transmission, and (2) the random sampling of organelle genomes during successive cell divisions. An issue of current concern is whether the common observation of homoplasmicity is consistent with reasonable models of organelle transmission dynamics. Clearly, a lineage in which a new mutant arises would be heteroplasmic for some period as the mutant drifts within the cell lineage, and yet this situation is rarely observed (see Hauswirth and Laipis, 1981, for a possible exception). Mathematical models of organelle transmission have been studied by Birky *et al.* (1983) and Chapman *et al.* (1982). These authors show that homoplasmicity is the predicted condition, even when moderate numbers of organelle genomes are sampled per cell generation. Birky *et al.* (1983) also conclude that the higher rate of mammalian mtDNA evolution is most reasonably accounted for by a higher mutation rate.

A second interesting theoretical problem concerns the two-locus behavior of a nuclear and cytoplasmic gene. Clark (1984) has analyzed viability selection models in which genes obeying these two different transmission processes interact. He has obtained the important result that general viability selection cannot maintain a two-locus polymorphism for diallelic models with random mating in which one locus is subject to nuclear transmission and the other is cytoplasmically transmitted. If sexual asymmetry in fertility is allowed, however, then such a polymorphism may exist (Gregorius and Ross, 1984).

The potentially most useful application of organelle genome restriction fragment data is exemplified by the work of Avise (Avise *et al.*, 1979, 1983). This work is based upon the observation that it is often possible to infer the sequence of mutational steps, separating different maternal lineages, from the analysis of restriction fragment patterns. Thus, in contrast to allozyme data, the restriction analysis of organelle genomes provides a natural distance metric. Because strict maternal transmission occurs in many organisms, the networks of relationships constructed from such analyses reflect the phylogenetic relationships among matriarchal lineages. Johnson *et al.* (1983) have applied these techniques to the study of the ethnic radiations of man. They present an mtDNA phylogeny that depicts the branch points in the evolution of human ethnic groups. Interestingly, these data require that a higher rate of mtDNA evolution be postulated for Bushmen.

Research on the population genetics of organelle transmission is expanding rapidly. The ease of study of organelle DNA and the novel problems presented by cytoplasmic transmission combine to make this area especially favorable for population genetic investigations.

III. Population Dynamics of Transposable Elements

The discovery and subsequent molecular characterization of transposable elements (TEs) have presented population geneticists with an exciting set of new theoretical problems. Despite the fact that genetic evidence for transposition has been available for more than three decades (McClintock, 1952, 1956), a clear demonstration of the ubiquity of these entities has only become available with the growth of recombinant DNA technology. It now appears that organisms ranging from *E. coli* to *Drosophila,* maize, and man carry families of transposable elements in their genome. Moreover, TEs have been implicated as causative agents in certain classes of mutation, some TEs may have oncogenic properties, TEs have been associated with the development of hybrid dysgenesis and may sometimes play a role in speciation, and TEs have been successfully exploited as transformation vectors to study problems of gene regulation and development. In short, TEs have manifold genetic effects.

The evolutionary implications of TEs are diverse. Questions of particular importance are, (1) what are the dynamics of spread of TEs within a population and within a genome? (2) What factors cause the mobilization of TEs and how might time honored notions about the randomness of mutation be affected? And, (3) to what extent are TEs involved in rapid genomic change and ultimately speciation? Below, we briefly consider the population genetic implications of each of these questions.

The hypothesis of "selfish" (or parasitic) DNA (Doolittle and Sapienza, 1980; Orgel and Crick, 1980) drew dramatic attention to the idea that TEs could obey a transmission dynamic that differed from ordinary genes. The actual details of TE replication and transposition are fairly well understood in prokaryotes (Campbell, 1983) but are still being clarified in eukaryotes. Nevertheless, certain general features are sufficiently clear to allow the construction of mathematical models of TE dynamics. The critical assumptions of these models are (1) that a fixed number of sites exist in the genome that can accept TEs; (2) an element can be lost from a given site with a fixed probability; (3) an element can transpose to a new site (while leaving a copy in the initial site) with a probability that depends on the number of sites occupied; and (4) because of deleterious mutational effects, an individual's fitness is a decreasing function of the number of sites occupied, so that the transposition process can be regulated by phenotypic selection. The variables of interest in finite populations are the mean number of copies per individual and the frequency distribution of copies per individual taken over the population. Recent analytical and simulation studies that incorporate some or all of these assumptions have been presented by Charlesworth and Charlesworth (1983), Ohta (1983), and Langley *et al.* (1983).

The results of these investigations are in general qualitative agreement with empirical studies of the distribution of TEs in populations of *Drosophila* (Montgomery and Langley, 1983). Specifically, the models show that in finite populations the number of occupied sites will vary among individuals so that the mean number of occupied sites is well below the maximum. Of course, these results depend upon the specific choice of loss and transposition probabilities (assumptions 2 and 3 above) and on the selection function (when selection is included in the model). The comparison of observed distributions to model expectations provides a basis for the estimation of loss and transposition probabilities (Kaplan and Brookfield, 1983). Montgomery and Langley (1983) conclude that both loss and transposition probabilities must be very high, because in a sample of 20 wild X chromosomes of *D. melanogaster,* relatively few occupied sites are shared among different chromosomes.

In addition to the distribution of occupied sites among individuals in a population, the other major theoretical problem is the rate of spread of TEs in populations. Because TEs can spread within a genome, an individual may transmit more copies of a TE to its offspring than it received from its parents. Hickey (1982) has shown that, assuming a zero loss probability and infinite population size, TEs can spread through a population despite very strong phenotypic selection. Charlesworth and Charlesworth (1983) have analyzed a more detailed model of infinite population dynamics with nonzero loss probabilities and give conditions for an equilibrium number of occupied sites. These considerations highlight the importance of site loss and phenotypic selection in checking the spread of TEs in populations.

The mutational properties of TEs may be important in providing novel genetic variants under conditions of environmental stress. McDonald (1983) has argued, in his "new wave" theory of evolution, that mutation can no longer be viewed as the genetically homogeneous, undirected process envisioned by elementary population genetic theory. Transposition-mediated genetic changes, which are specific in nature, can in some cases be induced by stressful environments. Environmental challenges may bring forth novel mutational variants that provide a substrate for natural selection and the origin of adaptive novelties. Similar arguments have been advanced by Campbell (1983), who notes a correspondence between Lamarckian theories of evolution and transposition-mediated genetic change in *E. coli*. We are clearly entering into a period of ferment that may bring forth significant modifications to the neoDarwinian paradigm. Evolutionary theory will have to accord mutation a less passive role, especially where transposition-induced mutations are concerned.

Evidence for rapid genetic change associated with TEs comes from the study of hybrid dysgenesis in *D. melanogaster*. The term "hybrid dysgenesis" was coined by Kidwell *et al.* (1977) to describe a syndrome of deleterious phenotypic effects observed in the progeny of certain crosses among *D. melanogaster* stocks. The phenotypic effects included high rates of male recombination, mutation, chromosomal aberrations, and gonadal sterility. Stocks of *D. melanogaster* could be grouped into different equivalence classes based upon a cytoplasm–chromosome interaction. In particular, P-strain males crossed to M-strain females (M × P) produce dysgenic progeny, whereas no effect is observed in P × M, P × P, or M × M crosses. A family of TEs, called the P-factor family, has been shown to be responsible for the dysgenic phenomena (Rubin *et al.*, 1982; Bingham *et al.*, 1982). P-strain individuals carry from 30 to 50 copies of the P element dispersed in their genome, while the P element is entirely absent from the genome of M-strain individuals.

Besides the P and M strains, a third Q-strain class has also been uncovered. Q-strain flies also carry 30–50 P elements per genome but do not exhibit gonadal sterility in M × Q crosses. In addition, a second entirely independent form of hybrid dysgenesis, called I–R dysgenesis, has been shown to occur in *D. melanogaster* (reviewed by Bregliano and Kidwell, 1983). In the following we confine our discussion to P–M dysgenesis.

Kidwell (1983a) has shown that M-strain X chromosomes can acquire P-strain characteristics from nonhomologous P-strain autosomes. This process, termed "chromosomal contamination," is apparently caused by P-factor transposition to the uninfected chromosomes in a single generation. There is some intriguing evidence for the very rapid spread (or loss) of TEs in *Drosophila* populations. Specifically, Kidwell (1983b) has surveyed *D. melanogaster* strains collected from the wild at different points in time, ranging from the 1920s to the present. She shows that the P type is absent from strains collected prior to 1950. After 1950, the frequency of the P type increases as the laboratory age of the strains decreases. Moreover, the global geographic distribution of P strains varies substantially among the major continents.

Two hypotheses are advanced to explain the change in the distribution of P-factor frequency. One hypothesis, termed the "rapid invasion" hypothesis, states that the P-factor family has recently invaded and spread through the *D. melanogaster* species. The second hypothesis, called the "recent loss" hypothesis, maintains that the P factor is subject to stochastic loss in small laboratory populations. While the present data are not sufficient to decide between these two hypotheses, they tend to favor the rapid invasion hypothesis.

The mere fact of hybrid dysgenesis is *prima facie* evidence for phenotypic selection against TEs. In addition, their mutational properties are also likely to be a cause of negative selection. That negative selection is not always the case is illustrated by experiments of Chao *et al.* (1983) with *E. coli* populations containing the Tn*10* transposable element. These investigators show that chemostat populations of Tn*10*-bearing strains out-compete strains lacking Tn*10*, provided the initial frequency of the Tn*10* strain exceeds 10^{-4}. Moreover, successful Tn*10* strains were always characterized by a transposition event into a particular region of the genome. In contrast, unsuccessful Tn*10* strains were not characterized by transposition events.

The fact that the competitive success of Tn*10* strains depends on initial frequency is interpreted in terms of the expected number of favorable transposition-induced mutational events. According to this explanation, when the initial frequency of Tn*10* is less than 10^{-4}, the expected number of favorable mutational events is low and the Tn*10* strains decline in frequency. If, however, the initial frequency exceeds

10^{-4}, the expected number of favorable mutational events is high and the Tn*10* clone carrying the favorable mutant wins in competition. The fact that winning Tn*10* strains experienced transposition events is consistent with the transposition-induced favorable mutation hypothesis.

Biel and Hartl (1983) have presented evidence that the transposon Tn*5* can confer a selective (growth rate) advantage in *E. coli* chemostat populations. The advantage depends upon the *E. coli* host strain but does not require transposition and occurs independent of Tn*5* chromosomal or episome location. The advantage is apparently not a consequence of mutational activity but instead may involve the transposase function (Hartl *et al.*, 1983). Although the underlying mechanisms of the growth rate advantage are unknown, this appears to constitute a case of positive phenotypic selection for a TE.

Clearly, TEs have manifold properties and their evolutionary consequences are equally diverse. Population genetic research in this area is still in a phenomenological phase. We must establish the major outlines of TE behavior before detailed theoretical analyses can be pursued.

IV. Quantitative Genetics

Natural selection must necessarily act on the individual phenotype. The phenotype is usually the manifestation of many loci, each of small overall effect. Recently developed quantitative genetic models represent an important effort to study the evolutionary dynamics of complex phenotypes. These models, which assume that the distribution of phenotypes is normal and therefore can be fully described by the mean and genetic covariance matrix, simplify the study of the evolutionary dynamics of important adaptive characters. In this section we examine polygenic models and their heuristic value for empirical and further theoretical studies.

A. Basic Models

A fundamental observation in evolutionary biology is that many major phenotypic traits exhibit continuous, normally distributed variation and a large component of this variation is heritable. These traits often have an optimal phenotype and extreme phenotypes are selected against. Stabilizing selection (also known as centripetal, normalizing, or noroptimal selection) has often been demonstrated, as for example, in the classical study of neonatal survivorship of human babies as a function of body weight at birth (Karn and Penrose, 1951). However, only recently have basic stabilizing selection models of polygenic characters been analyzed.

Stabilizing selection is expected to deplete genetic variability while mutation is required to replenish genetic variability. Turelli (1984) recently reviewed stabilizing selection–mutation models. Kimura (1965), Lande (1976), and Fleming (1979) analyzed a class of models (referred to as K–L–F models) in which it is assumed that

1. The distribution of phenotypes is normal.
2. Population size is effectively infinite.
3. There is a continuum of alleles at each locus.
4. Each allele at each (interchangeable) locus contributes an additive genetic effect (within and between loci) plus a normally distributed, genotype-independent, additive environmental effect.
5. The change in the effect of a gene by mutation has a normal distribution with mean zero.
6. The selective value of phenotypes has either a Gaussian or "quadratic" (in the Kimura model) distribtuion.
7. Linkage between the loci is "loose."

To insure a normal phenotypic distribution the K–L–F model makes further assumptions about the relative magnitudes of selection, mutability, and rate of mutation (Fleming, 1979; Turelli, 1984).

Bulmer (1980) analyzed a model that differs from the K–L–F model in that there are only two alleles per locus with mutation in both directions. Other models include the strictly phenotypic model of Slatkin (1970). Turelli (1984) recently developed a heuristic model that closely resembles the K–L–F model in terms of genetics and the Bulmer model in terms of assumptions regarding parameter values, yet differs from both in that "the distribution of effects associated with new mutants will be essentially independent of their premutation state."

The K–L–F, Bulmer, and Turelli models all predict large levels of genetic variability at equilibrium, which is consistent with most observations (Turelli, 1984). One important difference between the models is that for a given gametic mutation rate, $n\mu$ (where n equals the number of loci affecting the trait and μ is the per locus mutation rate), the K–L–F model counterintuitively predicts an increase in genetic variance, V_G, at equilibrium as n increases (and μ decreases) whereas the Bulmer and Turelli models are invariant. A critical difference between these models is that, for a reasonable range of the other parameter values, the K–L–F model must assume μ greater than 10^{-4}, and the Bulmer and Turelli models maintain normality when μ is less than 10^{-5}. (Although in some cases the Turelli model maintains normality with $\mu = 10^{-4}$.)

Clearly, mutation rate and the number of loci affecting a trait are critical parameters. Hence it is important to obtain sound empirical estimates of actual mutation rates, gametic mutation rates, mutability of loci, number of loci affecting the trait, and the strength of stabilizing selection acting on the trait, because all of these parameters are relevant to the validity of the various models and to the level of genetic variability maintained. Turelli (1984) reviews the extant empirical observations in detail. Estimates of mutation rates typically range from 10^{-7} to 10^{-5}; however, these may be underestimates because mutations of small effect may occur in high frequency but are not detected. For example, mutations may frequently occur in large, non-translated regions of DNA near structural loci and these mutations may affect the rate of production of the gene products and may be expected to have only slight effects on the phenotypic trait (Wright, 1977).

Turelli (1984) also reviews evidence for "unusual" mutational processes (transposons, unequal crossing over, and gene conversion) that can result in very high mutation rates, from 10^{-5} to 10^{-4}. He warns, however, that these mutation processes cannot simply be plugged into a model as a constant, because these processes can have different behaviors at the population level and some are genotype dependent. At present very little is known about the mutability of most loci, but some studies suggest that mutations with small effects may be much more frequent than mutations with larger effects (Gregory, 1966; Mukai, 1979). This result supports the normality assumption (with mean zero) made in the K–L–F model.

Gametic mutation rate estimates typically exceed 10^{-2} per gamete per generation for polygenic characters. The discrepancy with per-locus mutation rates may be due to the fact that (1) a large number of loci affect the trait, say 10^3–10^4; (2) per-locus mutation rates may be higher than 10^{-5}, as discussed above; or (3) gametic mutation rates are somehow biased (Turelli, 1984).

The number of loci affecting a trait is also important and can be estimated by observing the higher moments of progeny distributions from certain crosses (Mather and Jinks, 1971) and is facilitated where doubled haploid (plants) are available (Choo and Reinbergs, 1982). These studies probably greatly underestimate the actual number of loci due to inequality of per-locus effects and other factors. Estimates typically range from 5 to 10 loci (Lande, 1980a), although estimates of 100 or more loci for protein content in maize (Dudley, 1977) have been obtained. Turelli (1984) notes that more direct methods suggest that about 5000 loci can produce lethal mutations in *Drosophila melanogaster*. Spradling and Rubin (1981) estimate that 30,000 loci may produce protein products in *D. melanogaster*.

The approximate strength of stabilizing selection must also be known. The strength of selection has been estimated by observing the decrease in phenotypic variance within a cohort over time, or over age classes in age-structured populations (Lande, 1980b, 1982). This approach has been applied to a variety of characters in numerous species (see Lande and Arnold, 1983). The estimates obtained are consistent with the intensity of selection used by Turelli (1984) in simulations that were compared to the predicted results of the various models.

Pleiotropy can cause serious overestimation of the intensity of selection on a single trait (Lande and Arnold, 1983). Selection on phenotypically correlated traits and directional selection can also bias estimates of selection intensity upward. Furthermore, selection on phenotypic traits at different life cycle stages is rarely measured. This may lead to substantial underestimates or, conceivably, overestimates if selection is disruptive.

Neutral theory may at first appear to be inconsistent with the observation that major phenotypic traits are often under substantial natural selection. However, Kimura (1983) and Milkman (1982) have demonstrated that even with genetic loads as large as 50% (i.e., 50% selective mortality on the trait) only very weak selection is experienced at any locus (or nucleotide pair) if the number of loci is very large. For example, if the number of heterozygous nucleotide sites per individual (the number of loci multiplied by average heterozygosity) is 10^6, then selection is very weak and most genes will closely approximate a neutral dynamic (i.e., there is considerable random fixation of alleles). This calculation assumes that loci have an equal effect on fitness, which is perhaps unlikely.

B. Extensions of the Basic Models

Lande (1980a) extended the basic model to multiple characters that may share a common genetic basis, or *pleiotropy*. He makes the additional assumption that the joint distribution of phenotypes is multi-variate normal. Genes are still assumed to be additive in phenotypic effect for each trait both within and between loci.

The model gives rise to the beautifully simple relationship $\Delta z = G \nabla \ln w$, where Δz is the change in the vector representing the mean phenotypic values of the traits, G is the genetic variance–covariance matrix of the traits, and $\nabla \ln w$ is the "selection gradient" with respect to changes in z. The model assumes loose linkage between loci controlling the traits. If G remains constant the population evolves to an optimal fitness peak, although some traits may actually evolve mal-adaptively. M. Turelli (personal communication) reports very different behavior in an extension of his model to multiple characters with pleiotropy. In this context, Atchley et al. (1982) found that a variety of morphometric traits evolved unpredictably between replicates when weight (which did respond predictably) was selected for in rats. Lande (1979) applied his model to the problem of allometry (correlations in the development of phenotypic traits) with regard to trait similarity between species in time or across taxonomic units. The common observation of allometry has been discussed by many authors (Huxley, 1932; Gould, 1966), but previously, the relationship of allometry to the quantitiative genetic covariance structure of traits was ignored. Lande (1979) also discusses the importance of genetic covariance to the "ontogeny recapitulates phylogeny" concept.

Gillespie (1984) analyzed models similar to Bulmer's except that multiple alleles are allowed and loci contribute a pleiotropic effect on viability in addition to an additive contribution to the phenotypic trait under stabilizing selection. The pleiotropic effect is assumed to exhibit overdominance, and mutation is ignored. Gillespie (1984) found that large heritabilities can be maintained at equilibrium, and thus pleiotropic overdominance provides a viable alternative to the selection–mutation hypothesis favored by Lande (1976). The dynamics of the model are described and experimental procedures are suggested to test the two hypotheses.

Models other than the pleiotropic overdominance model of Gillespie (1984) can also account for observed heritabilities and may provide an alternative hypothesis to stabilizing selection. Alternatives include pleiotropic models with frequency–dependent viability selection, frequency-dependent phenotypic fitnesses, variable (either cyclic or stochastic) environmental selection (e.g., Slatkin and Lande, 1976), density-dependent selection, or genotype–environment interactions.

Many of these models may be mathematically intractable because the assumption of normality may be invalid.

Lande (1977) found that inbreeding or assortative mating has no effect on the amount of genetic variance at equilibrium, in contrast to traditional polygenic models (Fisher, 1918; Wright 1921). Lande (1980b) applied his multicharacter model to cases in which there is sexual dimorphism for the quantitative traits with or without sexual selection on the traits. Under weak natural selection, the mean phenotype of both sexes eventually evolves to locally optimum fitness peaks. Sexual selection, however, can cause maladaptive evolution in either sex and if strong enough, leads to extinction of the population. Sexual selection makes phenotypic fitnesses frequency dependent; thus this model represents an important attempt to incorporate frequency dependency into quantitative genetic models. Finally, Lande (1982) has analyzed the evolution of life history characters.

It is apparent that the study of quantitative characters has experienced a renaissance in recent years. Much of this renewed interest can be traced to the theoretical work initiated by Kimura and developed by Lande.

V. Genetic Analysis of Plant Mating Systems

The mating system of a population profoundly affects the structure of genetic variability. The mating system can be described in three ways: (1) by direct observation of matings; (2) by analysis of the reproductive biology; and (3) by observation of the distribution of genotypes in families. We will consider recent theoretical and empirical results based on observations of the third type. Three areas of contemporary research activity are discussed: (1) the use of multilocus data to measure the rate of self-fertilization and other mating system parameters, (2) modifications of the classical mixed mating model, and (3) the evolutionary dynamics of genes that modify plant mating systems.

A. Multilocus Estimation of Mating System Parameters

An important new approach to the study of plant mating systems has involved the use of multilocus gametic types. To see the advantages of this approach, it is first necessary to discuss tbe standard single locus estimator. Consider the progeny of a maternal parent of type A_2A_2. Assuming a diallelic locus with gene frequencies $f(A_1) = p$, $f(A_2) = q = 1 - p$, two progeny genotypes are possible, A_1A_2 and A_2A_2 with probabilities tp and $s + tq$, respectively, where $t(= 1 - s)$ denotes the probability of outcrossing. Clearly all A_1A_2 progeny result from outcross events; however, the A_2A_2 progeny class represents a mixture of self-fertilization and outcross events. If we could arrange it so that every gamete was uniquely marked, then all outcross events would be identifiable. Suppose we have m loci, each with k alleles, then there are k^m possible gametic types. It is easy to see that the joint multilocus gametic frequency distribution provides a means of marking most gametic types uniquely, depending on the number of polymorphic loci available in any particular population. This idea has been applied to the analysis of plant mating systems (Shaw et al., 1981; Ritland and Jain, 1981).

There are two areas in which multilocus mating system estimation has been employed. First, comparison of multilocus and single-locus estimates has been used to detect violations in the assumptions of single-locus estimation models. These models assume that no other forms of inbreeding are operating, selection is assumed to be absent, the probability of outcrossing is assumed to be the same for all individuals, and the outcross pollen is assumed to be randomized with respect to maternal parents (A. H. D. Brown and Allard, 1970; Clegg et al., 1978; Cheliak et al., 1983). Violations of these assumptions must be common because single locus estimates of t often vary strikingly for different loci within a sample from a single population (Allard et al., 1977; A. H. D. Brown et al., 1975; Moran and Brown, 1980). In self-compatible lodgepole pine, however, seven single-locus estimates of t were all near 1.0 (Epperson and Allard, 1984). Estimates of t based on multilocus family-structured data require assumptions similar to those of the basic one-locus estimator; however, they are more robust to violations of these assumptions (Shaw et al., 1981; Ritland and Jain, 1981). Violations can be identified by simply comparing single-locus estimates to the multilocus estimate (Shaw and Allard, 1982). We note in passing that the multilocus estimators do not utilize information on linkage of the marker loci, linkage disequilibrium between the marker loci, or the coupling–repulsion phase of multiple heterozygotes.

Ritland (1984) developed an elegant model that utilizes additional information on inbreeding in family-structured multilocus data sets to quantify the amount of inbreeding apart from selfing. Ritland (1985) has also developed procedures for estimation of population subdivision, particularly in the case of Wright's island model. The effect of population substructure on estimates of t has been shown to be large in experimental populations of the bee-pollinated plant *Ipomoea purpurea* (Ennos and Clegg, 1982). In addition, Ellstrand *et al.* (1978) suggested that subdivision caused low estimates of t, which ranged from 0.54 to 0.91, in populations of the self-incompatible plant *Helianthus annuus*.

The second major approach, based on multilocus gamete types, is appropriate for small, isolated populations that are highly polymorphic for several marker loci. This method uses multilocus family-structured data either to uniquely identify the paternal parent, or to estimate the most likely paternal parent (or parents). This has been done, for example, for completely censused small populations of *Raphanus sativus* (Ellstrand, 1984), prairie dogs (Foltz and Hoogland, 1981), and conifer seed orchards (Smith and Adams, 1983). In the absence of selection, this procedure allows a complete description of actual matings and results in theory-free estimates of the outcrossing rate. If the locations of the parents are known, mating by proximity can also be measured. However, without a measure of consanguinty between the putative paternal parent and the known maternal parent, we can not fully describe the amount of inbreeding other than by selfing (although this could be estimated indirectly by using the model of Ritland, 1984). Other important mating system attributes, such as variation in t between individuals, correlations in paternal parentage within families, and nonrandom distribution of the outcross pollen available to individuals, can also be measured. Finally, it is possible to accurately measure gene flow, which previously has proven extremely difficult, despite the fact that migration is one of the central parameters of population genetics theory.

B. Modifications of the Mixed Mating Model

A random distribution of the outcross pollen pool may be a reasonable assumption in some populations of wind-pollinated species, but this assumption is probably violated in many plant species. In insect-pollinated species, particularly if the number of pollinator visits per flower is small, the outcross pollen available to a single flower or individual may derive from a single male parent. Schoen and Clegg (1984) developed a single-locus estimator of t that assumes the pollen parent is the same for all progeny of a maternal family. They found, by computer simulations, that the standard one-locus estimator is substantially biased when correlated mating occurs, and conversely, that the correlated pollen parent model is biased when pollen is sampled independently within a family. Estimation models that posit an intermediate or variable degree of paternal correlation have yet to be developed. In addition, other aspects of reproductive biology, for example partial self-incompatibility, competition between gametes, and zygotic selection, all can play an important role in determining maternal–progeny arrays observed in a population sample.

C. Genetic Modification of the Mating System

It is becoming increasingly evident that there is considerable genetic variation for outcrossing both within and between populations (e.g., Rick *et al.,* 1979; Schoen, 1982). The nature of the genetic variability for outcrossing rate is wide ranging. For example, the narrow sense heritability of the distance between anthers and stigma, which affects the rate of selfing, is approximately 0.5 in the self-compatable species *Ipomoea purpurea* (Ennos, 1981). Moreover, a single gene responsible for white flower color is often in high frequency in populations of *I. purpurea*. This gene decreases the rate of pollinator visits to white flowers and thus increases the rate of selfing of white morphs (B. A. Brown and Clegg, 1984).

The existence of genetically determined variation in outcrossing rate establishes that mating systems can respond to evolutionary pressures. Moreover, genes that modify the mating process are *ipso facto* selected. To see this, consider the simplest model of the evolution of self-fertilization, which assumes that a single gene mutation increases selfing. In this model genes are transmitted to ovules in Mendelian ratios. By increasing the rate of selfing the gene biases the probability of being transmitted in the male gamete that fertilizes the ovule. As a result, a gene that increases selfing rate always increases in frequency to fixation (Fisher, 1941; Moran, 1962; Wells, 1979). In view of this argument, why aren't all species exclusively self-fertilizing?

Nagylaki (1976) analyzed a model in which the gene that increased selfing also causes less pollen to be contributed to the outcross pollen pool (i.e., pollen discounting). In this model the gene increases in frequency only if the pollen discounting is sufficiently small compared to the increase in rate of selfing. Actual pollen discounting depends on the reproductive biology of the species and the nature of the mutational change that causes increased selfing. For example, consider a monoecious, wind-pollinated species with partial self-incompatibility. A mutation that decreases self-incompatibility will increase selfing;

yet because of the large amounts of pollen being produced, there is negligible pollen discounting. In contrast, a mutation that causes a delay in flowering time (of both male and female flowers) may increase selfing and result in substantial or complete pollen discounting. Finally, mutations that increase selfing may affect seed set (e.g., in the above example if there was not enough self pollen available for 100% fertilization).

Any realistic model of the evolution of self-fertilization must include selection. Maynard-Smith (1978) analyzed a model in which selfed progeny are less fit. His model is subsumed in a model studied by Holsinger *et al.* (1984) that includes pollen discounting. These latter workers found that a mutant that increases selfing increases in frequency only if the selective disadvantage and pollen discounting are not too great relative to the increase in selfing. Various parameter values can lead to fixation, loss, or protected polymorphism for the selfing gene.

The model of Holsinger *et al.* (1984) assumes that the selective disadvantage of selfed progeny is constant. In reality, the immediate fitness of genotypes with increased selfing depends on selection at other loci and therefore depends on the nature of the selection (overdominance, partial dominance, epistasis, etc.), the distribution (particularly homozygosity) of multilocus genotypes for the selected loci, and linkage disequilbrium between the selfing locus and the selected loci. Models that include these factors have yet to be developed.

VI. Spatial Analysis of Genetic Variability

The distribution of genetic variability within and between populations has long been a primary interest in theoretical and empirical population genetics. In this section, we consider (1) the use of autocorrelation statistics to measure either the spatial arrangement of gene frequencies among discrete populations, or the spatial distribution of genetic variability within continuous populations; and (2) the development of a hypothesis-testing framework for processes such as clinal selection and isolation by distance, using autocorrelation statistics.

A. DESCRIPTION OF POPULATION SUBSTRUCTURE

Autocorrelation statistics measure the dependency of the character value, either nominal (e.g., genotypic), interval (e.g., gene frequency), or ranked data, at a location on the character value at other locations.

Here we consider gene frequencies and nominal genotypes. Although statistical procedures for simply comparing gene frequencies in discrete populations are well developed (e.g., Nei, 1973), autocorrelation statistics allow examination of gene frequency patterns in space. In this context, autocorrelation statistics measure the dependency of population samples of gene frequencies on population samples of gene frequencies at any arbitrary distance (or direction). If there is no pattern to the distribution of gene frequencies among populations, there is no spatial autocorrelation at any distance (or direction). A graph of autocorrelation statistics as a function of distance is known as a correlogram (Sokal and Oden, 1978b). Statistical tests have been developed for the null hypothesis that autocorrelation at a specified distance is zero (Sokal and Oden, 1978a) or that autocorrelation at several distances is zero (Oden, 1984). For discrete populations several empirical studies have made use of these statistics (e.g., Sokal and Oden, 1978b; Sokal and Menozzi, 1982).

Autocorrelation statistics may be especially useful in describing the genetic substructure of continuous populations. By measuring the autocorrelation between individual genotypes (or subsamples of gene frequencies), spatial structure can be described more precisely than has been possible in the past. However, the analysis of spatial autocorrelation patterns based on individual genotypes has not been widely employed in empirical population genetics. Sokal and Oden (1978b) report a study in which nearest neighbor genotypes were positively autocorrelated in a population of mice inhabiting a barn. In addition, rare heterozygotes showed positive autocorrelation on a small scale for allozymes in lodgepole pine (B. K. Epperson, unpublished data).

B. Tests of Patterns Caused by Various Evolutionary Factors

Sokal and colleagues have attempted to develop a framework for distinguishing spatial patterns of genetic variability caused by different evolutionary factors in terms of autocorrelation statistics (Sokal and Oden, 1978a,b; Sokal, 1979). For example, isolation by distance is expected to yield positive autocorrelation at short distances and negative autocorrelation at longer distances. If natural selection is acting in "patchy" environments, then positive autocorrelation occurs between genotypes within patches and negative autocorrelation occurs between genotypes from different patches. Circular gradients (i.e., an environmental gradient that radiates symmetrically from a single point) will cause positive autocorrelation at long distances relative to the patch area. Clinal selection results in short-distance positive and long-distance negative autocorrelation in the direction of the cline and long-distance positive autocorrelation in the direction perpendicular to the cline.

Once appropriate mathematical models or simulation realizations of various processes are available it may be possible to develop statistical hypotheses based on each process. The correlogram or analogous summary statistics could then be tested against various null hypotheses appropriate to the biological situation. If a stationary (or quasi-stationary) (Sokal, 1983) process is obtained, then it may be possible to find the form of the spectral density of the process and estimate the relevant parameters in corresponding real populations. These parameters may then be related to rates of migration or the strength of selection.

Real populations exist in a spatial context. Attempts to develop a statistical framework for the analysis and description of spatially determined genetic processes such as selection and migration are essential to the full interpretation of population genetic data.

VII. Conclusions

While the research areas discussed in this article illustrate the breadth of contemporary activity in population genetics, they are by no means exhaustive. For example, Cavalli-Sforza and Feldman (1981) have pioneered the use of population genetic models to investigate the processes of cultural evolution. This new area of research activity seeks to use the formal constructs of population genetics as a basis for the analysis of the transmission of nongenetic information. In so doing, Cavalli-Sforza and Feldman (1981) have moved one area of population genetic enquiry squarely into the arena of the social sciences. Another area of major research activity is kin selection theory (reviewed by Michod, 1982). This research program has attempted to account for the evolution of group behaviors or, more generally, group interactions. The dynamical behavior of multilocus genetic systems has also been a focus of recent research activity (reviewed by Clegg, 1984), and still other important areas of research activity could be listed. Taken together, these areas and those reviewed above provide substance to the contention that population genetics has been infused with a variety of new research directions.

ACKNOWLEDGMENTS

We thank Drs. Michael Turelli and S. E. Curtis for helpful discussions. This work was supported in part by National Science Foundation Grants BSR-8418382 and BSR-8418381.

REFERENCES

Allard, R. W., Kahler, A. L., and Clegg, M. T. (1977). Estimation of mating cycle components of selection in plants. *In* "Measuring Selection in Natural Populations" (F. B. Christiansen, and T. M. Fenchel, eds.), pp. 1–19. Springer-Verlag, Berlin and New York.

Anderson, S., Bankier, A. T., Barrell, B. G., de Bruijn, M. H. L., Coulson, A. R., Drouin, J., Eperson, I. C., Nierlich, D. P., Roe, B. A., Sanger, F., Schreier, P. H., Smith, A. J. H., Staden, R., and Young, I. G. (1981). Sequence and organization of the human mitochondrial genome. *Nature (London)* **290**, 457–465.

Anonymous (1983). Summarized proceedings of the workshop on population genetics June 8, 1983. National Institutes of Health Administrative Document.

Antonarakis, S. E., Orkin, S. H., Kazazian, H. H., Goff, S. C., Boehm, C. D., Waber, P. G., Sexton, J. P., Ostrer, H., Fairbanks, V. F., and Chakravarti, A. (1982). Evidence for multiple origins of the β^E-globin gene in Southeast Asia. *Proc. Natl. Acad. Sci. U.S.A.* **79**, 6608–6611.

Aquadro, C. F., and Greenberg, B. D. (1983). Human mitochondrial DNA variation and evolution: Analysis of nucleotide sequences from seven individuals. *Genetics* **103**, 287–312.

Arnheim, N., Krystal, M., Schmickel, R., Wilson, G., Ryder, O., and Zimmer, E. (1980). Molecular evidence for genetic exchanges among ribosomal chromosomes in man and apes. *Proc. Natl. Acad. Sci. U.S.A.* **77**, 7323–7327.

Asmussen, M. A. (1984). The use of incompletely linked markers in genetic counseling: accuracy versus linkage. *Hum. Hered.* (in press).

Asmussen, M. A., and Clegg, M. T. (1982). Use of restriction fragment length polymorphisms for genetic counseling: Population genetic considerations. *Am. J. Hum. Genet.* **34**, 369–380.

Asmussen, M. A., and Clegg, M. T. (1984). Multiallelic restriction fragment polymorphisms in genetic counseling: Population genetic considerations. *Hum. Hered.* (in press).

Atchley, W. R., Rutledge, J. J., and Cowley, D. E. (1982). A multivariate statistical analysis of direct and correlated response to selection in the rat. *Evolution* **36**, 677–698.

Avise, J. C., and Lansman, R. A. (1983). Polymorphism of mitochondrial DNA in populations of higher animals. *In* "Evolution of Genes and Proteins" (M. Nei and R. K. Koehn, eds.), pp. 147–164. Sinauer Associates, Sunderland, MA.

Avise, J. C., Giblin-Davidson, C., Laerm, J., Patton, J. C., and Lansman, R. A. (1979). Mitochondrial DNA clones and matriarchal phylogeny within and among geographic populations of the pocket gopher, *Geomys pinetis*. *Proc. Natl. Acad. Sci. U.S.A.* **76**, 6694–6698.

Avise, J. C., Shapira, J. F., Daniel, S. W., Aquadro, C. F., and Lansman, R. A. (1983). Mitochondrial DNA differentiation during the speciation process in *Peromyscus*. *Mol. Biol. Evol.* **1**, 38–56.

Biel, S. W., and Hartl, D. L. (1983). Evolution of transposons: Natural selection for Tn5 in *Escherichia coli* K12. *Genetics* **103**, 581–592.

Bingham, P. M., Kidwell, M. G., and Rubin, G. M. (1982). The molecular basis of P-M hybrid dysgenesis: The role of the P element, a P-strain-specific transposon family. *Cell* **29**, 995–1004.

Birky, C. W., Maruyama, T., and Fuerst, P. (1983). An approach to population and evolutionary genetic theory for genes in mitochondria and chloroplasts, and some results. *Genetics* **103**, 513–527.

Botstein, D., White, R. L., Skolnick, M., and Davis, R. W. (1980). Construction of a genetic linkage map in man using restriction fragment length polymorphisms. *Am. J. Hum. Genet.* **32**, 314–331.

Bregliano, J.-C., and Kidwell, M. G. (1983). Hybrid dysgenesis determinants. *In* "Mobile Genetic Elements" (J. A. Shapiro, ed.), pp. 363–410. Academic Press, New York.

Brown, A. H. D., and Allard, R. W. (1970). Estimation of the mating system in open-pollinated maize populations using isozyme polymorphisms. *Genetics* **66**, 133–145.

Brown, A. H. D., and Clegg, M. T. (1983). Analysis of variation in related DNA sequences. *In* "Statistical Analysis of DNA Sequence Data" (B. S. Weir, ed.), pp. 107–132. Dekker, New York.

Brown, A. H. D., Matheson, A. C., and Eldridge, K. G. (1975). Estimation of the mating system of *Eucalyptus obliqua* L'Herit. by using allozyme polymorphisms. *Aust. J. Bot.* **23**, 931–949.

Brown, B. A., and Clegg, M. T. (1984). Influence of flower color polymorphisms on genetic transmission in a natural population of the common morning glory *Ipomoea purpurea*. *Evolution* **38**, 796–803.

Brown, G. G., and Simpson, M. V. (1982). Novel features of animal mtDNA evolution as shown by sequences of two rat cytochrome oxidase subunit II genes. *Proc. Natl. Acad. Sci. U.S.A.* **79**, 3246–3250.

Brown, W. M. (1983). Evolution of animal mitochondrial DNA. *In* "Evolution of Genes and Proteins" (M. Nei and R. K. Koehn, eds.), pp. 62–88. Sinauer Associates, Sunderland, MA.

Bulmer, M. G. (1980). "The Mathematical Theory of Quantitative Genetics." Oxford Univ. Press (Clarendon), London and New York.

Campbell, A. (1983). Transposons and their evolutionary significance. *In* "Evolution of Genes and Proteins" (M. Nei and R. K. Koehn, eds.), pp. 258–279. Sinauer Associates, Sunderland, MA.

Cann, R. L., and Wilson, A. C. (1983). Length mutations in human mitochondrial DNA. *Genetics* **104**, 699–711.

Cavalli-Sforza, L. L., and Feldman, M. W. (1981). "Cultural Transmission and Evolution: A Quantitative Approach." Princeton Univ. Press, Princeton, NJ.

Chakravarti, A. (1983). Utility and efficiency of linked marker genes for genetic counseling. III. Proportion of informative families under linkage disequilibrium. *Am. J. Hum. Genet.* **35**, 592–610.

Chao, L., Vargas, C., Spear, B. B., and Cox, E. C. (1983). Transposable elements as mutator genes in evolution. *Nature (London)* **303**, 633–635.

Chapman, R. W., Stephens, J. C., Lansman, R. A., and Avise, J. C. (1982). Models of mitochondrial DNA transmission genetics and evolution in higher eucaryotes. *Genet. Res.* **40**, 41–57.

Charlesworth, B., and Charlesworth, D. (1983). The population dynamics of transposable elements. *Genet. Res.* **42**, 1–27.

Cheliak, W. M., Morgan, K., Strobeck, C., Yeh, F. C. H., and Dancik, B. P. (1983). Estimation of mating system parameters in plant populations using the EM algorithm. *Theor. Appl. Genet.* **65,** 157–161.

Choo, T. M., and Reinbergs, E. (1982). Estimation of the number of genes in doubled haploid populations of barley (*Hordeum vulgare*). *Can. J. Genet. Cytol.* **24,** 337–342.

Clark, A. G. (1984). Natural selection with nuclear and cytoplasmic transmission. I. A deterministic model. *Genetics* **107,** 679–701.

Clegg, M. T. (1984). The dynamics of multilocus genetic systems. *Oxford Surv. Evol. Biol.* **1,** 158–181.

Clegg, M. T., and Asmussen, M. A. (1983). Use of restriction fragment polymorphisms as genetic markers. *In* "Statistical Analysis of DNA Sequence Data" (B. S. Weir, ed.), pp. 201–229. Dekker, New York.

Clegg, M. T., Kahler, A. L., and Allard, R. W. (1978). Estimation of life cycle components of selection in an experimental plant population. *Genetics* **89,** 765–792.

Clegg, M. T., Brown, A. H. D., and Whitfeld, P. R. (1984). Chloroplast DNA diversity in wild and cultivated barley: Implications for genetic conservation. *Genet. Res.* **43,** 339–343.

Coen, E. S., and Dover, G. A. (1983). Unequal exchanges and the coevolution of X and Y rDNA arrays in *Drosophila melanogaster*. *Cell* **33,** 849–855.

Cullis, C. A. (1976). Environmentally induced changes in ribosomal RNA cistron number influx. *Heredity* **36,** 73–79.

Curtis, S. E., and Clegg, M. T. (1984). Molecular evolution of chloroplast DNA sequences. *Mol. Biol. Evol.* **1,** 291–301.

Doolittle, W. F., and Sapienza, C. (1980). Selfish genes, the phenotype paradigm and genome evolution. *Nature (London)* **272,** 123–124.

Dudley, J. W. (1977). 76 generations of selection for oil and protein percentage in maize. *In* "Quantitative Genetics" (E. Pollack, O. Kempthorne, and T. B. Bailey, eds.), pp. 459–473. Iowa State Univ. Press, Ames.

Ellstrand, N. C. (1984). Multiple paternity within the fruits of the wild radish *Raphanus sativus*. *Am. Nat.* **123,** 819–828.

Ellstrand, N. C., Torres, A. M., and Levin, D. A. (1978). Density and the rate of apparent outcrossing in *Helianthus annuus* (Asteraceae). *Syst. Bot.* **3,** 403–407.

Ennos, R. A. (1981). Quantitative studies of the mating system in two sympatric species of *Ipomoea* (Convolvulaceae). *Genetica* **57,** 93–98.

Ennos, R. A., and Clegg, M. T. (1982). Effect of population substructuring on estimates of outcrossing rate in plant populations. *Heredity* **48,** 283–292.

Epperson, B. K., and Allard, R. W. (1984). Allozyme analysis of the mating system in populations of lodgepole pine. *J. Hered.* **75,** 212–214.

Fauron, M.-R., and Wolstenholme, D. R. (1976). Structural heterogeneity of mitochondrial DNA molecules within the genus *Drosophila*. *Proc. Natl. Acad. Sci. U.S.A.* **73,** 3623–3627.

Fisher, R. A. (1918). The correlation between relatives on the supposition of Mendelian inheritance. *Trans. R. Soc. Edinburgh* **52,** 399–433.

Fisher, R. A. (1941). Average excess and average effect of gene substitution. *Ann. Eugen. (London)* **11,** 53–63.

Fleming, W. H. (1979). Equilibrium distribution of continuous polygenic traits. *SIAM J. Appl. Math.* **36,** 148–168.

Foltz, D. W., and Hoogland, J. L. (1981). Analysis of the mating system in the black-tailed prairie-dog *Cynomys ludovicianus* by likelihood of paternity. *J. Mammal.* **64,** 706–712.

Fox, T. D. (1984). Multiple forms of mitochondrial DNA in higher plants. *Nature (London)* **307**, 415.

Frankham, R., Briscoe, D. A., and Nurthen, R. K. (1980). Unequal crossing over at the rRNA tandon as a source of quantitative genetic variation in *Drosophila*. *Genetics* **95**, 727–742.

Gillespie, J. H. (1984). Pleiotropic overdominance and the maintenance of genetic variation in polygenic characters. *Genetics* **107**, 321–330.

Gould, S. J. (1966). Allometry and size in ontogeny and phylogeny. *Biol. Rev. Cambridge Philos. Soc.* **41**, 587–640.

Gregorius, H. R., and Ross, M. D. (1984). Selection with gene-cytoplasm interactions. I. Maintenance of cytoplasm polymorphisms. *Genetics* **107**, 165–178.

Gregory, W. C. (1966). Mutation breeding. *In* "Plant Breeding" (K. I. Frey, ed.), pp. 189–218. Iowa State Univ. Press, Ames.

Gusella, J. F., Wexler, N. S., Conneally, P. M., Naylor, S. L., Anderson, M. A., Tanzi, R. E., Watkins, P. C., Ottina, K., Wallace, M. R., Sakaguchi, A. Y., Young, A. B., Shoulson, I., Bonilla, E., and Martin, J. B. (1983). A polymorpbic DNA marker genetically linked to Huntington's disease. *Nature (London)* **306**, 234–238.

Hartl, D. L., Dykhuizen, D. E., Miller, R. D., Green, L., and de Framond, J. (1983). Transposable element IS50 improves growth rate of *E. coli* cells without transposition. *Cell* **35**, 503–510.

Hauswirth, W. M., and Laipis, P. J. (1981). Rapid variation in mammalian mitochondrial genotypes: Implications for the mechanism of maternal inheritance. *In* "Mitochondrial Genes" (P. Sloniminski, P. Borst, and G. Attardi, eds.), pp. 137–141. Cold Spring Harbor Lab., Cold Spring Harbor, NY.

Hickey, D. A. (1982). Selfish DNA: A sexually transmitted nuclear parasite. *Genetics* **101**, 519–531.

Holsinger, K. E., Feldman, M. W., and Christiansen, F. B. (1984). The evolution of self-pollination in plants. *Am. Nat.* **124**, 446–453.

Huxley, J. S. (1932). "Problems of Relative Growth." Methuen, London.

Jeffreys, A. (1979). DNA sequence variants in the $^G\gamma$-, $^A\gamma$-, δ- and β-globin genes of man. *Cell* **18**, 1–10.

Johns, M. A., Strommer, J. N., and Freeling, M. (1983). Exceptionally high levels of restriction site polymorphism in DNA near the *Adh*1 gene. *Genetics* **105**, 733–743.

Johnson, M. J., Wallace, D. C., Ferris, S. D., Rattazzi, M. C., and Cavalli-Sforza, L. L. (1983). Radiation of human mitochondria DNA types analyzed by restriction endonuclease cleavage patterns. *J. Mol. Evol.* **19**, 255–271.

Kan, Y. W., and Dozy, A. M. (1978a). Polymorphism of DNA sequence adjacent to the human β globin structural gene: Relationship to sickle cell mutation. *Proc. Natl. Acad. Sci. U.S.A.* **75**, 5631–5635.

Kan, Y. W., and Dozy, A. M. (1978b). Antenatal diagnosis of sickle cell anemia by DNA analysis of aminiotic fluid cells. *Lancet* 1(ii), 910–911.

Kan, Y. W., and Dozy, A. M. (1980). Evolution of hemoglobin S and C genes in world populations. *Science* **209**, 388–391.

Kaplan, N. L., and Brookfield, J. F. Y. (1983). Transposable elements in mendelian populations. III. Statistical results. *Genetics* **104**, 485–495.

Karn, M. N., and Penrose, L. S. (1951). Birth weight and gestation time in relation to maternal age, parity, and infant survival. *Ann. Eugen. (London)* **16**, 147–164.

Kazazian, H. H., Phillips, J. A., Boehm, C. D., Vik, T. A., Mahoney, M. J., and Ritchey, A. K. (1980). Prenatal diagnosis of β-thalassemias by amniocentesis: Linkage analysis using multiple polymorphic restriction endonuclease sites. *Blood* **56**, 926–930.

Kazazien, H. H., Chakravarti, A., Orkin, S. H., and Antonarakis, S. E. (1983). DNA polymorphism in the human β globin gene. *In* "Evolution of Genes and Proteins." (M. Nei and R. K. Koehn, eds.), pp. 137–146. Sinauer Associates, Sunderland, Massachusetts.

Kidwell, M. G. (1983a). Hybrid dysgenesis in *Drosophila melanogaster:* Factors affecting chromosomal contamination in the P-M system. *Genetics* **104,** 317–341.

Kidwell, M. G. (1983b). Evolution of hybrid dysgenesis determinants in *Drosophila melanogaster. Proc. Natl. Acad. Sci. U.S.A.* **80,** 1655–1659.

Kidwell, M. G., Kidwell, J. F., and Sved, J. A. (1977). Hybrid dysgenesis in *Drosophila melanogaster:* A syndrome of aberrant traits including mutation, sterility and male recombination. *Genetics* **86,** 813–833.

Kimura, M. (1965). A stochastic model concerning the maintenance of genetic variability in quantitative characters. *Proc. Natl. Acad. Sci. U.S.A.* **54,** 731–736.

Kimura, M. (1983). The neutral theory of evolution. *In* "Evolution of Genes and Proteins" (M. Nei and R. K. Koehn, eds.), pp. 208–233. Sinauer Associates, Sunderland, MA.

Kreitman, M. (1983). Nucleotide polymorphism at the alcohol dehydrogenase locus of *Drosophila melanogaster. Nature* **304,** 412–417.

Lande, R. (1976). The maintenance of genetic variability by mutation in a polygenic character with linked loci. *Genet. Res.* **26,** 221–235.

Lande, R. (1977). The influence of the mating system on the maintenance of genetic variability in polygenic characters. *Genetics* **86,** 485–498.

Lande, R. (1979). Quantitative genetic analysis of multivariate evolution, applied to brain:body size allometry. *Evolution* **33,** 402–416.

Lande, R. (1980a). The genetic covariance between characters maintained by pleiotropic mutations. *Genetics* **94,** 203–215.

Lande, R. (1980b). Sexual dimorphism, sexual selection and adaptation in polygenic characters. *Evolution* **34,** 292–305.

Lande, R. (1982). A quantitative genetic theory of life history evolution. *Ecology* **63,** 607–615.

Lande, R., and Arnold, S. J. (1983). The measurement of selection on correlated characters. *Evolution* **37,** 1210–1226.

Langley, C. H., Montgomery, E., and Quattlebaum, W. F. (1982). Restriction map variation in the *Adh* region of *Drosophila. Proc. Natl. Acad. Sci. U.S.A.* **79,** 5631–5635.

Langley, C. H., Brookfield, J. F. Y., and Kaplan, N. (1983). Transposable elements in mendelian populations. I. A theory. *Genetics* **104,** 457–471.

Levings, C. S. (1983). The plant mitochondrial genome and its mutants. *Cell* **32,** 659–661.

Lewontin, R. C. (1974). "The Genetic Basis of Evolutionary Change." Columbia Univ. Press, NY.

Li, W.-H. (1983). Evolution of duplicate genes and pseudogenes. *In* "Evolution of Genes and Proteins" (M. Nei and R. K. Koehn, eds.), pp. 14–37. Sinauer Associates, Sunderland, MA.

Long, E. D., and Dawid, I. B. (1980). Repeated genes in eukaryotes. *Annu. Rev. Biochem.* **49,** 727–764.

McClintock, B. (1952). Chromosome organization and gene expression. *Cold Spring Harbor Symp. Quant. Biol.* **16,** 13–47.

McClintock, B. (1956). Controlling elements and the gene. *Cold Spring Harbor Symp. Quant. Biol.* **21,** 197–216.

McDonald, J. F. (1983). The molecular basis of adaptation: A critical review of relevant ideas and observations. *Annu. Rev. Ecol. Syst.* **14,** 77–102.

Mather, K. (1943). Polygenic inheritance and natural selection. *Biol. Rev. Cambridge Philos. Soc.* **18,** 32–64.

Maynard-Smith, J. (1978). "The Evolution of Sex." Cambridge Univ. Press, London and New York.

Michod, R. E. (1982). The theory of kin selection. *Annu. Rev. Ecol. Syst.* **13,** 23–55.

Milkman, R. D. (1982). Toward a unified selection theory. *In* "Perspectives in Evolution" (R. D. Milkman, ed.), pp. 105–118. Sinauer Associates, Sunderland, MA.

Montgomery, E. A., and Langley, C. H. (1983). Transposable elements in Mendelian populations. II. Distribution of three copia-like elements in a natural population of *Drosophila melanogaster.* **104,** 473–483.

Moran, G. F., and Brown, A. H. D. (1980). Temporal heterogeneity of outcrossing rates in alpine ash (*Eucalyptus delegatensis,* R. T. Bak.). *Theor. Appl. Genet.* **57,** 101–105.

Moran, P. A. P. (1962). "The Statistical Processes of Evolutionary Theory." Oxford Univ. Press (Clarendon), London and New York.

Mukai, T. (1979). Polygenic mutation. *In* "Quantitative Genetic Variation" (J. N. Thompson and J. M. Thoday, eds.), pp. 177–196. Academic Press, New York.

Nagylaki, T. (1976). A model for the evolution of self-fertilization and vegetative reproduction. *J. Theor. Biol.* **58,** 55–58.

Nagylaki, T. (1984). The evolution of multigene families under intrachromosomal gene conversion. *Genetics* **106,** 529–548.

Nagylaki, T., and Petes, T. D. (1982). Intrachromosomal gene conversion and the maintenance of sequence homogeneity among repeated genes. *Genetics* **100,** 315–337.

Nei, M. (1973). Analysis of gene diversity in subdivided populations. *Proc. Natl. Acad. Sci. U.S.A.* **70,** 3321–3323.

Nei, M. (1979). Proportion of informative families for genetic counseling with linked marker genes. *Jpn. J. Hum. Genet.* **24,** 131–142.

Nei, M. (1983). Genetic polymorphism and the role of mutation in evolution. *In* "Evolution of Genes and Proteins" (M. Nei and R. K. Koehn, eds.), pp. 165–190. Sinauer Associates, Sunderland, MA.

Oden, N. L. (1984). Assessing the significance of a spatial correlogram. *Geogr. Anal.* **16,** 1–16.

Ohta, T. (1980). "Evolution and Variation of Multigene Families." Springer-Verlag, Berlin and New York.

Ohta, T. (1983). Theoretical study on the accumulation of selfish DNA. *Genet. Res.* **41,** 1–15.

Ohta, T. (1984). Some models of gene conversion for treating the evolution of multigene families. *Genetics* **106,** 577–582.

Olsen, G. J., and Sogin, M. L. (1982). Nucleotide sequence of *Dictyostelium discoideum* 5.8S ribosomal ribonucleic acid: Evolutionary and secondary structural implications. *Biochemistry* **21,** 2335–2343.

Orgel, L. E., and Crick, F. H. C. (1980). Selfish DNA: The ultimate parasite. *Nature (London)* **284,** 606–607.

Palmer, J. D., and Shields, C. R. (1984). Tripartite structure of the *Brassica campestris* mitochondrial genome. *Nature (London)* **307,** 437–440.

Panny, S. R., Scott, A. F., Smith, K. D., Phillips, J. A., Kazazian, H. H., Talbot, C. C., and Boehm, C. D. (1981). Population heterogeneity of the *Hpa*I restriction site associated with the β globin gene: Implications for prenatal diagnosis. *Am. J. Hum. Genet.* **33,** 25–35.

Rick, C. M., Fobes, J. F., and Tanksley, S. D. (1979). Evolution of mating system in *Lycopersicon hirsutum* as deduced from genetic variation in electrophoretic and morphological characters. *Plant Syst. Evol.* **132,** 279–298.

Ritland, K. (1984). The effective proportions of self-fertilization with consanguineous matings in inbred populations. *Genetics* **106,** 139–152.

Ritland, K. (1985). The genetic mating structure of subdivided populations. I. Open-mating model. *Theor. Pop. Biol.* (in press).

Ritland, K., and Jain, S. (1981). A model for the estimation of outcrossing rate and gene frequencies using *n* independent loci. *Heredity* **47,** 35–52.

Ritossa, F. (1976). The *bobbed* locus. *In* "The Genetics and Biology of Drosophila" (M. Ashburner and E. Novitski, eds.), Vol. 1B, pp. 801–846. Academic Press, London.

Rotwein, P., Chyn, R., Chirgwin, J., Cordell, B., Goodman, H. M., and Permutt, M. A. (1981). Polymorphism in the 5'-flanking region of the human insulin gene and its possible relation to type 2 diabetes. *Science* **213,** 1117–1120.

Rubin, G. M., Kidwell, M. G., and Bingham, P. M. (1982). The molecular basis of P-M hybrid dysgenesis: The nature of induced mutations. *Cell* **29,** 987–994.

Schoen, D. J. (1982). The breeding system of *Gilia achilleifolia:* Variation in floral characteristics and outcrossing rate. *Evolution* **36,** 352–360.

Schoen, D. J., and Clegg, M. T. (1984). Estimation of mating system parameters when outcrossing events are correlated. *Proc. Natl. Acad. Sci. U.S.A.* **81,** 5258–5262.

Sederhoff, R. R., Levings, C. S., Timothy, D. H., and Hu, W. W. L. (1981). Evolution of DNA sequence organization in mitochondrial genomes of *Zea. Proc. Natl. Acad. Sci. U.S.A.* **78,** 5953–5957.

Sederhoff, R. R. (1984). Structural variation in mitochondrial DNA. *Adv. Genetics* **22,** 1–108.

Shah, D. M., Hightower, R. C., and Meagher, R. B. (1983). Genes encoding actin in higher plants: Intron positions are highly conserved but the coding sequences are not. *J. Mol. Appl. Genet.* **2,** 111–126.

Shaw, D. V., and Allard, R. W. (1982). Estimation of outcrossing rates in Douglas-fir using isozyme markers. *Theor. Appl. Genet.* **62,** 113–120.

Shaw, D. V., Kahler, A. L., and Allard, R. W. (1981). A multilocus estimator of mating system parameters in plant populations. *Proc. Natl. Acad. Sci. U.S.A.* **78,** 1298–1302.

Slatkin, M. (1970). Selection and polygenic characters. *Proc. Natl. Acad. Sci. U.S.A.* **66,** 87–93.

Slatkin, M., and Lande, R. (1976). Niche width in a fluctuating environment—density independent models. *Am. Nat.* **110,** 31–55.

Smith, D. B., and Adams, W. T. (1983). Measuring pollen contamination in clonal seed orchards with the aid of genetic markers. *Proc. 17th South. Forest Tree Improve. Conf.,* Athens, Georgia, pp. 69–77.

Sokal, R. R. (1979). Ecological parameters inferred from spatial correlograms. *In* "Contemporary Quantitative Ecology and Related Ecometrics" (G. P. Patil and M. Rosenzweig, eds.), pp. 167–196. Int. Coop. Publ., Burtonsville, Maryland.

Sokal, R. R. (1983). Analyzing character variation in geographic space. *In* "Numerical Taxonomy" (J. Felsenstein, ed.), pp. 384–403. Springer-Verlag, Berlin and New York.

Sokal, R. R., and Menozzi, P. (1982). Spatial autocorrelation of HLA frequencies in Europe support demic diffusion of early farmers. *Am. Nat.* **119,** 1–17.

Sokal, R. R., and Oden, N. L. (1978a). Spatial autocorrelation in biology. 1. Methodology. *Biol. J. Linn. Soc.* **10,** 199–228.

Sokal, R. R., and Oden, N. L. (1978b). Spatial autocorrelation in biology. 2. Some biological implications and four applications of evolutionary and ecological interest. *Biol. J. Linn. Soc.* **10,** 229–249.

Spradling, A. C., and Rubin, G. M. (1981). *Drosophila* genome organization: Conserved and dynamic aspects. *Annu. Rev. Genet.* **15,** 219–264.

Stern, D. B., and Lonsdale, D. M. (1982). Mitochondrial and chloroplast genomes of maize have a 12-kilobase DNA sequence in common. *Nature (London)* **299,** 698–702.

Turelli, M. (1984). Heritable genetic variation via mutation-selection balance: Lerch's zeta meets the abdominal bristle. *Theor. Popul. Biol.* **25,** 138–193.

Wells, H. (1979). Self-fertilization: Advantageous or deleterious. *Evolution* **33,** 252–255.

Woo, S. L. C., Lidsky, A. S., Guttler, F., Chandra, T., and Robson, K. J. H. (1983). Cloned human phenylalanine hydroxylase gene allows prenatal diagnosis and carrier detection of classical phenylketonuria. *Nature (London)* **306,** 151–155.

Wright, S. (1921). Systems of mating. V. General considerations. *Genetics* **6,** 167–178.

Wright, S. (1977). "Evolution and the Genetics of Populations," Vol. 3. Univ. of Chicago Press, Chicago.

Wyman, A. R., and White, R. (1980). A highly polymorphic locus in human DNA. *Proc. Natl. Acad. Sci. U.S.A.* **77,** 6754–6758.

Zurawski, G., and Clegg, M. T. (1984). The barley chloroplast DNA *atpBE, trnM2,* and *trnV1* loci. *Nucleic Acids Res.* **12,** 2549–2559.

GENETICS, CYTOLOGY, AND EVOLUTION OF
Gossypium

J. E. Endrizzi,* E. L. Turcotte,† and R. J. Kohel‡

*Department of Plant Sciences, University of Arizona, Tucson, Arizona

†USDA, ARS, Cotton Research Center, Phoenix, Arizona

‡USDA, ARS, Cotton and Grain Crops Genetics Research
College Station, Texas

ADVANCES IN GENETICS, Vol. 23

1. Introduction

Cotton was one of the first crop plants to which the rediscovered Mendelian principles of genetics were applied. The results of research on the inheritance of lint color in *Gossypium barbadense* L. were reported by Balls in 1906. In *G. hirsutum* L. the inheritance of Okra leaf shape was published by Shoemaker in 1908, and the inheritance of red leaf color was reported by McLendon in 1912. Certainly, the widespread growing of cotton in both the Old and New World and its economic importance have created the interest, need, and financial resources for significant studies resulting in a continuous flow of publications from genetic research.

As research on cotton genetics advanced and genetic relations among species became better understood, researchers became aware that the cultivated Asiatic species, *G. arboreum* L. and *G. herbaceum* L., were diploids with 26 chromosomes and the cultivated New World species, *G. hirsutum* and *G. barbadense,* were allotetraploids with 52 chromosomes. These scientific revelations along with concurrent changes in cotton culture had an important impact on research. Cultivation of the diploid Asiatic species has declined over the years until now they are grown to only a limited extent in China, India, and Pakistan. On the other hand, cultivation of Upland cotton, *G. hirsutum,* increased and it has become, by far, the predominant type grown worldwide.

One effect of these cultural changes has been a near cessation of qualitative genetic research on the Asiatic diploid species. It is indeed unfortunate that, to a large extent, inheritance information and genetic traits of the Asiatic diploids have not been related effectively and transferred to the allotetraploid species because a significant amount of genetic information has been lost. Difficulties associated with interspecific hybridization have been one reason for this. Efforts to preserve Asiatic species germ plasm dwindled also as research on the genetics of the Asiatic species decreased. The published records are, by and large, all that remain of this material (Knight, 1954a; Sikka and Joshi, 1960).

The shift in cotton culture to varieties of *G. hirsutum* did not result in a large upsurge of genetic research in the allotetraploid species. The complexity of the genetics of the $2n = 52$ chromosome allotetraploids and the uniqueness of the allotetraploid genomic relationships shifted emphasis to studies of relations of genomes in the cultivated allotetraploids and the wild relatives of cotton. This emphasis created increased interest in cotton cytology and cytogenetics, centering primarily on studies of chromosome and genome manipulation, with the evolution of species taking prominence.

These cytogenetic investigations produced a wealth of knowledge about the cytological and genetical relationships of plant forms between and within species of *Gossypium*; however, they generated only limited knowledge of the genetic effects of individual chromosomes on growth and development of the cotton plant. The lack of major research efforts to detect these genetic effects was due perhaps to two factors: (1) cytogenetics and qualitative genetics, although generally recognized as important for genetic improvement, were presumed to not result in immediate or short-term economic returns; and (2) in the early days, the development of proper breeding, selection, and testing techniques could be relied upon to increase yields of cotton.

It has been just in recent times that research was directed more toward basic studies in genetics and cytogenetics at the individual chromosome level. The fact that germ plasm making up the bulk of United States Upland cultivars, and varieties grown in many other countries, are descendents of about a dozen introductions had an important influence on research emphasis (Richmond, 1951; Ramey, 1966). In view of the present status of varietal improvement, it is agreed generally that fundamental knowledge of the specific genes each chromosome carries, of the linkage relationships of these genes, and of the genetic interactions affecting agronomic traits is of paramount importance for further improvement of cotton cultivars.

An extended review of the cytology, genetics, and evolution of *Gossypium* has not been presented since the reviews of Hutchinson *et al.* (1947), Stephens (1947), Knight (1954a), and Phillips (1974), which are treatises that encompassed rather specific areas of investigations or were limited in their extent of coverage. The purpose of the present article is to present a comprehensive review of the published information on the cytology, genetics, and evolution of *Gossypium*. In addition, we present recent data and information on genome organization with which we form a hypothesis for the origin of the allotetraploid species that is different from that generally assumed.

II. Genetic Mutants, Gene Symbols, and Genetic Linkages

A. Cultivated Species

The genus *Gossypium* consists of 35 diploid species that are divided into seven genome groups and six allotetraploid species, each with the

same two subgenomes (see Table 5 in Section III). The genome relationships are discussed in Section III. Research in qualitative genetics
has been conducted primarily with the cultivated Asiatic diploid species with the A genome, *G. arboreum* and *G. herbaceum,* and with the
cultivated allotetraploids, *G. barbadense* and *G. hirsutum,* which have
the A and D subgenomes. The asiatic diploids and the allotetraploids
have the A genome in common, thus they are expected to have a
number of similar loci. In addition, because the allotetraploids have
both the A and D subgenomes, duplicate factor inheritance and the
existence of duplicate linkage groups occur in these cottons.

B. ASIATIC DIPLOID SPECIES

During the early worldwide expansion of commercial cotton production, the Asiatic cultivated diploids, *G. arboreum* and *G. herbaceum,*
were of major economic importance. However, the change in world
cotton production and research emphasis to primarily *G. hirsutum*
cultivars has resulted in a near cessation of reported qualitative genetic research on the Asiatic species. The information abstracted by
Knight (1954a) represents basically the current state of knowledge of
Asiatic diploid cotton genetics. Knight (1954a) prepared an abstract
bibliography that included Asiatic cotton genetics, and Sikka and
Joshi (1960) included Asiatic cottons in their review of cotton genetics.
Table 1 summarizes the gene list published by Knight, but gene sym-
bolization and genetic homologies are not current with the contemporary information on the cultivated New World cottons. Included in
Table 1 are the gene symbol, linkage groups where known, name,
origin, and reference of the mutant genes.

The last summary of linkage tests in Asiatic cottons also was made
by Knight (1954a). At that time seven linkage groups were identified
and several linkage or pleiotropic relations between mutant and morphological or agronomic traits were described (Table 2). Of these seven
linkage groups, two were identifiable with linkage groups I and IV of
the allotetraploid species. Other individual loci have suspected homoeology with the allotetraploids, but specific linkage relations with mutants are not known.

C. NEW WORLD ALLOTETRAPLOID SPECIES

The cultivated New World allotetraploid cotton species, *G. barbadense* and *G. hirsutum,* now dominate world cotton production. *G.*

barbadense is grown for extra long, strong, and fine fibers. Its cultivars include Sea Island, Egyptian, and Pima cottons as well as extra long staple cultivars grown in Russia, Sudan, Peru, India, and other countries. Cultivated *G. hirsutum,* Upland cotton, dominates world production and accounts for the worldwide spread of cotton cultivation.

The Technical Committee of United States Regional Project S-77, concerned with cotton genetics and cytogenetics research, has formed a subcommittee to establish rules for genetic nomenclature and preservation of cotton mutant and cytogenetic germ plasm. The Technical Committee adopted the subcommittee proposal based on the guidelines of the International Committee on Genetic Symbols and Nomenclature (Anonymous, 1957), and the lists of nomenclature rules and genetic mutants were published (Kohel, 1973b) to provide permanence and widespread distribution.

The mutants described included leaf shape and texture, plant hairs, plant stature and color, bacterial blight resistance, male and female sterility, flowering response, seed fibers, and various other traits. Most of the mutants were spontaneous in origin. A few mutants have been induced by mutagenic agents, and one new mutant in *G. barbadense* and one in *G. hirsutum* were identified following irradiation. Several mutants have been transferred to the allotetrapoloids from diploid species, and from one to another allotetraploid species. A comprehensive listing of allotetraploid mutants, gene symbols, linkage groups, origins, and references is presented in Table 3.

The linkage map for the cultivated allotetraploids is given in Table 4 It combines the linkage group designation, linkages among mutants, centromere location, and chromosome when known. All existing evidence indicates that this linkage map is valid for both *G. barbadense* and *G. hirsutum;* however, most of the linkage analyses and chromosome associations were determined in *G. hirsutum* and all of the mutants have not been transferred and tested in both genomes.

Research during the past 25 years has resulted in significant progress in the genetic identification of the allotetraploid chromosomes. Stephens (1955) identified four linkage groups with 12 mutant loci, but one of these loci, N_1, has been eliminated from its proposed linkage group (Endrizzi and Taylor, 1968).

There are currently 17 linkage groups identified with 63 mutant loci (Table 4). Thirteen of these linkage groups have been associated with specific chromosomes, two have been associated with specific subgenomes, and two remain unassociated. The active study of genetic mutant germ plasm with linkage analyses, monosome tests, and other

TABLE 1

List of Gene Symbols, Linkage Group, Name, Origin, and Reference of Asiatic Genes

Gene symbol[a]	Linkage group	Name	Origin (Gossypium)	Reference
B_4		Blackarm resistance	aboreum	Knight (1948a,b)
B_{6m}		Blackarm resistance intensifier	arboreum	Knight (1953b)
b_8	VII	blackarm resistance	arboreum	Knight (1954a)
b_9		blackarm resistance	herbaceum	Knight (1956)
chl_1		chlorophyll deficient	herbaceum	Balasubrahmanyan (1947); Ramanathan and Balasubrahmanyan (1938)
chl_2	VII	chlorophyll deficient	arboreum	Balasubrahmanyan (1947); Hutchinson and Nath (1938); Ramanathan and Balasubrahmanyan (1938); Yu (1939a)
cl	VII	short branch	herbaceum	Patel et al. (1947)
Cp_a, Cp_b		Crumpled (complementary)	arboreum, herbaceum	Hutchinson (1932a)
cr	I	crinkled	arboreum	Balasubrahmanyan and Santhanam (1951b)
cu		curly	arboreum, herbaceum	Yu (1939b)
d		female-sterile dwarf	arboreum	Khadilkar (1946)
d_1		Anakapalle dwarf	arboreum	Balasubrahmanyan and Santhanam (1950a)
d_2		Cocanada dwarf	arboreum	Balasubrahmanyan and Santhanam (1950a)
d_3		dwarf bushy	herbaceum	Venkoba Rao and Ramachandran (1943)
de_1, de_2		incomplete boll dehiscence	herbaceum, arboreum	Balasubrahmanyan et al. (1949)
de_3		closed boll	herbaceum	Kokuev (1935)
Fz, fz		Tufted seed, fuzzy	arboreum	Hutchinson (1935)
g		no ovary	arboreum, herbaceum	Balasubrahmanyan (1950)
H_1		Hairy leaves	herbaceum	Harland (1944); Knight (1952b, 1954b)
Hv^1		Super stellate hairs	herbaceum	Ramiah and Paranjpe (1944)

Symbol	Group	Description	Species	References
h_a	II	glabrous, lintless	*arboreum*	Afzel and Hutchinson (1933); Hutchinson and Gadkari (1937); Kottur (1927)
h_b		glabrous, lintless	*arboreum, herbaceum*	Afzal and Hutchinson (1933); Balasubrahmanyan (1951); Hutchinson and Gadkari (1937); Silow (1939)
L	I	Narrow leaves	*arboreum*	Balasubrahmanyan (1951); Hutchinson (1934); Silow (1939)
L^B		Mutant broad leaves	*arboreum*	Hutchinson (1934); Silow (1939)
L^J		Mutant intermediate leaves	*arboreum*	Hutchinson (1934); Silow (1939)
L^L		Laciniate leaves	*arboreum*	Hutchinson (1934); Silow (1939)
L^N		Narrow leaves	*arboreum*	Hutchinson (1934); Silow (1939)
$Ld_1{}^K$		Khaki lint	*arboreum, herbaceum*	Balasubrahmanyan et al. (1959); Hutchinson (1934); Ramanathan and Balasubrahmanyan (1933b)
Lc_1	I	Khaki lint	*arboreum, herbaceum*	Balasubrahmanyan et al. (1950)
lc_2	II	white lint	*arboreum, herbaceum*	Balasubrahmanyan et al. (1950); Ramanathan and Balasubrahmanyan (1933b); Silow (1944)
$Lc_2{}^B$		Light brown lint	*arboreum, herbaceum*	Balasubrahmanyan et al. (1950); Hutchinson (1935); Silow (1944)
$Lc_2{}^K$		Khaki lint	*arboreum, herbaceum*	Balasubrahmanyan et al. (1950); Ramanathan and Balasubrahmanyan (1933b); Silow (1944)
$Lc_2{}^M$		Medium brown lint	*arboreum, herbaceum*	Silow (1944)
$Lc_2{}^V$		Very light brown lint	*arboreum, herbaceum*	Balasubrahmanyan et al. (1950); Silow (1945)
$Lc_3{}^B$	III	Light brown lint	*arboreum, herbaceum*	Hutchinson (1935); Silow (1944)
$Lc_4{}^K$		Khaki lint	*arboreum*	Silow (1945)
li_a		hairy, lintless	*herbaceum*	Govande (1944a); Hutchinson and Gadkari (1937)

(*continued*)

TABLE 1 (*Continued*)

Gene symbol[a]	Linkage group	Name	Origin (*Gossypium*)	Reference
li_b	I	hairy, lintless	*herbaceum*	Hutchinson and Gadkari (1937)
li_c		hairy, lintless (lethal)	*arboreum*	Afzel and Hutchinson (1933); Hutchinson and Gadkari (1937)
li_d		hairy, lintless (lethal)	*herbaceum*	Govande (1944b)
li_e		hairy, lintless (lethal)	*arboreum*	Balasubrahmanyan and Santhanam (1952)
$li_{(sh)}$		short lint	*arboreum*	Balasubrahmanyan and Santhanam (1952)
li_{sp}		sparse lint	*arboreum*	Balasubrahmanyan and Santhanam (1950b)
lm		immature lint	*arboreum*	Balasubrahmanyan et al. (1949)
ls		single lobed leaf	*arboreum*	Ramiah and Nath (1943)
m_1		increased number of floral parts	*arboreum*	Balasubrahmanyan (1950); Ramanathan and Balasubrahmanyan (1938)
m_2		multibracteolate	*herbaceum*	Govande (1946)
ne	IV	nectaries absent	*arboreum, herbaceum*	Leake (1911)
P_a, P_b	V, IV	Pollen color: $P_a\ P_b$ yellow, $p_a\ P_b$ cream, $P_a\ p_b$ pale, $p_a\ p_b$ cream	*herbaceum*	Ramanathan and Balasubrahmanyan (1933a); Silow (1941)
Ppf		Petalody	*arboreum*	Afzel and Singh (1939); Balasubrahmanyan (1950); Hutchinson and Ghose (1937b); Ramanathan and Sankaran (1934)
pte		pistillate	*arboreum, herbaceum*	Balasubrahmanyan and Santhanam (1951a)
R_2^{ASA}	VII	Sun red spotted	*arboreum*	Balasubrahmanyan et al. (1949); Ramiah (1945)
R_2^{BO}	VII	Sun red spotless	*arboreum*	Balasubrahmanyan et al. (1949); Ramiah (1945)
R_2^{CS}	VII	Red calyx, spotted	*herbaceum*	Balasubrahmanyan et al. (1949); Ramiah (1945)
R_2^{DS}	VII	Thumbnail red, spotted	*arboreum*	Balasubrahmanyan et al. (1949); Ramiah (1945)
R_2^{DO}	VII	Thumbnail red, spotless	*herbaceum*	Balasubrahmanyan et al. (1949); Ramiah (1945)
R_2^{FO}	VII	Greenstem spotless	*arboreum*	Balasubrahmanyan et al. (1949); Ramiah (1945)

$R_2{}^{GS}$	VII	Tinged petal weak thumbnail red, spotted	*arboreum*	Ramiah (1945); Ramiah and Nath (1944)
$R_2{}^{HO}$	VII	Green stem spotless, untinged petal	*arboreum*	Balasubrahmanyan *et al.* (1949); Ramiah (1945); Ramiah and Nath (1944)
$R_2{}^{LO}$	VII	Red leaf, spotless	*arboreum, herbaceum*	Hutchinson and Ghose (1937a)
R^{LS}	VII	Red leaf, spotted	*herbaceum*	Ramiah (1945)
$R_2{}^{MS}$	VII	Red margin, spotted	*arboreum*	Ramiah (1945); Ramiah and Nath (1944)
$R_2{}^{MM}$	VII	Red margin, spotless	*arboreum*	Silow and Yu (1942)
$R_2{}^{OS}$	VII	Green stem, ghost spot	*arboreum, herbaceum*	Balasubrahmanyan *et al.* (1949); Ramiah (1945); Silow (1941); Silow and Yu (1942)
$R_2{}^{RS}$	VII	Full red, spotted	*arboreum*	Harland (1935a); Hutchinson (1932b); Knight (1945)
$R_2{}^{TS}$	VII	Green stem, tinged, ghost spot	*aboreum*	Ramiah (1945)
$R_2{}^{VS}$	VII	Red vein, spotted	*arboreum*	Ramiah (1945)
$R_2{}^{VO}$	VII	Red vein, spotless	*arboreum*	Balasubrahmanyan *et al.* (1949); Ramiah (1945)
$r_3{}^{oo}$	VII	green, spotless	*arboreum*	Silow (1941)
Rl_a	VI	Red lethal, complementary with factor from *G. hirsutum*	*arboreum*	Gerstel (1953a, 1954)
Sr		Spot reducer	*arboreum*	Hutchinson and Ghose (1937a)
stg		female sterile	*herbaceum*	Balasubrahmanyan (1950); Vijayaraghavan *et al.* (1936)
stp		male sterile	*arboreum, herbaceum*	Hutchinson and Gadkari (1935)
v_1		virescent yellow	*arboreum, herbaceum*	Yu (1939b, 1941)
v_2		virescent yellow	*arboreum, herbaceum*	Yu (1941)
v_3		virescent yellow	*arboreum, herbaceum*	Yu (1941)

(continued)

279

TABLE 1 (*Continued*)

Gene symbol[a]	Linkage group	Name	Origin (*Gossypium*)	Reference
v_4		virescent yellow	*arboreum, herbaceum*	Yu (1941)
vc		few loculed bolls	*arboreum*	Ramiah and Nath (1947)
W_1		*Fusarium* resistance	*arboreum, herbaceum*	Kelkar *et al.* (1947)
W_2		*Fusarium* resistance	*arboreum, herbaceum*	Kelkar *et al.* (1947)
W_3		*Fusarium* resistance	*arboreum, herbaceum*	Kelkar *et al.* (1947)
X		Lint color modifier	*arboreum, herbaceum*	Ramanathan and Balasubrahmanyan (1933a)
Y_a	III	Yellow petal	*arboreum, herbaceum*	Nath (1942); Hutchinson (1931); Silow (1941)
Y_a^P	III	Pale petal	*arboreum, herbaceum*	Hutchinson (1931); Leake (1911); Silow (1941)
y_a		white petal	*arboreum, herbaceum*	Hutchinson (1931)
Y_b	VI	Chinese yellow petal	*arboreum, herbaceum*	Nath (1942); Silow (1941)
Y_b^P	VI	Chinese pale petal	*arboreum, herbaceum*	Silow (1941)
Ydp	V	Yellow depressor	*anomilum*	Silow (1941)

[a] Identical subscripts indicate alleles; different subscripts indicate different genes.

TABLE 2
Asiatic Linkage Map

Linkage group	Gene order				
I	L	15	cu	19	Lc_1
	L	17.1	Li_d	20.5	Lc_1
II	Ha	7.1	Lc_2		
III	Y_a	24	Lc_3		
IV	P_b	16	Ne		
V	P_a	29.7	Ydp		
VI	R_3	1.2	Y_b		
VII	Chl_2	9	R_2		
	cl	30	R_2		
	b_8	1.4	R_2		

cytological methods for chromosome identification have combined to produce these results.

Of the 19 chromosomes that have chromosome or genome identity, five homoeologous pairs have been identified by mutant similarity and monosomic phenotypes. Silow (1946) identified the homoeologies between the loci for anthocyanin pigmentation (R_1 and R_2) and cluster fruiting habit (cl_1 and cl_2) of linkage groups I and III, and Rhyne (1958, 1960) extended homoeology to include yellow-green plant color (yg_1 and yg_2) and brown lint color (Lc_1 and Dw). The proposed homoeology of the leaf shape loci, L_1^L and L_2^O, was established with the chromosome association of lp_1 with L_1^L and lp_2 with L_2^O (Endrizzi and Stein, 1975; Endrizzi and Stith, 1970), and with the linkage of v_5 with L_1^L and v_6 with L_2^O (Kohel, 1973c). Homoeology of linkage groups V and XIII on chromosome 12 with linkage group IX on chromosome 26 is well established with the identification of subgenome location (Endrizzi and Ramsay, 1980) and the identification of six duplicate loci, gl_2gl_3, bw_1,bw_2, ne_1,ne_2, ms_8ms_9, N_1n_2, and Le_1Le_2 (Holder *et al.*, 1968; Rhyne, 1962a,b, 1965a; Rhyne and Rhyne, 1972; Endrizzi and Ramsay, 1980; Lee, 1965, 1982; Kohel, 1979; Turcotte and Feaster, 1979). The fourth homoeologous pair, chromosome 6 and 25, is identified by the similar phenotypes of the monosomes (Endrizzi and Ramsay, 1979). Chromosome 6 has linkage group IV, containing Lc_2 and H_2, and chromosome 25 has been associated with Sm_1^{sl} (J. E. Endrizzi, unpublished).

Of the three homoeologous chromosomes that are marked genetically with 14 homoeologous gene pairs, six pairs are expressed as

TABLE 3
List of Gene Symbols, Linkage Group, Name, Origin, and Reference of Allotetraploid Genes

Gene symbol[a]	Linkage group	Name	Origin	Reference
ac	III	curly cotton	*hirsutum*	Tiranti (1967)
as_1, as_2		asynapsis	*hirsutum*	Beasley and Brown (1942)
av_1		albivirescent	*barbadense*	Turcotte and Feaster (1978)
av_2		albivirescent	*barbadense*	Turcotte and Feaster (1978)
B_1		Blight resistance	*hirsutum*	Knight (1947); Knight and Clouston (1939)
B_2		Blight resistance	*hirsutum*	Knight and Clouston (1939)
B_3		Blight resistance	*hirsutum*	Knight (1944)
B_4	XI	Blight resistance	*arboreum*	Knight (1944, 1948b)
B_5		Blight resistance	*barbadense*	Knight (1950)
B_6		Blight resistance	*arboreum*	Knight (1953a)
B_7		Blight resistance	*hirsutum*	Knight (1953b)
B_8		Blight resistance	*anomalum*	Knight (1954c)
B_{9K}		Blight resistance	*herbaceum*	Knight (1963)
B_{9L}		Blight resistance		Lagiere (1959)
B_{10K}		Blight resistance		Knight (1957)
B_{10L}		Blight resistance		Lagiere (1959)
B_{11}		Blight resistance	*herbaceum*	Knight (1963)
B_{In}		Blight resistance	*hirsutum*	Green and Brinkerhoff (1956)
B_N		Blight resistance	*hirsutum*	Green and Brinkerhoff (1956)
B_S		Blight resistance	*hirsutum*	Green and Brinkerhoff (1956)
bp		smooth boll	*barbadense*	Knight (1954a); Smith (1942)
bw_1, bw_2	V, IX	withering bracts	*hirsutum*	Knight (1952a); Rhyne (1965a)
cg_1, cg_2		cleistogamy	*hirsutum*, *barbadense*	Hau *et al.* (1980a)
chl_1, chl_2		chlorophyll deficient	*barbadense*, *hirsutum*	Harland (1932, 1934); Hutchinson (1946b)

ck^x, ck^y, ck^o		corky	barbadense, hirsutum	Stephens (1950b)
cl_1	III	cluster	hirsutum	Hutchinson and Silow (1939)
cl_2	I	cluster	barbadense	Silow (1946)
cl_3	III	cluster	hirsutum	Hau et al. (1980b)
cr	II	crinkled leaf	barbadense	Harland (1917); Hutchinson (1946a)
Crp		Crumpled	hirsutum	Kohel (1973d)
cu		cup leaf	hirsutum	Lewis (1954)
cy		crinkle yellow	hirsutum	Kohel (1982)
de		depauperate	hirsutum	Kohel (1973e)
Dw	III	Dirty white lint	raimondii	Rhyne (1957)
E		Enhancer, fertility	barbadense	Sheetz and Weaver (1980)
F	VI	Flowering	barbadense	Lewis and Richmond (1960)
fg, Fg^h		frego bract	hirsutum	Green (1955); Kohel (1967)
$fng\text{-}a, fng\text{-}b$		fringeless nectaries	barbadense	Stephens (1974b)
fs_1		female sterile	hirsutum	Justus and Meyer (1963)
gl_1, gl_6, gl_1^y		glandless boll	hirsutum, barbadense	McMichael (1954, 1970); Murray (1965)
$gl_2, gl_3, Gl_2^s, GL_3^r, GL_3^{dav}$		glandless and glanded plant	hirsutum, barbadense	Barrow and Davis (1974); Lee (1962, 1978); McMichael (1960)
(gl_4, gl_5)		(modifiers)		
G^o, G^x, G^y		Tumorigenesis	hirsutum, gossypioides	Phillips (1976)
Gv		Golden veins	barbadense, hirsutum, tomentosum	Turcotte and Feaster (1983)
H_1, H_2, Sm_2^b	IV	Pubescent, Pilose, Smooth		Harland (1944); Saunders (1961b, 1963, 1965a,b); Simpson (1947); Lee (1968)
(H_3, H_4, H_5)		(modifiers)		
H_6		Pubescent	raimondii	Saunders (1965c)

(continued)

283

TABLE 3 (Continued)

Gene symbol[a]	Linkage group	Name	Origin	Reference
hb_1		hairy boll	*hirsutum, barbadense*	Hau *et al.* (1980a)
ia	VI	accessory involucre	*hirsutum*	Kohel (1965)
L_1^L	VII	Laciniate leaf	*arboreum*	Hutchinson (1934)
L_2^O, L_2^s, L_2^u	II	Okra leaf	*barbadense, hirsutum*	Shoemaker (1908); Stephens (1945)
Lc_1	I	Brown lint	*hirsutum*	Harland (1935b)
Lc_2	IV	Brown lint	*barbadense, darwinii, tomentosum*	Harland (1935b)
Le^{dav}, Le_1, Le_2	V, IX	Lethal	*barbadense, davidsonii*	Lee (1981a)
Lf	V	Leaf fleck	*hirsutum*	Kohel *et al.* (1977b)
Lf		Filiform	*hirsutum*	Hau *et al.* (1981)
Lg	II	Green lint	*hirsutum*	Harland (1929b)
Li	XV	Ligon lintless	*hirsutum*	Kohel (1972)
Lk		Leaky glands	*darwinii*	Stephens (1974a)
lp_1, lp_2	VII, II	abnormal palisades	*hirsutum*	Kohel (1964)
ltg		light green	*barbadense*	Turcotte and Feaster (1978)
ml	VIII	mosaic leaf	*hirsutum*	Lewis (1958)
ms_1		male sterile	*hirsutum*	Justus and Leinweber (1960)
ms_2		male sterile	*hirsutum*	Richmond and Kohel (1961)
ms_3	III	male sterile	*hirsutum*	Justus *et al.* (1963)
Ms_4		Male sterile	*hirsutum*	Allison and Fisher (1964)
ms_5, ms_6		male sterile	*hirsutum*	Weaver (1968)
Ms_7		Male sterile	*hirsutum*	Weaver and Ashley (1971)
ms_8, ms_9	V, IX	indehiscent anthers	*hirsutum*	Rhyne (1971)
Ms_{10}		Male sterile	*hirsutum*	Bowman and Weaver (1979)

Symbol	Linkage group	Character	Species	Reference
Ms_{11}	V	Male sterile	barbadense	Turcotte and Feaster (1979)
mt	V	mottled leaf	hirsutum	Lewis (1960)
N_1		Naked seed	barbadense, hirsutum	Carver (1929); Griffee and Ligon (1929); Harland (1929c); Kearney and Harrison (1927); Thadani (1923); Ware (1929)
n_2	IX	naked seed	barbadense, hirsutum	Harland (1929c); Ware et al. (1947)
ne_1,ne_2	V, IX	nectariless	tomentosum	Meyer and Meyer (1961)
ob	XVI	Open bud	hirsutum	Kohel (1973a)
P_1	XI	Pollen color	hirsutum	Harland (1929b); Stephens (1954a)
P_2		Pollen color	barbadense	Turcotte and Feaster (1966)
pg		pale green	hirsutum	Murray and Brinkerhoff (1966)
R_1, R_1^{dar}	III	Red plant	hirsutum	McLendon (1912); Harland (1935a); Stephens (1974a)
R_2, R_2^{v}	I	Petal spot	barbadense, darwinii, hirsutum	Harland (1929a); Stephens (1974a)
Rc	XIX	Cracked root	harknessii	Weaver and Weaver (1979)
Rd	XIV	Dwarf red	hirsutum	McMichael (1942); Johnson (1949)
Rf		fertility restorer	harknessii	Weaver and Weaver (1977); daSilva et al. (1981)
Rg	X	Ragged leaf	hirsutum	Kohel and Lewis (1962a)
rl_1	X	round leaf	hirsutum	Brown and Cotton (1937)
Rl_2	XIX	Round leaf	hirsutum	Percival et al. (1976)
rs	XVII	rudimentary stigma	barbadense	Turcotte and Feaster (1964)
Ru	X	Rugate	barbadense	Turcotte and Feaster (1965)
rx^c		unnamed	hirsutum	Kohel (1972)
s,s^r,s^l,S^h	II	strap leaf	hirsutum	Dilday and Waddle (1974); Dilday et al. (1975)
Se		Semigamy	barbadense	Turcotte and Feaster (1975b)
Sm_1,Sm_1^{sl}		Smooth stem, leaf	barbadense, amourianum	Lee (1966, 1968, 1976); Meyer (1957)

(continued)

285

TABLE 3 (Continued)

Gene symbol[a]	Linkage group	Name	Origin	Reference
Sm_3, sm_3		Smooth leaf	hirsutum	Lee (1968)
st_1	VIII	club stigma–style	hirsutum	McMichael (1965)
st_2	XIV	mini stigma–style	hirsutum	McMichael (1965)
st_3	XVII	invert stigma–style	hirsutum	McMichael (1965)
v_1		virescent	hirsutum	Killough and Horlacher (1933)
v_2		virescent (golden crown)	hirsutum	Duncan and Pate (1967)
v_3	XV	virescent	hirsutum	Percival and Kohel (1974)
v_4		virescent	hirsutum	Quisenberry and Kohel (1970)
v_5, v_6	VII, II	virescent	hirsutum (irradiation)	Kohel (1973c)
v_7		virescent	barbadense	Turcotte and Feaster (1973)
v_8	XIV	virescent	hirsutum	Kohel (1978b)
v_9		virescent	hirsutum	Percival and Kohel (1974)
v_{10}	XII	virescent	hirsutum	Percival and Kohel (1976)
v_{11}	XI	virescent	hirsutum	Percival and Kohel (1974)
v_{12}		virescent	hirsutum	Kohel (1983a)
v_{13}		virescent	hirsutum	Kohel (1983a)
v_{14}		virescent	hirsutum	Kohel (1983a)
v_{15}		virescent	hirsutum	Kohel (1983a)
v_{16}, v_{17}	VII, II	virescent	hirsutum	Kohel (1983a)
vf, Vf^h	II	veins-fused	hirsutum	Kohel and Lewis (1962b); Dilday et al. (1975)
wr		wrinkled leaf	barbadense	Turcotte and Feaster (1980)

Y_1	XII	Yellow petals	*barbadense, hirsutum*	Harland (1920, 1936a); Turcotte and Feaster (1963); Stephens (1954b)
Y_2	XVI	Yellow petals	*darwinii*	Harland (1936a); Hutchinson and Silow (1939); Stephens (1954b)
yg_1, yg_2	III, I	yellow-green	*barbadense, hirsutum*	Rhyne (1955)
yv	XVII	yellow veins	*hirsutum*	Kohel (1983b)

Working designation		
Lc_4	Light brown lint	
$mtBrown = Lc_y, Lc_z$	Brown lint (from *mt* stock, two factors)	
$TTLc = Lc_w$	Light brown (from translocation stock)	
$stpl$	stipple leaf	
RSL	Recessive seedling lethal	
im	immature fiber	
sd	spindly dwarf	
an	annual stem	
ax	(possibly as_2)	

[a] Identical subscripts indicate alleles; different subscripts indicate different genes.
[b] H_1, H_2, and Sm_2 reported to be alleles (Endrizzi and Ramsay, 1983a).
[c] Temporary gene symbol.

287

TABLE 4
Allotetraploid Linkage Map

Linkage group	Gene order	Chromosome	References
I	R_2 16 cl_2 4 yg_2 32 Lc_1	7	Endrizzi and Taylor (1968); Harland (1935a); Kammacher (1968); Poisson (1968); Rhyne (1957); Silow (1946); Stephens (1955)
II	v_6 0 L_2O3 sxl^a 44 Lg 5 vf 1 s .2 cr ? lp_2 ———51——— ———6———	15	Dilday et al. (1975); J. E. Endrizzi (unpublished); Endrizzi and Kohel (1966); Endrizzi and Stith (1970); Stephens (1955); Wilson and Kohel (1970); R. J. Kohel (unpublished)
III	v_{17} 28 L_2O sxl^a cl_3 24 cl_1 17 R_1 19 yg_1 5 ms_3 33 ac 17 Dw ———30——— ———26——— ———34———	16	Endrizzi and Kohel (1966); J. E. Endrizzi (unpublished); Harland (1935a); Hau et al. (1980b); Kammacher et al. (1967); R. J. Kohel (unpublished); Lefort (1970); Rhyne (1957)
IV	Lc_2 10 sxl^a 4 H_2 (Sm_2,H_1) ———22———	6	Endrizzi (1975); Endrizzi and Kohel (1966); Stephens (1955); Endrizzi and Ramsay (1983a)
V, XIII	gl_2 20 bw_1 39 ne_1 ? ms_8 ———40——— gl_2 8 Ms_{11} gl_2 27 Le_1 N_1 14 Ms_{11}	12	J. E. Endrizzi (unpublished); Endrizzi and Ramsay (1980); Holder et al. (1968); Kohel et al. (1977b); Lee (1965, 1982); Rhyne (1962a, 1965a,b); Rhyne and Rhyne (1972); Turcotte and Feaster (1979); R. J. Kohel and J. A. Lee (unpublished)

Group	No.	Map	References
		N_1 7 Lf	
		gl_2 32 N_1	
		sxl^a 11 N_1	
VI	3	fg 30 ia 30 sxl^a	Endrizzi and Ramsay (1980, 1983b); Kohel et al. (1965)
VII	1	v_5 0 L_1^L 38 sxl^a 44 lp_1 v_{16} 25 L_1^L	J. E. Endrizzi (unpublished); Endrizzi and Stein (1975); Endrizzi and Stith (1970); White and Endrizzi (1965); R. J. Kohel (unpublished)
VIII	4	st_1 32 sxl^a 23 ml	R. J. Endrizzi (unpublished); Endrizzi and Bray (1980); White and Endrizzi (1965)
IX	26	bw_2 5 gl_3 35 ne_2 16 ms_9 ? n_2 gl_3 26 Le_3 Gl_3^{dav} 26 Le_2^{dav}	Endrizzi and Ramsay (1980); Holder et al. (1968); Kohel (1979); Lee (1965, 1981b, 1982); Rhyne (1962a, 1965a,b); Rhyne and Rhyne (1972)
X		rl_1 20 Rg 15 rx ⟵ 31 ⟶	Kohel (1972); A. E. Percival and R. J. Kohel (unpublished)
XI	5	P_1 4 B_4 ? v_{11} ⟵ 36 ⟶	Endrizzi and Ramsay (1980); A. E. Percival and R. J. Kohel (unpublished); Tayel et al. (1973)
XII	A	v_{10} 4 Y_1	Percival and Kohel (1976); Stephens (1954b)
XIV	D	v_8 13 Rd 33 st_3 ⟵ 37 ⟶	Gerstel and Phillips (1958); R. J. Kohel (unpublished); Phillips (1962)
XV		v_3 12 Li	Kohel (1978b)
XVI	18	ob_1 sxl^a 18 Y_2	Endrizzi (1975, unpublished); Kohel (1983a); Weaver and Weaver (1979)
XVII		Ru 37 yv 30 v_1	
XVIII		Rc 14 Rf	

^{a} The symbols sxl are, respectively, short arm, centromere, long arm.

completely recessive duplicates (yg_1yg_2, bw_1bw_2, ms_8ms_9, lp_1lp_2, v_5v_6, and $v_{16}v_{17}$). Gene expression of known or suspected homoeologous loci in the allotetraploid species argues for an intermediate level of differentiation. Patterns of gene expression range from completely recessive duplicates to fully diploidized, e.g., R_1R_2, cl_1cl_2, Lc_1Dw.

Isogenic lines involving several allotetraploid mutant genes have been developed and evaluated (Kohel *et al.*, 1967; 1974; Kohel and Richmond, 1971; Turcotte and Feaster, 1975a). In addition, several mutant genes have potential economic value including *glandless* for improved food and feed, and *nectariless, plant hairs, frego bract,* and *okra leaf* for the control of insect damage and boll rot (Andries *et al.*, 1969, 1970; Jenkins *et al.*, 1966; Jones and Andries, 1969; Knight, 1952b; Knight and Sadd, 1953; Meredith *et al.*, 1973; Wannamaker, 1957; Wilson and Wilson, 1975).

In addition to mutations of nuclear genes affecting chloroplast development, there is a class of mutations that is confined to the chloroplast genome. These *plastome* mutants are maternally inherited and, like some of the *plastid* mutants determined by nuclear genes, produce a variegated phenotype. They occur routinely as spontaneous mutants in cotton plantings, appearing most commonly as sectorial chimeras in leaves or parts of plants (Endrizzi *et al.*, 1974; Katterman and Endrizzi, 1973; Kestler *et al.*, 1977; Kohel, 1967; Kohel and Benedict, 1971).

The cytoplasm of several diploid species, when transferred to the allotetraploids, interacts with the nuclear genes of the latter to produce male sterility (Meyer, 1973, 1975).

III. Genomes of Gossypium—Their Differentiation and Identification

The first accurate chromosome counts that demonstrated there were two chromosome numbers in the genus *Gossypium* were made by Nikolajeva (1923) and Denham (1924a,b), who observed $2n = 26$ for the Old World cottons *G. arboreum* and *G. herbaceum* and $2n = 52$ for the New World cottons *G. barbadense* and *G. hirsutum*.

In the late 1920s and early 1930s, the collection and cytological analysis of species and types of *Gossypium* were being actively pursued by a number of investigators. During this period, chromosome numbers of $2n = 26$ and 52 were reported for many species including wild and cultivated forms occurring in America, Asia, Africa, and Australia (Youngman and Pande, 1927; Harland, 1928; Banerji, 1929; Baranov,

1930; Kearney, 1930; Longley, 1933; Skovsted, 1933, 1934a,b, 1935b,c; Webber, 1934a,b, 1935, 1939). There are now 35 diploid and six allotetraploid species in the genus (Fryxell, 1979; Valiček, 1981). All are wild species except for two diploids and two allotetraploids, which are commercially grown for their seed and fiber. Most of the diploid species have been placed into one of seven genome groups (A–G) based on cytological and taxonomic affinities, and the allotetraploids have the A and D genomes (Table 5). It has been suggested that the diploid number of 26 may be a secondary polyploid in origin (Davie, 1933; Skovsted, 1933; Abraham, 1940a).

Beal (1928) noted that the chromosomes of the allotetraploid species varied in size at metaphase I, and Davie (1933) recorded that the somatic chromosome lengths of *G. hirsutum* ranged between 0.8 and 1.8 μm and those of *G. herbaceum* were 1.5–2.5 μm. Genome size differences of additional species were reported by Skovsted (1934a,b, 1935a). He noted that the Old World and Australian diploids, *G. arboreum, G. herbaceum, G. stocksii,* and *G. sturtianum,* had 13 large chromosomes whereas the New World diploids, *G. thurberi, G. davidsonii, G. aridium,* and *G. harknessii (G. armourianum,* Skovsted, 1937), had 13 small chromosomes. He observed that the chromosomes of *G. arboreum* and *G. herbaceum* were homologous and equal in size, and that the mean length of the somatic chromosomes of *G. arboreum* was about 2.5 μm. Skovsted made the important observation that the New World tetraploid cottons had 13 large chromosomes with a mean length of 2.25 to 2.36 μm and 13 small chromosomes with a mean length of 1.25–1.45 μm.

In meiotic studies of species, their hybrids, and synthesized hexaploids, Beasley (1940b, 1942) emphasized the comparative genome size differences of the chromosomes. In these reports he defines a cytological classification of genomes that is closely related to taxonomic affinities. Species with similar genomes exhibit highly regular pairing of the chromosomes in their hybrids, whereas species with dissimilar genomes show highly irregular chromosome pairing in their hybrids, indicating major chromosome differentiation.

Symbols were devised by Beasley (1940b) to designate the genomes of species: Similar genomes are designated by the same capital letter and closely related genomes are distinguished by a numerical subscript after each letter of that class. Seven diploid and one tetraploid genome groups are now recognized in *Gossypium.* The A to E genomes were assigned by Beasley (1940b) and F and G were assigned respec-

TABLE 5

The Species of *Gossypium*

Species	Genomic group[a]	Distribution
Diploids ($2n = 26$)		
G. herbaceum L.	A_1	Old World cultigen
G. h. var. *africanum* (Watt) Mauer	A_1	Africa
G. arboreum L.	A_2	Old World cultigen
G. anomalum Wawr. and Peyr.	B_1	Africa
G. triphyllum (Harv. and Sand.) Hochr.	B_2	Africa
G. capitis-viridis Mauer	B_3	Cape Verde Islands
G. sturtianum J. H. Willis	C_1	Australia
G. sturtianum var. *nandewarense* (Derera) Fryx.	C_{1-n}	Australia
G. robinsonii F. Muell.	C_2	Australia
G. australe F. Muell.	—	Australia
G. costulatum Tod.	—	Australia
G. cunninghamii Tod.	—	Australia
G. nelsonii Fryx.	—	Australia
G. pilosum Fryx.	—	Australia
G. populifolium (Benth.) Tod.	—	Australia
G. pulchellum (C.A. Gardn.) Fryx.	—	Australia
G. thurberi Tod.	D_1	Mexico, Arizona
G. armourianum Kearn.	D_{2-1}	Mexico
G. harknessii Brandg.	D_{2-2}	Mexico
G. klotzschianum Anderss.	D_{3-k}	Galapagos Islands
G. davidsonii Kell.	D_{3-d}	Mexico
G. aridum (Rose & Standl.) Skov.	D_4	Mexico
G. raimondii Ulbr.	D_5	Peru
G. gossypioides (Ulbr.) Standl.	D_6	Mexico
G. lobatum Gentry	D_7	Mexico
G. laxum Phillips	D_8	Mexico
G. trilobum (DC.) Skov.	D_9	Mexico
G. turneri Fryx.	—	Mexico
G. stocksii Mast.ex. Hook.	E_1	Arabia
G. somalense (Gürke) Hutch.	E_2	Africa
G. areysianum (Defl.) Hutch.	E_3	Arabia
G. incanum (Schwartz) Hillc.	E_4	Arabia
G. ellenbeckii (Gürke) Mauer	—	Africa
G. longicalyx Hutch. and Lee	F_1	Africa
G. bickii Prokh.	G_1	Australia
Allotetraploids ($2n = 4x = 52$)		
G. hirsutum L.	$(AD)_1$	Central America
G. barbadense L.	$(AD)_2$	South America
G. tomentosum Nutt. ex Seem.	$(AD)_3$	Hawaii
G. mustelinum Miers ex Watt	$(AD)_4$	Brazil
G. darwinii Watt	$(AD)_5$	Galapagos Islands
G. lanceolatum Tod.[b]	(AD)	Mexico

[a] A dash (—) indicates that the genome designation has not been determined.

[b] Species status of *G. lanceolatum* needs experimental verification.

tively by Phillips and Strickland (1966) and Edwards and Mirza (1979). The species in the genus, genome designation, and natural geographical distribution are listed in Table 5.

The relative chromosome size for the genomic groups has been classified by Stephens (1947) and Katterman and Ergle (1970), and their interpretation along with additional information is as follows: (1) The C genomes have very large chromosomes. (2) The E and F genomes have large chromosomes which are slightly larger than those of the A and B genomes. (3) The B genomes have large chromosomes, some of which are slightly larger than those of the A genomes (in Stephens' classification, the B and A genome groups are switched in relation to one another). (4) The A genomes have moderately large chromosomes. (5) The G genomes have chromosomes which are moderately large but smaller than those of A genomes (Wilson and Fryxell, 1970; Kadir, 1976; Edwards and Mirza, 1979. (6) The D genomes have small chromosomes.

Saunders (1961a) speculates that the center of origin of the genus *Gossypium* is in Central Africa because four of the seven diploid genome groups occur on the African Continent. It is generally assumed that divergence of the genus into the different genome groups occurred before or during the separation of the continents in the Cretaceous period (Skovsted, 1934a; Hutchinson *et al.*, 1947; Prokhanov, 1947; Mauer, 1954; Saunders, 1961a; Phillips, 1963; Hawkes and Smith, 1965; Fryxell, 1965a, 1979). A tertiary differentiation of genomes within each of the genome groups was advanced by Fryxell (1965a) and Phillips (1966).

In the analysis of species relationships to determine which genome is most likely to be the progenitor genome of the genus, workers have stressed the signal importance of chromosome homologies (Phillips, 1966), plant morphological characters (Fryxell, 1971; Valiček, 1978), and seed protein patterns (Johnson and Thein, 1970).

Phillips (1966, 1974) summarized the data on chromosome pairing in intra- and intergenomic hybrids of the diploid species known at that time. In determining genome relationships and level of divergence, the frequency of unpaired or univalent chromosomes was emphasized by Phillips since it was assumed to be a better indicator of differentiation than the frequency of paired associations. The average univalent frequencies for the intergenomic combinations are given in Table 6. The reduction of chromosome pairing and chiasma frequency between two genomes was attributed to structural changes, the degree of which was assumed to be directly related to the time elapsed since their diver-

TABLE 6

Summary of Average Univalent Frequency in Intergenomic Hybrids
of Diploid *Gossypium* Species[a]

Intergenomic hybrid	Univalents per cell	Chiasmata per bivalent
A × B	2.82	1.44
A × C	8.50	1.19
A × D	13.98	1.10
A × E	17.13	1.08
B × C	11.17	1.11
B × D	18.19	1.07
B × E	22.35	—
C × D	13.10	1.10
C × E	24.68	1.01
D × E	25.15	1.00
D × F	21.60	1.02
G × C[b]	3.84	1.48

[a] All averages except the A × D and C × D intergenomic combination are from
Phillips (1974); the A × D average includes data from Bahavandoss and Veluswamy
(1969); Bahavandoss *et al.* (1973); and Jayaraman *et al.* (1973). C × D includes mean
frequencies from Tiranti (1966).

[b] Involvement of the G genome is indirect; see text.

gence from a common ancestor. Thus, based on chromosomal affinities,
the general pattern of intergenomic relationships indicated that the
A–E genomes diverged sequentially from a basic generic lineage.

The intergenomic hybrids involving the E genome species had the
highest frequency of univalents, indicating that the E genome is ancient in origin and therefore closest to the ancestral genome of the
genus. Genomic hybrids of A and B species exhibited a very low univalent frequency, suggesting that these two genomes represent a
rather recent divergence in which A evolved from the B genome. In the
hybrids involving C or D species crossed with the A or B species, the
univalent frequency indicated that the C and D genomes represent
evolutionary lines of intermediate age. Divergence of the D genome
was assumed to have occurred somewhat prior to the divergence of the
C genome since the univalent frequencies in A or B × C hybrids are
lower than those of A or B × D (Phillips, 1963, 1966).

Univalent frequency of two additional intergenomic combinations, D
× F and G(?) × C, has been reported more recently (Table 6; Phillips,
1974). Phillips and Strickland (1966) have shown that the F genome is
closely related to the A genome and therefore to the B genome. The

frequency of univalents in the D × F combination is very high and similar to that in B × D. These analyses indicate that the F genome is of more recent origin similar to that of the A and B genomes. Schwendiman *et al.* (1980) surmised that the F genome is a relic of a common ancestor of the Old World genomes, a conclusion based primarily on chromosome pairing in the F_1 hybrids between hexaploids of *G. hirsutum* × *G. longicalyx* (F genome) and *G. hirsutum* × *G. stocksii*.

Morphologically and cytologically, *G. australe* is closely related to the G genome species *G. bickii* (Fryxell, 1965b; Phillips, 1974; Valiček, 1978), which implies that it likewise has a G genome. Chromosome pairing data between *G. australe* and *G. robinsonii* (C genome) show that the C and G genomes are rather closely related (Phillips, 1974) and that the two diverged at a later date, most likely following the isolation of the prototype genome with the Australian land mass.

Based on intergenomic chromosome affinities, as interpreted by Phillips and discussed above, the data suggest that the original ancestral generic genome was composed of large chromosomes like those of the E genome and the later divergence of the A, B, D, F, and G genomes with smaller chromosomes involved a downward shift in nuclear content. An increase in nuclear DNA would have had to occur for the origin of the C genome since its chromosomes are the largest in the genus.

Fryxell (1971) reported a comparative study of plant morphological characters among the diploid species by the Wagner Groundplan/Divergence Method. With some significant exceptions, the phylogenetic branching sequence confirms in broad outline Phillips' view derived from chromosome affinity. However, species with the D genome were found to have the least amount of evolutionary advancement, thus suggesting that the small D genomes are closest to the ancestral or prototype genome. If the D genome is most similar to the progenitor genome of the genus, then the plant morphological data indicate that the other six genomes must have risen by an upward shift in nuclear DNA content.

Valiček (1978) combined his phylogenetic observations with part of those of Fryxell (1971) in evaluating by numerical taxonomy the phenetic relationships of the species of the A–F genomes. He arrived at a genome phylogeny similar to that of Johnson and Thein (1970) discussed below. The B genomic group of species was found to be the most homogeneous of all groups, and its phylogenetic characteristics were noted to be significantly correlated with those of the A, E, and D genomic groups. Valiček interprets these relationships as indicating that the B genome is the prototype of the diploid species of the genus. The characteristics of the C and F genomes showed a low correlation with those of other genomic groups, indicating that they may be of more recent origin.

Seed protein banding patterns were utilized by Johnson and Thein (1970) to ascertain the evolutionary affinities of the genomes. Based on the estimated correlations of the similarities of the banding patterns, they concluded that the B genome was the closest survivor of the prototype genome, which gave rise to the A genome, a group of D species (D $_\xi$), and the C and E genomes. At a much later date, the remaining D species group (D$_\beta$) evolved from a B-like genome.

The interpretations of Johnson and Thein and Valiček of the phylogeny of the genomes are for the most part opposite to that of Phillips and Fryxell, but like Phillips', necessitates a two-way shift in nuclear DNA content in their evolution.

Qualitative and/or quantitative differences in amino acids, fatty acids, and A + T/C − 5MC + G nucleotide ratios of both DNA and RNA did not prove for the most part to be adequate as a means for estimating evolutionary affinities of the species (Ergle *et al.*, 1964; Sarvella and Stojanovic, 1968; El-Nockrashy *et al.*, 1969).

IV. Translocations

A. Origin and Identification

Many of the earlier cytologists noticed the presence of multivalent chromosome configurations at metaphase I in species hybrids, primarily between the Asiatic species and between the Asiatics and the New World allotetraploid species. Beasley (1942) referred to the multivalents as translocations without determining the number involved.

Gerstel (1953a) reexamined chromosome pairing in hybrids of *G. herbaceum, G. arboreum,, G. anomalum* and the allotetraploids and found that the genome of G. arboreum and the A subgenome of the allotetraploid differ from each other and the genomes of *G. herbaceum* and *G. anomalum* by chromosomal interchanges.

Prior to Gerstel's publication, cotton cytologists at the Beasley Laboratory in Texas noticed a high frequency of multivalents at metaphase I in trispecies hybrids of *G. hirsutum* × Asiatic–New World diploid F_1 and decided to reexamine the available information and to reanalyze cytologically hybrids between *G. hirsutum* and the Asiatic and New

World diploid species (Brown and Menzel, 1950; Menzel and Brown, 1954a). They found that all the viable hybrids of *G. hirsutum*–New World diploids almost regularly formed 13 small bivalents and 13 large univalents. The *G. hirsutum*–Asiatic hybrids, however, gave entirely different modal pairing behavior. Like Gerstel (1953a), they observed chromosome structural changes that differentiate the A genomes of *G. arboreum* and *G. hirsutum* from *G. herbaceum* and they diagrammed chromosome end arrangements of the first five chromosomes of the three A genomes that would account for the multivalent formation in each hybrid combination (Fig. 1). *Gossypium herbaceum* has the standard or primitive chromosome end arrangement, and *G. arboreum* differs from it by a single interchange between chromosomes 1 and 2. The A subgenome of the allotetraploids differ from the genome of *G. herbaceum* by two independent reciprocal interchanges involving, respectively, chromosomes 2 and 3 (IV_1) and 4 and 5 (IV_2). The chromosomes of *G. arboreum* and the allotetraploids differ by three interchanges, which are apparent in their hybrids by the occurrence of a ring-of-

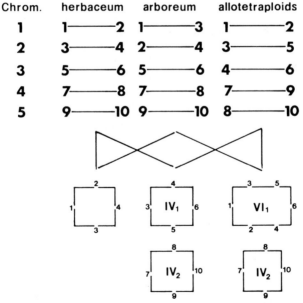

FIG. 1. Chromosome end arrangement of the first five chromosomes of the A genomes of *G. herbaceum, G. arboreum,* and the allotetraploids, and the multivalent pairing configuration(s) observed in their F_1 hybrids. (Modification of Menzel and Brown, 1954a.)

six (VI_1) and a ring-of-four (IV_2). The ring-of-six forms from combining the interchange between chromosomes 1 and 2 of *G. arboreum* with the interchange between chromosomes 2 and 3 of the allotetraploids. The ring-of-six contains a differential segment of sufficient length for an occasional formation of a chiasma. From exchange in this segment and self-fertilization, it is theoretically possible to recover four types of rings-of-four (Endrizzi and Brown, 1962). In outcrossing to the allotetraploid standard, it is possible to recover two of the four types of rings-of-four, one of which is a new interchange designated IV_a, as demonstrated by Tiranti and Endrizzi (1963).

Through backcrossing, the IV_1, IV_2, and VI_1 chromosome translocations were transferred to *G. hirsutum*, providing the first set of chromosomally identified translocations in *Gossypium* (Brown, 1980). These three sets of translocations were used as the starting point for numbering the remaining eight A chromosomes in translocations that were later isolated in *G. hirsutum*.

Until the 13 homoeologous chromosome pairs of the allotetraploids can be identified, cotton workers have agreed to provisionally number the A subgenome chromosomes A1–A13 (H1–H13) and the D subgenome chromosomes D1–D13 (H14–H26) (Kohel, 1973b).

The first induced translocations were from seed exposed in the atom bomb test at Bikini (Brown, 1950). Following this, additional seed were exposed to radiation (X, γ, and fast neutron rays) to induce translocations for isolation and identification in *G. hirsutum* (Menzel and Brown, 1954b; Kammacher, 1958; Brown, 1980). Sixty-two translocations have been isolated, made homozygous, and identified as to the chromosomes involved (Brown, 1980) (Table 7), of which 53 were radiation induced. The remaining nine consist of the four that came from the *G. hirsutum*–Asiatic hybrids and five that were spontaneous in origin. The 62 translocations identify chromosomes 1 through 25 of the 26 chromosomes in the allotetraploid. Chromosome 26 has not been isolated in a translocation, but it is identified as a telocentric chromosome (see Table 8, Section V,A). Fifty-eight of the translocations are simple reciprocal interchanges, i.e., involving only two nonhomologous chromosomes, whereas the remaining four are multiple associations involving interchanges among three or four nonhomologous chromosomes. The 58 reciprocal translocations comprise the most complete set of cytological markers for the A and D genomes of cotton. A tester set selected from this group can be used to identify any one of the 26 chromosomes occurring in a structural or numerical chromosome change (Ray, 1982).

Of the 62 homozygous translocations, 26 are between Ah chromosomes, 26 are exchanges between Ah and Dh chromosomes, and 10 are exchanges between two Dh chromosomes (Table 7). The cytological procedure for assigning chromosomes in translocations to their respective genomes was described by Menzel (1955). This can be done by crossing the homozygous translocations to diploid species with either the A genome or the D genome. In practice, however, hybridization has usually been limited to the (AD) × D combination since it was much easier to accomplish prior to recently developed ovule culture methods (Stewart and Hsu, 1978; Stewart, 1979). Pairing of the standard chromosomes in the F_1 of an (AD) × D hybrid is primarily 13 small bivalents plus 13 large univalents, with a very low frequency of trivalents. Therefore, when a homozygous simple reciprocal translocation is crossed to the D species, there are three modal pairing possibilities: (1) an AA translocation will have 13 bivalents and 13 univalents; (2) an AD translocation will have 12 bivalents, one trivalent, and 12 univalents; and (3) a DD translocation will have 11 bivalents, one quadrivalent, and 13 univalents (Brown, 1967; Menzel *et al.*, 1978). Interchanges involving more than two nonhomologous chromosomes will give a more complex modal pairing combination. Since most of the A and D chromosomes differ appreciably in size, the size of chromosomes in the pairing configuration of the heterozygote in *G. hirsutum* provides an additional criterion for determining genome identity.

The identity of the chromosomes in each of the 62 translocations was determined in most cases by intercrossing the homozygous translocations and determining whether the same or different chromosomes were involved (Brown, 1980). The results are given in Table IV of Brown's publication and the method is illustrated with three reciprocal interchanges in Fig. 2. When two translocations (A and C in Fig. 2) involve different chromosomes, their F_1 hybrid will show 22 bivalents and 2 quadrivalent configurations at metaphase I; however, when two translocations (B and C) have one chromosome in common, 23 bivalents and a hexavalent configuration will be observed.

In some cases, stocks that were monosomic for known chromosomes were used to identify chromosomes in the translocations and vice versa (Fig. 2). If the monosome is not one of the chromosomes involved in the translocation, the monosomic F_1 from crossing the two cytological types will show 23 bivalents, one quadrivalent, and a univalent chromosome at metaphase I. Whenever the monosome is one of the same chromosomes as in the translocation, pairing in the monosomic F_1 will consist of 24 bivalents and a trivalent configuration.

Not illustrated is a case in which two different translocations are interchanges between the same two chromosomes. When this occurs,

TABLE 7
Homozygous Translocation Lines in *G. hirsutum*[a]

Translocation line[b]		Genetic markers	Rings and/or chains	One or two interstitial chiasma	Frequency dp–df[c]
AA Translocations					
IVa	T1L–2Ra[d]	R_1 or r_1 Y_1	+		0.36
DP30	T1L–2Lb	cr^D	+		0.37
2935	T1L–3L		+		0.41
5-4c	T1L–7L		+		0.57
2775b	T1L–8L		+		0.30
IV$_1$	T2–3a	R_1 N_1	+		0.39
1059	T2–3b	L_2^O N_1	+		0.37
7–2b	T2–6		+		0.33
1039	T2–8Ra	R_1	+		0.35
1058b	T2–8Rb		+		0.30
8B-3	T2L–9R		+		0.46
8-5Gb	T3R–5			+	0.51
4010	T3L–6L		+		0.43
8-30-5	T3R–9R			+	0.47
IV$_2$	T4L–5	H_1 or h_1	+		0.34
C14-3	T5L–9L	R_1 P_1	+		0.42
SL18	T5–12R	Gl_2 or gl_2	+		0.39
1048	T6–7L	R_1 cu fg N_1 Lc_1 or lc_1 ne_1 Ne_2	+		
Z9-9	T6L–10R	R_1 R_2 Lc_1	+		0.51
1052	T7R–11R			+	0.42
1043	T7L–12R	L_2^O N_1	+		0.55
2778	T8–12L	*crumpled-mottled*		+	0.43
2785	T10R–11R		+		0.52

Line	Translocation	Genes			Value
6-5M	T11R–12L		+	+	0.45
10-5Kb	T11R–13L	$R_1\ Rd$	+	+	0.50
AD Translocations					
2780	T1L–14L			+	0.40
2B-1	T2R–14R			+	0.40
AZ-7	T6L–14L		+		0.51
1040	T4R–15L	$L_2^O\ N_1$	+	+	0.48
1058a	T11L–15L	$L_2^O\ Lc_1$ or lc_1	+	+	0.43
2770	T1R–16Ra				0.49
4672	T1R–16Rb			+	0.43
1036	T9R–17Ra		+	+	0.45
6340	T9R–17Lb		+	+	0.54
1316	T11R–17R	$R_1\ L_2^O\ N_1\ Lc_1$ or lc_1	+	+	0.58
4659	T7–18R	N_1	+	+	0.53
E20-7	T3L–19L	$R_1\ cu\ v_1$		+	0.43
10-5ka	T4L–19R	$R_1\ N_1\ Lc_1$ or lc_1	+	+	0.46
5-5B	T8R–19R		+	+	0.59
1626	T10R–19R	$R_1\ R_2\ v_1\ N_1$ or n_1	+	+	0.55
9-5H	T12R–19R				0.55
2925	T13R–19R		+		0.49
4669	T1L–20R	P_1		+	0.59
2772	T9R–20L	V_1 or $v_1\ cu\ gl_1\ cl_1$	+		0.47
2790	T7L–21R		+		0.50
4675	T10L–21L		+		0.39
2775a	T5–23			+	0.42
2870	T9R–25			+	0.43
DD Translocations					
2777	T14L–23		+	+	0.48
2781	T14R–24R				0.53

(continued)

TABLE 7 (Continued)

Translocation line[b]		Genetic markers	Most common MI configuration		Frequency dp–df[c]
			Rings and/or chains	One or two interstitial chiasma	
8-5Ga	T15R–16La	cu	+	+	0.50
2767	T15L–16Lb		+		0.55
SL15	T15R–20R	$L_2^O R_2 cl_1 Lc_1$	+		0.54
E22-13	T19–21R		+	+	0.52
2786	T19R–24R		+		0.57
7-3F	T20R–21L	R_1 or $r_1 N_1$ or n_1	+	+	0.42
DP4	T20L–22R	L_2^O or l_2^O	+		9.54
2791	T20R–25		+		0.68
Complex translocations					
VI$_1$	T1–2–3		+		
DP6	T2–19–23	$R_1 rl_1 Lg$	+		
2779	T5–11–21				+
4655	T2–11–13–24				

[a] Data from Seyam and Brown (1973); Menzel and Brown (1978a,b); Brown (1980); Ray (1982); Brown et al. (1981); M. Y. Menzel (unpublished).

[b] R and L, Right and left arms. R arbitrarily assigned to arm with largest number of breaks, not necessarily the longer arm cytologically.

[c] dp–df, Duplication–deficiency.

[d] Chromosome designation 1–13, A subgenome; 14–26, D subgenome.

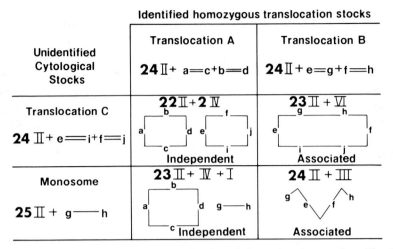

FIG. 2. Cytological procedure for using the identified translocations stocks to identify the chromosomes occurring in translocations or as aneuploids. In the same kind of crossing scheme, an identified monosome can be used to identify chromosomes in translocations.

the expected chromosome association in their F_1 hybrid will consist of either 26 bivalents and/or 24 bivalents plus one quadrivalent, depending on whether the exchange segments in the two translocations involve the same or different arms of the two chromosomes.

Menzel and Brown (1978a,b) and Brown *et al.* (1981) have determined the arm location of breakpoints in the translocated chromosomes of many translocations. For those translocations in which it has not been determined whether the breaks are in the long or short arms, the break positions are given provisionally in Table 7 as occurring in the "right" or "left" arms, in which "right" is arbitrarily assigned to the arm involved in the larger number of breaks.

Given in Table 7 for each of the translocation lines are the arm location of breakpoints, most common type(s) of M1 configuration of the heterozygote, frequency of recovered duplication–deficiencies, and, if present, marker genes occurring in each line.

B. Uses of Translocations

1. Genetic Analysis with Duplications and Deficiencies

Since *G. hirsutum* is an allotetraploid, it is not surprising that viable duplications and deficiencies for segments of chromosomes can be recovered as a result of adjacent disjunction in a heterozygous interchange (Table 7) (Brown, 1950; Brown *et al.,* 1981; Menzel and Brown, 1952, 1954b, 1978b). From data in Table 7, it can be assumed that duplication–deficiency types can be recovered for any interchange in cotton. Therefore, as demonstrated by Menzel and Brown (1954b), it is very likely that the translocations can be used to assign genes to chromosomes as described in the following. For locating a recessive gene, a translocation heterozygote carrying the normal alleles would be crossed as female to a recessive mutant line and the F_1 generation scored for the presence of the recessive (hemizygous $-/a$) phenotype. Cytological analysis of these plants should reveal the presence of a duplication–deficiency type of chromosome configuration.

For dominant mutant genes in which the allelic heterozygotes have an intermediate phenotype, it would be desirable for the translocation heterozygote to be homozygous for the dominant alleles. The translocation heterozygote would then be crossed to the normal or standard line and the F_1 hybrids classified for the presence of two genotypic and phenotypic classes, the deficient or hemizygous normal phenotype ($-/+$) and the duplication ($DD+$) appearing as a DD homozygote. An F_1 with the normal phenotype indicates that the chromosome segment carrying the $+$ allele is hemizygous deficient, whereas an F_1 with the homozygous dominant phenotype indicates that the chromosome segment carrying the D allele is duplicated. Cytological analysis would confirm that the two phenotypes are associated with duplication–deficiency types of interchanges (Menzel and Brown, 1954b). Normally, a translocation line will not carry a particular dominant mutant gene. Nevertheless, the duplication–deficiency interchanges can be used for assigning these loci to chromosomes by first crossing that homozygous translocation to a normal standard line. The heterozygous translocation is then crossed as female to the mutant dominant line and a number of the F_1 hybrids are self-pollinated. The F_2 populations are then scored for those that do not segregate for a dominant mutant allele. Nonsegregation in the F_2 indicates that the parental F_1 was a duplication–deficiency interchange and that the gene is located on one of the chromosomes in the interchange. Plants having the duplication–deficiency interchanges can often be identified morphologically by resemblance to corresponding monosomes or monotelodisomes (Menzel, 1982).

Homozygous duplication–deficiencies which are tetra-nullo combinations for short terminal chromosome segments have been recovered from the translocations (Menzel and Brown, 1978b). Combinations of this type should prove useful for locating genes on chromosomes and for examining the diploidization phenomenon that has occurred in the genomes of the allotetraploids.

2. Analysis for Incipient Genome Differentiation

The translocations are useful for detecting incipient genome differentiation within the A and D subgenomes of the allotetraploid species and between the subgenomes of *G. hirsutum* and the A and D genomes of the diploid species (Menzel *et al.*, 1978, 1982; Hasenkampf and Menzel, 1980). Seven AD translocations were crossed with *G. raimondii* and *G. lobatum* and the chiasma frequency in specific D genome regions was analyzed. Since the translocated chromosomes are *G. hirsutum* AD interchanges, the (AD)D hybrid will show a trivalent consisting of the two AD interchanged chromosomes with the normal Dx chromosome between the two. Therefore, in crosses with two or more D genome species, the chiasma frequencies observed in the Dx chromosome of the different hybrids can be compared and any significant differences would indicate that chromosome differentiation had taken place (Menzel *et al.*, 1978). In the same manner, the Asiatics and the wild allotetraploid species were crossed to the translocations of *G. hirsutum* for analysis of chromosome differentiation between their genomes (Hasenkampf and Menzel, 1980; Menzel *et al.*, 1982). This is a cytological method that can be coupled with the cytogenetic method used by others (Gerstel, 1956; Gerstel and Phillips, 1957, 1958; Phillips and Gerstel, 1959; Rhyne, 1958, 1960, 1962b, 1965a,b; Stephens, 1961; Saunders, 1969; Lee, 1972) for studying chromosome differentiation within and between the genomes of the diploid and the allotetraploid species.

3. Production of Aneuploids

As the consequence of unequal disjunction of an interchange heterozygote, offspring can be recovered with an extra chromosome or deficient for a chromosome. The extra chromosome and the deficient chromosome can be a standard or an interchange chromosome. The modal pairing configuration at M1 in the duplicate or deficient plants will depend on whether the parental heterozygous interchange was self-pollinated or outcrossed with a standard line. In the chromosome deficient plants, which have a modal pairing of 24 bivalents and a trivalent, the trivalent association can be either of two possible types, one consisting of two interchanged chromosomes with a standard chromosome between them (*pseudomonosome*), and the other consisting of two standard chromosomes connected by an interchange chromosome (*tertiary monosome*). Both chromosome-deficient types are useful for locating genes on specific chromosomes (Menzel and Brown, 1954b; Endrizzi, 1963).

4. *Study of Chromosome Orientation*

The cotton translocations are useful for studying the orientation of the four chromosomes on the spindle apparatus at metaphase I. One of the transferred interspecific translocations and two of the AD translocations, because of their special cytomorphological features, were used to demonstrate the occurrence of two kinds of alternate orientation in addition to the two commonly observed adjacent orientation types (Endrizzi, 1974). A third type of alternate orientation, as well as total randomness of centromere orientation of quadrivalents, was also observed by Endrizzi *et al.* (1983).

C. CHROMOSOME STRUCTURAL CHANGES AND GENOME STABILITY

Besides the chromosomal translocations differentiating the A genomes of *G. arboreum* and the allotetraploids from that of *G. herbaceum* (Fig. 1), structural alterations have been observed in some of the hybrids between races of *G. hirsutum* (Endrizzi, 1966a) and in a few species hybrids of *G. hirsutum* with the diploid species (Sarvella, 1958; Kammacher, 1959; Douglas and Brown, 1971) and with *G. mustelinum* (Hasenkampf and Menzel, 1980). In some cases it has been postulated that these structural changes are examples of additional genome differentiation in *Gossypium*. It is doubtful that this is the case, but rather it is highly likely that they represent cases of transitory spontaneous chromosomal mutations.

Five of the 58 simple reciprocal translocations are spontaneous in origin (Brown, 1980) and were isolated by chance. For many years, we have been cytologically analyzing phenotypically aberrant plants that were found in commercial fields and breeding plots, and in a majority of cases these "off-types" were due to chromosomal structural changes, apparently arising spontaneously. For example, during a 4-year period, 45 "off-type" plants were analyzed and 20 plants (44%) had 21 translocations. The cytological configurations of 15 of 21 translocations indicated the interchanges were between A and D subgenome chromosomes, suggesting possibly an occasional pairing and exchange between A and D homoeologs.

Thus, it is apparent that spontaneous chromosome alterations are not uncommon in genomes of *Gossypium* (see also Barrow and Dunford, 1974), at least in the allotetraploid and that they have been occurring for a long time, at least since the origin of the allotetraploids. However, except for the reciprocal translocations that are permanently fixed in the A genomes of *G. arboreum* and the allotetraploids, there is no good evidence that similar or related types of structural changes differentiate the genomes of *Gossypium*.

Excluding for the moment the natural interchanges in the A genomes, we assume that the reason for the lack of fixation of structural changes in an established species is that they normally upset the highly stable relationship of the molecular organization and arm sizes of the chromosomes, a relationship that was established in the chromosomes at least at the time of the origin of the generic genome of *Gossypium,* and are therefore selected against.

One might speculate that changes in chromosome end arrangements in *G. arboreum* and in the allotetraploid species from that of the ancestral arrangement played a significant role in the derivation of these two plant forms as species. Those in the allotetraploid could have been established simultaneously in the early polyploid by a saltational event involving multiple breaks and rearrangements, as assumed for the origin of species in several genera of plants (Lewis, 1966; Carr, 1980). The exchanges in cotton involve almost the entire arms, which apparently at the time of their origin did not cause a disruption in the interdependence of the chromosomal molecular organization and arm size relationships but instead generated a chromosome reorganization having a selective advantage in that particular genomic background. In the ancestral forms of *G. arboreum* and the allotetraploid, it is likely that the gross structural changes were part of the genetic stabilization process that occurred during their initial stages of speciation.

These results suggest that the genomes have remained structurally stable in their evolution, a viewpoint that is supported by observations to be presented later (Section XII).

V. Monosomes

A. Origin and Identification

Allopolyploids such as *G. hirsutum* can tolerate the loss of whole chromosomes and transmit the haplodeficiency to the following generations. Plants that are deficient for one chromosome of a pair are referred to as *monosomics* and have a $2n-1$ chromosome number. Chromosome pairing at metaphase I will normally consist of 25 bivalents plus one univalent (see Fig. 3, Section V,C). Since $n = 26$ in the allotetraploids, it is theoretically possible to recover monosomic chromosomes for each of the 26 chromosomes.

The first plant identified as a monosomic in the allotetraploid cottons was found in an Upland cultivar at the Texas Agricultural Experiment Station by J. O. Beasley prior to 1942; however, 175 progeny plants grown over a period of several years from self- and cross-pollination of the monosomic plant failed to include any monosomic plants. At about the same time, a second monosomic plant was recovered from a different Upland cotton and it likewise failed to produce any monosomic progeny (Brown and Endrizzi, 1964).

These observations at first tended to discourage efforts to isolate monosomic plants and to encourage the isolation of trisomic ($2n + 1$) plants for cytogenetic studies. However, in 1946, a third monosomic plant, found among the progeny produced by crossing two trisomic lines, did produce monosomic progeny. The monosome of this line was later designated as chromosome 17 of the D subgenome. The finding that $n-1$ gametes can be recovered in succeeding generations from monosomic plants, coupled with the significant advantage of monosomics over trisomics in cytogenetic studies, renewed interest at the Beasley Laboratory of the Texas Agricultural Experiment Station in identifying plants with transmissible monosomics for other chromosomes of *G. hirsutum*.

Fifty-one monosomic plants of different origins were isolated during this period (1946–1962) at the Beasley Laboratory, and the monosome chromosome of 23 of the lines was identified in cytogenetic tests (Brown and Endrizzi, 1964; Endrizzi and Brown, 1964). Twenty of these involved chromosomes 1, 2, 4, and 6 of the A subgenome and 17 and 18 of the D subgenome of *G. hirsutum*. The remaining three consisted of one line with chromosome 1 from *G. arboreum* and two lines with chromosome 2 of *G. herbaceum*.

Following 1962, over 225 additional monosomic plants have been recovered as phenotypically aberrant plants in commercial fields, breeding blocks, and lines carrying different kinds of cytological aberrations isolated from species hybrids or from irradiation of seed or pollen of normal lines. A vast majority of these involved monosomes for chromosomes 2, 4, and 6 of the A subgenome (Endrizzi and Ramsay, 1979; Edwards *et al.*, 1980).

Monosomes for 15 of the 26 chromosomes of *G. hirsutum* were identified in the above group of monosomic plants (Endrizzi and Ramsay, 1979, 1980). These identified chromosomes are 1, 2, 3, 4, 6, 7, 9, 10, and 12 of the A subgenome and 16, 17, 18, 20, 22, and 25 of the D subgenome. The translocations were used to identify and number the monosomes as illustrated in Fig. 2.

As more of the chromosomes of *G. hirsutum* are identified as monosomes, the chance isolation of a monosome for a chromosome not already identified as a monosome becomes increasingly unlikely; therefore, the isolation of monosomes for the remaining 11 chromosomes will require a more direct approach. Pollinating plants that are recessive for mutant genes with irradiated pollen of the dominant marker line has been demonstrated as a reliable method for isolating monosomics and telosomics for specific chromosomes in cotton (Galen and Endrizzi, 1968; Endrizzi and Ramsay, 1979). With the increase in development of genetic marker lines in *G. hirsutum* and *G. barbadense,* radiation will provide the shortest route for completing the monosomic series of *G. hirsutum.* The translocations may be used for isolating monosomes for those chromosomes that remain unmarked genetically.

Haploids were successfully used in wheat and oats for isolation of monosomes, but haploids of cotton proved to be a disappointing source for monosomes. *G. hirsutum* haploids are normally completely sterile and are therefore not useful, but haploids of *G. barbadense* do produce a few seeds. In this species, a source of haploids produced by the genetic phenomenon of semigamy (Turcotte and Feaster, 1967, 1969) was tested and found to be of limited value for the production of a high frequency of monosomics (Endrizzi, 1966b).

All of the monosomic lines, including the monotelodisomics to be discussed below, are being transferred to the genetic background of Texas Marker 1 (TM 1), a highly inbred line of "Deltapine 14" that serves as the standard reference for plant morphological characters in genetic and cytogenetic studies (Kohel *et al.,* 1970).

Although *G. hirsutum* is an allotetraploid consisting of two homoeologous sets of chromosomes, one may nevertheless expect that the genetic imbalance owing to the monosomic condition will alter normal plant development in one form or another depending on the chromosome involved. Consequently, a syndrome of plant morphological characters is associated with each monosomic type (Table 8), which can be used to easily identify the monosomic plants from their disomic sibs. Several of the monosomics can be recognized in the seedling stage (given in parentheses in the table) with a high degree of success. Douglas (1968) noted that some monosomic plants showed amounts of pollen

TABLE 8

Plant Description and Origin of the Monosomics and Monotelodisomics of Cotton[a]

Chromosome	Plant description	Origin/Reference
A subgenome		
Mono 1	Smaller plant; small narrow leaf; narrow or twisted bracteole; plants usually have yellowish appearance and tend to be decumbent. Small oblong boll. Homoeolog of telo 15L (very narrow cotyledons)	$2n - 1 - 1$/Brown and Endrizzi (1964); Endrizzi and Brown (1964)
Telo 1L	Smaller, narrower leaf; narrow or twisted bracteole; generally lighter green	Mono 1
Telo 1S	Smaller, narrower leaf; narrow or twisted bracteole; generally lighter green; small oblong boll. Telo S resembles the mono more than telo L	Mono 1/Endrizzi and Stein (1975)
Mono 2	Smaller plant; smaller, narrower and cupping leaf; terminal leaves generally pointing upward; shorter sympodia; distinct midlock furrow in ovary and boll; bolls smaller and rounded; open bolls have storm-resistant shape. May be homoeolog of telo 14L (smaller cotyledons)	$2n + 1$/Brown and Endrizzi (1964); Endrizzi and Brown (1964)
Telo 2L	Smaller plant; smaller, narrower leaf; rounded boll that may have midlock furrow	γ-Radiation
Telo 2S	Smaller plant; smaller, narrower leaf; shorter sympodia; round boll, usually with midlock furrow. Telo S resembles the mono more than does telo L	Mono 2
Mono 3	Rounded leaf that is flatter and slightly crinkled; some plants have convex leaf; plant usually glabrate, resulting in leaves having a glossy appearance	γ-Radiation/Endrizzi (1975)
Telo 3L	Narrow, flatter leaf that is semirounded with some crinkling	Mono 3
Telo 3S	Narrow to rounded leaf, which tends to be glabrate	γ-Radiation

TABLE 8 *(Continued)*

Chromosome	Plant description	Origin/Reference
Mono 4	Shorter, bushy plant; leaf lobes are longer and leaves usually with wavy margins; longer and narrower bracteole; long peduncle and boll; bolls usually with midlock furrow. May be homoeolog of mono 22	X-Radiation/Brown and Endrizzi (1964); Endrizzi and Brown (1964)
Telo 4L	Bushy plant; long peduncle and bracteole	Mono 4
Telo 4S	Bushy plant; long peduncle, bracteole and bolls	Monos 3 and 7
Telo 5L	Narrow leaf with shallow base	γ-Radiation/Endrizzi and Ramsay (1980)
Mono 6	Short sympodia; glabrate; larger leaf; small bolls. May be homoeolog of mono 25	Endrizzi (1963)
Telo 6L	Short sympodia; glabrate; larger leaf. Telo L resembles the mono more than does telo S	Mono 6/Endrizzi and Kohel (1966)
Telo 6S	Glabrate; shorter sympodia; smaller, rounded bolls	Mono 6/Endrizzi and Kohel (1966)
Mono 7	Darker green plant; larger and thicker leaf, generally with secondary lobing; larger, thicker and contorted bracteole with more teeth; contorted calyx; shorter sympodia; smaller boll. Homoeolog of mono 16	Strain A/Endrizzi and Taylor (1968)
Telo 7L	Generally darker green plant; generally larger leaf; narrow bracteole with some cortortion or twisting	γ- and X-Radiation
Telo 7S	Darker green plant; larger leaf and bracteole. Telo S resembles the mono more than does telo L	$2n$ + centric fragment; Mono 7/Endrizzi and Taylor (1968)
Mono 9	Slow growing; short and rather stiff plant; larger, rounded, leathery leaf; large style and stigma, longer calyx teeth; pitted boll	γ-Radiation/Endrizzi and Ramsay (1980)
Telo 9L	Larger, flatter leaf; semi-short sympodia	Monos 3 and 9

(continued)

TABLE 8 (*Continued*)

Chromosome	Plant description	Origin/Reference
Telo 9S	Rounded, leathery, distorted leaf with shallow lobes	Mono 9
Mono 10	Large, round, and generally flatter leaf; generally darker green; short sympodia; large, abnormal stigma; longer boll. May be homeolog of mono 20	γ-Radiation and T10–19 translocation/Endrizzi and Ramsay (1980)
Telo 10L	Larger leaf; short sympodia; large, abnormal stigma; longer boll	Mono 10
Telo 10S	Larger, narrower leaf; short style	Mono 10
Mono 12	Smaller, narrower, crinkly leaf with redder veins; leaf shape and texture resembles heterozygous cup leaf (*Cu cu*); shorter sympodia; narrower bracteole; longer style and stigma; shorter filaments; semi-pollen sterile; inner involucral nectaries rarely present; smaller, rounded boll. Homoeolog of telo 26S (smaller cotyledons, generally lighter green and undulating)	25″ + t1″/Endrizzi and Ramsay (1980)
Telo 12L	Narrow bract; longer style and stigma; shorter filaments; semi-pollen sterile	Mono 12
D subgenome		
Telo 14L	Generally shorter plant; smaller leaf; semi-short sympodia; boll with midlock furrow	γ-Radiation/Endrizzi and Ramsay (1980)
Telo 15L	Narrow, twisted bracteole. Homoeolog of mono 1	$2n - 1Ah \times$ T4–5/Endrizzi and Kohel (1966)
Mono 16	Light green plant, stems and petioles almost devoid of anthocyanin; pulvinus spot light red; shorter sympodia; prolific flowering; smaller boll. Homoeolog of mono 7 (light green, less anthocyanin)	M25 and M39 of Brown and Endrizzi (1964); White and Endrizzi (1965)
Telo 16L	Indistinguishable from disomic sibs	Mono 16 and Acala 1517D
Mono 17	Smaller plant; smaller, narrower and more pointed leaf; longer style and stigma; small, round-	$2n + 1 \times 2n + 1$/Brown and Endrizzi (1964); Endrizzi and Brown (1964)

TABLE 8 (*Continued*)

Chromosome	Plant description	Origin/Reference
	ed boll; shorter lint; less seed fuzz (small cotyledons)	
Telo 17S	Smaller plant; smaller, narrower leaf, small boll	γ-Radiation
Mono 18	Smaller plant; smaller, narrower and more pointed leaf with wavy or ruffled margins; slightly longer style and stigma; longer peduncle; less glanded boll; shorter lint; longer and denser seed fuzz (small cotyledons)	γ-Radiation/Brown and Endrizzi (1964); Endrizzi and Brown (1964)
Telo 18L	Smaller semi-crinkled leaf	Mono 18
Telo 18S	Smaller leaf; longer style and peduncle	Monos 16 and 18
Mono 20	Darker green, very large leaf; short sympodia; sparse flowering; short but large stigma and style; larger rounded boll; dense seed fuzz. May be homoeolog of mono 10	Acala 6016/Endrizzi and Ramsay (1980)
Telo 20L	Usually darker green; larger leaf but not as large as mono; semi-short sympodia; short but large stigma and style; larger rounded boll	Mono 20/Endrizzi and Ramsay (1980)
Telo 20S	Very large leaf; short sympodia. Telo S resembles the mono more than does telo L	Telo 20L/Endrizzi and Ramsay (1980)
Mono 22	Leaf lobes are longer and leaves usually with wavy margins; plant darker green; longer narrow bracteole; long peduncle; bolls smaller, longer with mid-lock furrow. May be homoeolog of mono 4	Stoneville 7A-164
Telo 22L	Long style and stigma; peduncle, boll and bracteole teeth tend to be longer	Mono 22
Telo 22S	Shorter, bushy plant, darker green; leaves with wavy margins; narrow bracts; longer peduncle	Iso 22

(*continued*)

TABLE 8 *(Continued)*

Chromosome	Plant description	Origin/Reference
Mono 25	Short sympodia; generally gla-brate. May be homoeolog of mono 6	γ-Radiation/Endrizzi and Ramsay (1980)
Telo 25L	Short sympodia, generally glabrate	Mono 25
Telo 26S	Smaller, narrower, crinkled leaf with redder veins and petioles; leaf shape and texture resemble *Cu cu*; short sympodia, prolific flowering; inner involucral nectaries frequently absent. Homoeolog of mono 12	γ-Radiation/Endrizzi and Ramsay (1980)

^a Comparison of the morphological characters is with the disomic sibs. Notations in brackets are for seedlings.

abortion significantly different from the standard TM 1 and suggested that such monosomics may be selected in this manner. This, however, is unnecessary since each of the monosomics can be easily identified by their associated plant morphological characters.

Results so far have shown plants monosomic for a homoeologous set of chromosomes have for the most part similar syndromes of morphological characters. The matching of these morphological characters, along with the association of duplicate genetic factors, have enabled us to tentatively assign the following A and D subgenome chromosomes as homoeologous sets: 1–15, 4–22, 6–25, 7–16, 10–20, and 12–26 (Table 8). Following the isolation of monosomes for all 26 chromosomes, the chromosomes will be renumbered to identify homoeologous pairs.

B. BREEDING BEHAVIOR

Monosomic plants when self-pollinated or outcrossed as female normally produce two kinds of progenies: disomics and monosomics. The frequency of the monosomic progeny is dependent on the rate of transmission of the haplo-deficient gametes.

When using the monosomics for cytogenetic studies, the usual procedure with unselected seed of most monosomic lines is to transplant in the field 10–20 seedlings. The frequency of monosomic plants in this progeny size will generally range between 20 and 50%. However, the monosomics for chromosome 3, 9, and 16 usually occur in much lower frequencies and require larger populations to insure their recovery.

Table 9 shows the frequency of aneuploid plants in a single planting consisting of more than 100 progeny in each of 13 monosomics and two monotelodisomic lines (Malek-Hedayat, 1981). In this test, the rate of transmission of the haplo-deficient gamete ranges from a low of 4% for monosome 9 to a high of 49% for monosome 4. There appears to be no relationship between the genome of a monosomic chromosome and its rate of transmission.

Douglas (1972) noted in a preliminary study with six monosomes that within a single boll, the lightweight seed were predominantly monosomic. These observations were verified by Malek-Hedayat

TABLE 9

Frequency of Aneuploids in Unselected and Selected Seed in 13 Monosomic and Two Monotelodisomic Lines[a]

Chromosome	Unselected seed		Selected seed—Percentage of aneuploids in seed less than 1 SD from mean weight
	Plants (No.)	Aneuploids (%)	
A subgenome			
1	104	35 (33)[b]	67
2	121	45 (40)	83
3	139	7	30
4	173	49 (38)	68
6	138	32 (37)	85
7	128	21	52
9	137	4	18
10	119	23	45
12	128	30	44
D subgenome			
16	133	7	19
17	137	37 (21)	81
18	127	43 (37)	76
20L[c]	130	41	82
20S[c]	133	23	82
25	128	32	45

[a] Data of Malek-Hedayat (1981). Aneuploids are the respective monosomic or telosomic chromosome listed in the table.

[b] Values in parenthesis from Brown and Endrizzi (1964).

[c] Telocentrics for the long (L) and short (S) arms of chromosome 20.

(1981) with the seed producing the progenies given in Table 9. She found that the mean weight of the aneuploid seed was significantly less than the mean weight of the disomic seed. To increase the frequency of aneuploids in bulk harvested seed, she recommends the selection of those seeds weighing less than one standard deviation from the mean weight. The last column of Table 9 shows the percentage of aneuploids that would have been recovered if this method had been used.

Selection of minimum weight seed is especially recommended for monosomes 3, 9, and 16, which are normally transmitted in low frequencies. The low rate of transmission of mono 9 appears to be due in part, if not wholly, to the effect the deficiency has on chromosome segregation, presumably during the formation of the female gametophyte. In addition to the normal disomes and a few monosomic 9 plants from selfing or outcrossing mono 9 plants, the progeny include plants that are multiple monosomics, monosomic–trisomics, and combinations involving isosomic and/or telosomic chromosomes with monosomic and/or trisomic chromosomes. The abnormal cytological behavior occurs also in monotelodisomic 9S plants, but not in monotelodisomic 9L plants, establishing that the gene(s) regulating normal cytological divisions occur in the long arm. The cause of the erratic cytological behavior of monosome 9 plants is being investigated. All other monosomics are essentially stable and only occasionally a univalent shift is noted in the progeny.

No systematic attempt has been made to determine the rate of male transmission of $n-1$ gametes for each of the monosomes, but in studies requiring the use of several as the male parent in crosses showed male transmission of the haplo-deficiency in only two cases (Table 10). Although the populations for some of the monosomes are perhaps too small for an adequate test, the transmission of the haplo-deficiencies only for chromosomes 4 and 6 has been observed. These are the two chromosomes that occur most frequently as spontaneous monosomes in natural populations of cotton.

Most of the monosomic chromosomes are fairly stable and misdivide infrequently to give telocentric or isochromosomes (Brown, 1958). However, as measured by the recovery of centric fragments in progeny of plants monosomic for chromosomes 12, 22, and 25, the misdivision rates of these three chromosomes appear to be very high. Consequently, the monosome of these three chromosomes is generally accompanied by a centric fragment ($2n-1$ + fragment) of variable size, which so far has not interfered in the use of either monosome in cytogenetic studies. Except for the centric fragment of chromosome 12, the

TABLE 10
Male Transmission of $n - 1$ Gametes in Crosses
of $2n \times 2n - 1^a$

Chromosome	No. of plants	No. of $2n - 1$ plants
H1	115	0
H2	35	0
H4	66	4
H6	191	2
H7	44	0
H16	28	0
H17	143	0
H18	35	0

[a] From Brown and Endrizzi (1964, and unpublished data).

centric fragments of chromosomes 22 and 25 do not undergo exchange pairing with their normal homologs. This suggests that they consist entirely of heterochromatic chromatin and, therefore, do not contain structural genes.

C. Uses of Monosomes

In studies involving the monosomes, the monosomes must be used as the female parent for their recovery in the F_1 generation, which will consist of two kinds, the disomics and monosomics.

Monosomes provide a quick and simple method of identifying the chromosomes in translocations, as illustrated in Fig. 2 and discussed in Section IV,A.

The monosomic stocks have been used extensively to assign genetic factors to specific chromosomes. The protocol for locating recessive and dominant marker genes on monosomic chromosomes was outlined by Endrizzi (1963) and is illustrated in part in Figs. 3 and 4 and described more fully in the following section.

1. Identifying Recessive Mutant Genes Associated with a Monosome

The procedure for locating recessive mutant alleles on a specific chromosome is illustrated in Fig. 3 with two marker genes, one independent of and the other associated with the monosome chromosome. The monosomic parent produces two kinds of gametes, one with a full complement (n) of chromosomes and normal A and B alleles, and the other deficient ($n-1$) for the monosome chromosome carrying the nor-

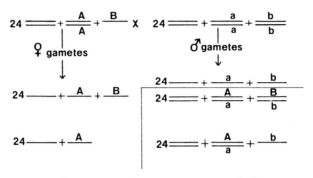

FIG. 3. Procedure for locating recessive genes on monosomic chromosomes. The monosomic line carries the normal or dominant alleles *A* and *B*. Note that the disomic F_1 progeny have the normal phenotype of *A* and *B* alleles, whereas the monosomic F_1 progeny have the normal phenotype of the *A* allele, but the recessive phenotype of the *b* allele.

mal *B* allele. Consequently, in a cross with the recessive *aa bb* line, two F_1 cytotypes are recovered: the noncritical disomics, which are heterozygous for both loci and have the normal *A* and *B* phenotypes, and the monosomics, which are the critical F_1 hybrids for determining the presence or absence of association with a gene. Nonassociation of a gene with the monosome is illustrated in Fig. 3 with the *A* locus; it can be seen that the monosomic F_1 progeny, like the disomes, have the normal *A* phenotype. Association is shown with alleles at the *B* locus; the monosomics F_1 progeny have the hemizygous 0/*b* recessive genotype and phenotype rather than the normal *B* phenotype. Recovery of the recessive phenotype in the monosomic F_1 plants establishes that the *B* locus but not the *A* locus is located on the monosomic chromosome.

The above results can be verified by scoring test cross or F_2 populations of the two cytotypes. It is most desirable to use the $2n-1$ F_1 offspring as the male parent in a test cross since $n-1$ gametes are rarely transmitted through pollen.

In a test cross, the disomic F_1 progeny will produce a population segregating in typical disomic ratios of one normal: one mutant for both loci. The test cross population of the $2n-1$ F_1 offspring, however, will consist of only the recessive *bb* phenotypes, but segregate in a ratio of one normal: one mutant phenotype for the nonassociated *Aa* allelic pair. In the F_2, the disomics, as expected, will produce populations showing the normal 3:1 phenotypic segregation ratios for alleles at the *A* and *B* loci; whereas the monosomics will produce populations showing a 3:1 phenotypic segregation ratio for alleles at the A locus but only the mutant phenotype of the b allele.

Some recessive genes in cotton, however, behave like antimorphs in that the phenotype expressed by the hemizygous recessive genotype of the F_1 is intermediate and similar to that of the heterozygote. In such cases, determining whether association exists will require either a test cross or an F_2 population. If the locus and the monosomic chromosomes are independent, the test cross population for the disomic and monosomic F_1 progeny will show a 1:1 ratio of intermediates (heterozygotes) to mutants, whereas the F_2 population of the two cytotypes will produce a 1:2:1 ratio of phenotypes for that marker locus. If association occurs, the monosomic F_1 offspring will show only mutant phenotypes in the test cross and two phenotypic types in the F_2 generation: mutants and "intermediates" (hemizygous recessive), in which the individuals with the mutant phenotype are disomics and those with the "intermediate" phenotype are monosomics, the frequency of which is dependent on the transmission rate of $n-1$ gametes.

2. Identifying Dominant Mutant Genes Associated with a Monosome

The genotypes of the disomic and monosomic F_1 offspring from crossing a monosomic plant carrying normal or wild-type alleles with a dominant marker line are given in Fig. 4. Test cross and F_2 populations are always required for determining whether a dominant mutant is associated with the monosome. When association occurs, the monosomic F_1, when either selfed or test crossed as male to a standard line, will not segregate for the marker located on the monosome and therefore will produce progenies expressing only the dominant phenotype.

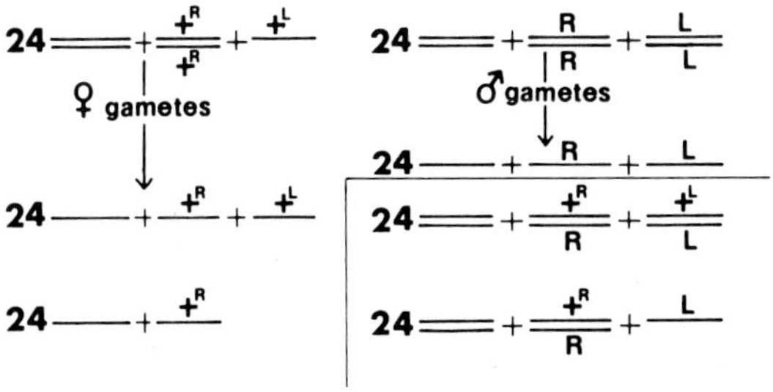

FIG. 4. Procedure for locating dominant mutant genes on monosomic chromosomes. The monosomic line carries the normal or wild-type alleles. Note that the disomic and monosomic F_1 progeny have the dominant phenotype of the R and L alleles, but that the monosomic F_1 is hemizygous for the L allele.

TABLE 11

Summary of Tests for Association of Mutant Genes with
Monosomes and Telosomes for 19 of the 26 Chromosomes
of *G. hirsutum*

Mutant loci	Chromosome location	No. of monosomes and telesomes found to be independent
ac	16	5
as_1	2^a	4
$as_2{}^b$		6
ax		6
B_2, B_3		6
B_4		4
bw_1	12	4
chl_1, chl_2		6
cl_1	16	7
cr	15	5
Crp		8
cu		11
cy		8
de		7
fg	3	11
gl_1		12
gl_2	12	6
gl_3	26	6
$H_1{}^c$	6	4
$H_2{}^c$	6	11
ia	3	6
$L_1{}^L$	1	3
$L_2{}^O$	15	13
Lc_1	7	12
Lc_2	6	
Lc_4		11
Lc_w		3
Lc_y, Lc_z		9
lethal		1
Lg	15	3
Li		7
lp_1	1	8
lp_2	15^a	1
ml	4	
ms_1		7
ms_2		6
ms_3	16	

TABLE 11 (*Continued*)

Mutant loci	Chromosome location	No. of monosomes and telesomes found to be independent
N_1	12	12
n_2	26	7
ne_1	12	
ne_2	26	
ob	18	6
P_1	5	12
P_2		8
pg		5
R_1	16	11
R_2	7	12
Rd		11
Rf		18
Rg		4
rl_1		8
Rl_2		5
Ru		2
RSL		6
rx		7
Sm_1^{sl}	25	2
Sm_2^c	6	7
st_1	4	5
st_2		6
st_3		7
$stpl$		3
v_1	20	11
v_2		8
v_3		6
v_4		5
v_8		8
v_9		5
v_{12}		2
vf		2
Y_1		13
Y_2	18	8
yv		8

[a] Needs retesting.

[b] A desynaptic mutant found in 1517 Acala C or D; believed to be a mutation at the A_2 locus.

[c] Recently determined to be alleles (Endrizzi and Ramsay, 1983a).

The nonassociated marker loci, which are heterozygous, will segregate into dominant and recessive phenotypic classes in both the test cross and the F_2.

If the monosomic F_1 is used as the female parent in a test cross, the disomic progeny will have the dominant phenotype and the monosomic progeny will have the normal phenotype. Nonassociated markers will segregate independently of the two cytological classes.

Following the methods outlined above, 487 different test combinations involving 75 different marker genes and one or more of the monosomes or monotelodisomes have been carried out (Endrizzi, 1963, 1975; Endrizzi and Kohel, 1966; Endrizzi and Taylor, 1968; Poisson, 1968; Endrizzi and Stein, 1975; White and Endrizzi, 1965; Kohel and Douglas, 1974; Kohel, 1978a; Endrizzi and Bray, 1980; Anonymous, 1981). The results of these studies are given in Table 11.

Use of the monosomic initially to assign genes to specific chromosomes eliminates the laborous procedure of randomly crossing different mutant genes in testing for linkage combinations.

Pseudo- and tertiary monosomes, recovered from unequal disjunction of heterozygous translocations, were discussed in Section IV,B,3.

3. Separating Duplicate Linkage Groups

Monosomes are ideal for separating duplicate linkage groups for determining the linkage values in each of the duplicate linkage groups. Separating the two linkage groups for independent analysis avoids the confounding effect normally encountered in the segregation of the duplicate loci in an F_2 or a testcross population. This can be accomplished simply by crossing the two appropriate homoeologous A and D monosomics as female to a line with the duplicate recessive linkage groups. The F_1 plants that are monosomic for the A subgenome chromosome will be heterozygous (disomic) for only the markers of the D subgenome linkage group; the monosome chromosome in these plants will carry the duplicate recessive markers of the A subgenome linkage group. Conversely, the F_1 plants that are monosomic for the D subgenome chromosome will be heterozygous for the markers of the A subgenome linkage group and recessive for the duplicate set of the D subgenome. These two types of monosomic F_1 progeny can be used to produce test crosses of F_2 populations for analysis of recombination in the two unimeric linkage groups.

4. Synthesis of Chromosome Substitution Lines

Monosomics can also be used in the synthesis of chromosome substitution lines as outlined by Endrizzi *et al.* (1963b) and White *et al.*

(1967). Individual chromosomes can be transferred intact from cultigens, primitive races, and wild species. Such disomic substitution stocks can be used for the study of the genetic effects of individual chromosomes on plant traits and for estimating the number of genes and their interactions and linkage relationships controlling plant characters of economic importance. A study of this kind was conducted by Kohel *et al.* (1977a) in backcross-derived disomic substitution lines with 25 pairs of *G. hirsutum* chromosomes and either one pair of chromosome 6 or one pair of chromosome 17 of *G. barbadense*. They found that higher lint percentage, finer fiber, and later flowering were significantly associated with substitution line 6, and short fiber length was significantly associated with substitution line 17. Schwendiman (1975) also developed a *G. hirsutum* line with the disomic substitution of chromosome 6 of *G. barbadense*. The disomic substitution line did not differ from *G. hirsutum* in plant morphological characters, but it did show an improvement in fiber length, uniformity, and micronaire. The yield of seed cotton, however, was lower.

Following the localization of genetic factors controlling agronomic traits on one or more chromosomes in the substitution lines, their position within the chromosome can be determined with the telosomics and in linkage analysis with the genetic marker loci located on that chromosome.

The alien substitution lines are also useful for studying chromosomal differentiation between two species. This can be accomplished cytologically by combining in an F_1 hybrid the substituted chromosome with their homologous telocentric chromosome of the receptor line. Chromosome pairing in the monotelodisomic pair of the F_1 is then compared with pairing in the standard parental monotelodisomic. A genetic approach for studying chromosome differentiation would involve combining a multiple genetically marked chromosome with the substituted chromosome and noting shifts in recombination values between the marker loci as demonstrated by Rhyne (1958, 1960, 1962b, 1965a,b,) Stephens (1961), Saunders (1969), and Lee (1972).

With the development of the monosomic series in two allotetraploid species, more than one alien chromosome may be substituted into one or the other species (White *et al.*, 1967).

VI. Monotelodisomes and Their Use

A monotelodisomic plant in cotton has 25 bivalents plus a heteromorphic bivalent, in which the heteromorphic or tl pair consists of a telocentric chromosome plus a complete homologous chromosome (Fig. 5). Telocentric chromosomes arise by misdivision of univalents at the first or second meiotic division. The first telocentric chromosomes in cotton were reported by Brown (1958) who recovered telocentrics and other heteromorphic bivalents from several sources including the monosomics. One of these involved chromosome 17, the first chromosome identified as a telocentric. In addition to their recovery from monosomics and other types of cytological aberrant lines, telocentrics have been recovered in irradiation experiments (Galen and Endrizzi, 1968; Endrizzi and Ramsay, 1979).

Table 8 lists the 30 monotelodisomic stocks that mark one or both arms of 19 chromosomes that have been identified. With the exception of monotelodisomic 16L, all other monotelodisomics exhibit a syndrome of morphological characters closely paralleling in one form or another that which is associated with their monosomic counterparts. Most monotelodisomics were first recognized and isolated by these associated plant characters. Cytological analysis and later tests with translocations confirmed their identity.

Like monosomics, monotelodisomics produce two kinds of progenies, disomes and monotelodisomes, when either self- or cross-pollinated. For most, the female transmission frequency of the telocentric chromosome is generally about 50%. For those monotelodisomics we have used as the male parent in crosses, the telocentrics were transmitted through the pollen, ranging from a few percent for most to a high of 33% for telo 16L. Only the ditelocentric for the long arm of chromosome 16 has been recovered from self-pollination of monotelodisomics.

Once a marker gene has been assigned to a specific chromosome by the monosome test, the monotelodisomics can then be used for determining the arm location of the marker as well as its linkage relationship with the centromere. The procedure employed for determining whether a marker is associated with the monosomic arm is identical to that described above for locating genes on monosomic chromosomes and has been described in several reports (Endrizzi and Kohel, 1966; Endrizzi and Taylor, 1968; Endrizzi and Stein, 1975; Endrizzi and Bray, 1980).

In mapping the centromere and the marker locus, it is desirable to use the monotelodisomic (tl) F_1 as the pollen parent in test crossing with the standard recessive parent (Fig. 5), since the telocentric chromosome is infrequently transmitted through the pollen. The tl pair occurs primarily as a bivalent at MI, thus it is not expected that the linkage value between two markers would differ significantly whether

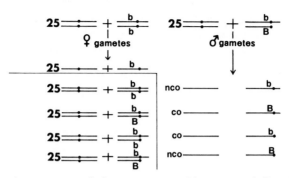

FIG. 5. Mapping a gene and the centromere with a monotelodisomic. The mono-telodisomic F_1 can produce four kinds of gametes; their frequencies are dependent on the distance of the gene locus from the centromere. If the monotelodisomic F_1 is used as the male parent, a majority of the test cross progeny will be disomic. Nco and co are, respectively, noncrossover and crossover gametes.

the monotelodisomic F_1 is used as male or female in the testcross. However, this has not been critically tested. If monotelodisomic plants are recovered in the test cross progeny, they can be identified by their diagnostic morphological characters. Thus, the recombinant and non-recombinant types in both disomic and monotelodisomic testcross plants can be readily identified. Through the use of these procedures, the position of the centromere in the genetic map of eight of the chromosomes has been determined (Table 4).

Like monosomes, the monotelodisomics can be employed to identify the individual chromosomes in translocations in a minimal number of crosses. Also, for translocations having chiasmata forming in an interstitial segment, the telocentric chromosomes can be used to determine the arm in which the translocation had occurred.

VII. Monoisodisomes

Misdivision of univalent and telocentric chromosomes can give rise to isochromosomes, which are chromosomes with two homologous or identical arms. Plants with an isochromosome are referred to as mono-isodisomics since they have 25 normal bivalents plus a heteromorphic bivalent consisting of a normal chromosome and an isochromosome. The heteromorphic bivalent is monosomic for one arm and trisomic for the other.

Isochromosomes can be used in the same manner as the telocentric chromosomes for positioning genes in arms of the chromosomes; however, they may not be as useful as the telocentrics for mapping studies since they may give rates of recombination different from that of the monotelodisomic as the consequence of exchanges in the trisomic arms (Endrizzi and Bray, 1980).

VIII. Trisomes, Tetrasomes, and Alien Chromosome Additions

Prior to the demonstration that monosomic plants produce functional haplo-deficient gametes, a program was under way at the Texas Agricultural Experiment Station to isolate and identify primary trisomic, tetrasomics, and alien chromosome addition lines in *G. hirsutum* (Beasley and Brown, 1943; Brown, 1966). However, the program was abandoned in favor of the monosomic program since the latter are far more useful for cytogenetic studies.

Over the years many primary and a few secondary and tertiary trisomics, either spontaneous in origin or in a number of cases from unequal disjuention of heterozygous translocations, have been found in the cytogenetics nursery, but then have generally been ignored. Primary trisomic and tetrasomic individuals occur occasionally in cultivars of cotton (Endrizzi *et al.*, 1963a).

More recently, alien chromosome addition lines have been bred and studied for their practical use (Kammacher and Poisson, 1964; Poisson, 1967, 1970, 1971; Kammacher and Schwendiman, 1969; Vieira da Silva and Poisson, 1969; Sharif and Islam, 1970; Schwendiman, 1977; Hau, 1981a,b, 1982).

Disomic addition lines involving alien chromosomes provide a method for identifying homoeologous chromosomes of different species, as demonstrated by Poisson *et al.* (1969) and Schwendiman (1974). The latter developed 14 disomic addition lines in which eight different chromosomes of *G. anomalum* and six different chromosomes of *G. stocksii* were added to the normal complement of *G. hirsutum*. By comparing the modified plant morphological characters associated with each disomic addition line, he demonstrated that four, possibly five, chromosomally different *G. anomalum* additions and the same number of *G. stocksii* additions exhibited parallel plant modifications. Matching similarities in phenotypic modification indicated that in each case the *G. anomalum* and *G. stocksii* disomic ($2n + 2$) additions are homoeologs and that they therefore carry the same set of structural genes performing identical functions. The lack of complete chromosome pairing between the homoeologs in their F_1 hybrids is due largely to their differences in quantities and qualities of repetitive DNA.

IX. Origin of the Allotetraploids

The discovery of tetraploid species that were confined to the New World raised the question of whether they were auto- or allotetraploid and of the time and place of their origin.

Longley (1933), Webber (1934a), and Gates (1938) assumed that they were autotetraploids. Amphidiploidy was proposed by Davie (1933) who speculated that these forms arose from two differentiated wild American 13-chromosome species, which themselves had arisen from a hybrid between two species with six and seven chromosomes. The six-chromosome species were assumed to have originated by the fusion of two of the chromosomes in a seven-chromosome species. The reduction of seven to six chromosomes is based on his and Denham's (1924a) observations that one pair of chromosomes in the tetraploids was distinctly larger than the rest. However, cotton cytologists agree that such a size difference in one pair of chromosomes does not exist in the genus *Gossypium* (see also Beal, 1928; Skovsted, 1934a; Webber, 1934a).

Chromosome pairing in the F_1 hybrids of Asiatic species (*G. arboreum* and *G. herbaceum*) × American tetraploids was reported by Baranov (1930), Zhurbin (1930), and Nakatomi (1931) to consist of 13 bivalents and 13 univalents, but it was Skovsted (1934a,b) who noted that the 13 large chromosomes of the Asiatics had paired with the 13 large chromosomes of the American tetraploids and that the remaining 13 small chromosomes were left as univalents. Based on the meiotic behavior of a number of interspecific hybrids, including the one above, and on the comparative size of the chromosomes of species in the genus, Skovsted (1934a) concluded that the New World cottons were amphidiploids that originated by doubling of the nonhomologous chromosomes of two species with $n = 13$, one of which was cytologically similar to the Asiatic cottons with large A chromosomes and the other of which was probably an American diploid with small D chromosomes. Confirmation of the hypothesis was provided by the cytological observations of Skovsted (1934b) and Webber (1934b) on chromosome pairing in hybrids between American wild diploids and the tetraploid cottons. Moreover, Skovsted noted that the 13 small chromosomes of the American diploid species (D genome) paired with the 13 small chromosomes of the American tetraploid species, leaving the 13 large chromosomes of New World cottons as univalents.

Following the cytological demonstration that the New World allotetraploids consisted of A and D subgenomes, efforts were made to produce a synthetic hybrid between an A and D species. This feat was accomplished independently in 1940 by Beasley and Harland, who crossed *G. arboreum* with *G. thurberi* and doubled the chromosomes of the F_1 hybrid (Beasley, 1940a; Harland, 1940). Beasley (1940a) showed that less than one-half of the chromosomes paired in the diploid F_1, but that the chromosomes of the synthetic amphidiploid showed regular bivalent pairing just as one would expect according to the allotetraploid hypothesis of Skovsted. Furthermore, the synthetic amphidiploid, although partly male sterile, was female fertile, setting a full complement of seed when pollinated with either *G. hirsutum* or *G. barbadense*. Beasley's (1942) analysis of genome affinities not only confirmed Skovsted's proposal that the amphidiploid species consisted of A and D genomes, but also eliminated all other genomic combinations as possible progenitors of the amphidiploids.

Genetic support for the allopolyploid origin was demonstrated by Harland (1937a) and Harland and Atteck (1941). They showed that (1) the genome of the diploid New World wild species carries the normal allele of the crinkle dwarf (*cr*) mutant of New World tetraploid cottons; and (2) for the duplicate R_1 and R_2 anthocyanin loci in the New World cottons, the R gene of the New World wild diploids is allelic with R_1, whereas the R allelic series of the Asiatics is allelic with R_2. Allopolyploid origin was more vividly demonstrated by Silow (1946) in reporting the existence of duplicate linkage groups R_1-cl_1 and R_2-cl_2. These studies showed that the loci of the allotetraploids were related (homologous) to those of the diploids by common origin.

Thus, by the early 1940s, it was firmly established that the New World cottons were allotetraploids that originated from combining A and D genomes, but which genomes of the two Asiatic cottons and of the several American wild diploids were most similar to the progenitor forms were unknown. Stephens (1944a,b) provided the first clue as to which species of the American wild diploids has a D genome most closely related with the D subgenome of the allotetraploids in a study of leaf developmental patterns in hybrids of Asiatics with various alleles for leaf shape and *G. raimondii*, one of the four American D species, with entire leaves. The general morphology of leaves and bracteoles of the hybrids suggested that the D subgenome of the allotetraploids is most closely related to either *G. raimondii* or *G. klotzschianum* and its closely related form *G. davidsonii.*

Additional information supporting Stephens' proposal was provided by Hutchinson *et al.* (1945), who discussed a number of factors that should be considered in formation of a natural allotetraploid. These included the kinds of seed hairs occurring in the diploid species and the vigor of hybrid forms. The vigor of the diploid F_1 should be at least equal to that of the parents, and the vigor of the synthetic F_1 allotetraploid should be more than that of the diploid F_1. Among the then known six D genome species, *G. thurberi* and *G. aridum* were ruled out because they gave hybrids of low vigor with the Asiatic cottons, and the seed hairs of these species are not of the right type. *G. armourianum, G. harknessii, G. klotzschianum,* and *G. davidsonii,* though not successfully crossed with the Asiatic cottons, were ruled out on the basis of their seed hairs. This left *G. raimondii* as the most likely candidate as donor of the D subgenome of the allotetraploids, which in combination with the genome of an Asiatic cotton would provide the genetic system including convoluted seed hairs, characteristic of the allotetraploids.

Finally, published data on chromosome pairing in hexaploid hybrids showed that the hexaploid combinations of *G. hirsutum–raimondii* and *G. hirsutum–*Asiatics had a higher frequency of multivalents than other genomic combinations, indicating a greater affinity of the chromosomes of *G. raimondii* and the Asiatics with their "homologous" counterparts in the New World allotetraploids (Hutchinson *et al.,* 1947).

Thus, by the late 1940s, it was evident from cytogenetic and morphological studies that (1) the ancestral A genome of the allotetraploids was similar to the A genome of the Asiatic species and (2) though not firmly established, the ancestral D genome was more similar to the D genome of *G. raimondii* than to the D genomes of other diploids.

The first critical evidence of which of the two Asiatic species has an A genome more closely related to the A subgenome of the allotetraploids was provided by Gerstel (1953a), who reexamined chromosome pairing among hybrids between *G. arboreum* × *G. herbaceum, G. hirsutum* × *G. herbaceum,* and *G. hirsutum* × *G. arboreum* and concluded that the genomes of the two Asiatic species differ by one reciprocal interchange, *G. herbaceum* and *G. hirsutum* differ by two interchanges, and *G. arboreum* and *G. hirsutum* differ by three interchanges (Fig. 1). The pairing data of the hybrids with *G. hirsutum* showed that the genome of *G. herbaceum* is less structurally differentiated from that of the allotetraploids. In addition, the chromosomes of *G. herbaceum* were shown to have the structural arrangement of the ancestral A genome since the hybrid between it and the closely related wild diploid *G. anomalum* had 13 bivalents at metaphase I, whereas the hybrid between *G. arboreum* × *G. anomalum* had 11 bivalents and a quadrivalent. The B genome is assumed to be the progenitor of the A genome (Hutchinson *et al.*, 1947; Phillips, 1966). These results firmly established that the genome of *G. herbaceum* rather than *G. arboreum* is more closely related to the A subgenome of the allotetraploids, even though the A subgenome differs from the *G. herbaceum* genome by two independent reciprocal translocations. Gerstel's observations were confirmed by Menzel and Brown (1954a).

Stephens (1942, 1947, 1949a, 1950a) suggested cytogenetic tests involving synthetic allopolyploids that would not only provide a more sensitive test for determining the relationship of genomes but would also be valuable in determining whether the natural allotetraploids are ancient or recent in origin. Utilizing this information, Gerstel, Sarvella, and Phillips synthesized amphidiploids and allohexaploids involving Asiatic, New World wild, and the natural allotetraploid species and analyzed them cytologically for multivalent frequencies and genetically for segregation of mutant marker genes located in the A and D genomes (Gerstel, 1953b, 1956, 1963; Sarvella, 1958; Gerstel and Phillips, 1957, 1958; Phillips and Gerstel, 1959; Phillips, 1960, 1962, 1963, 1964). The data for the hexaploid hybrids are summarized in Table 12 along with the expected ratios for an autotetraploid for comparison.

Cytological and genetic ratios observed in the analysis of the hexaploid hybrids involving the D species and the allotetraploids showed that those involving *G. raimondii* most closely approached autotetraploid behavior, confirming that its genome is indeed most closely related to the contributor of the D subgenome of the tetraploids. It is now a firmly established cytogenetic fact that the progenitor genomes of the allotetraploid species were most similar to the genomes existing in *G. herbaceum* and *G. raimondii*. Additional new diploid species have been identified, particularly in the D genome group, but their taxonomic status does not question the current concept of the origin of the allotetraploids.

The data in Table 12 also shows that the A genome chromosomes of the Asiatics and the allotetraploid species in hexaploid combinations

TABLE 12

Relationship between the Frequencies of Multivalents and Segregation of Marker
Genes in (AD)A and (AD)D Hexaploid Hybrids of *Gossypium*

Hexaploid	Average multivalents	Average test cross segregation	No. of loci
Expected for autotetraploid[a]	8.70	5:1	All
G. hirsutum × *G. arboreum*	8.68	5.1:1	4
G. barbadense × *G. arboreum*	7.80	6.1:1	5
G. hirsutum × *G. raimondii*	6.16	9.5:1	10
G. barbadense × *G. raimondii*	—	11.5:1	3
G. barbadense × *G. raimondii*[b]	6.33	—	—
G. barbadense × *G. raimondii*[b]	3.30	—	—
G. hirsutum × *G. harknessii*	3.65	17.1:1	5
G. hirsutum × *G. armourianum*	3.96	17.4:1	5
G. hirsutum × *G. davidsonii*	3.96	19.8:1	—
G. hirsutum × *G. aridum*	3.48	21.3:1	4
G. hirsutum × *G. lobatum*	3.66	23.7:1	4
G. hirsutum × *thurberi*	3.61	32.9:1	3
G. barbadense × *G. gossypioides*	1.13	71.6:1	4

[a] Theoretical expected multivalent frequency and segregation ratio for random pairing of four homologous chromosomes and random chromosome segregation.

[b] From Sarvella (1958); the remaining data from Phillips (1962, 1964, 1974).

behave cytologically and genetically as newly synthesized autotetraploids, establishing that the A genomes of the three species remained essentially completely homologous since the time of the origin
of the allotetraploid. However, in a cytological study of hybrids between the Asiatics and the translocation lines of *G. hirsutum*, Menzel
et al. (1982) found that overall chiasma frequencies were slightly but
significantly lower than in the controls, indicating some divergence in
the A genome chromosomes. However, most of the reduction appeared
to be localized in chromosomes 2–5, perhaps primarily in chromosomes
4 and 5, which are the chromosomes involved in the natural interchanges. Obviously, the differentiation they observed does not hinder random chromosome association and segregation in hexaploid combinations.

Even though the D genomes of *G. raimondii* and the allotetraploids
are highly homologous, they do show some divergence in that multivalent frequencies, and segregation ratios are not quite like those expected for an autotetraploid.

X. Time and Place of Origin of the Allotetraploids

It is generally presumed that divergence of the diploid species is very ancient, occurring no later than the Cretaceous period (Skovsted, 1934a; Hutchinson *et al.*, 1947; Prokhnov, 1947; Mauer, 1954; Saunders, 1961a; Phillips, 1963; Hawkes and Smith, 1965; Fryxell, 1965a, 1979).

Skovsted (1934a) and Saunders (1961a) assumed that continental drift provided the means of distribution or separation into two major chromosomal size groups due to the spatial relationships of the progenitors, the large chromosomes of the diploid species of Asia, Africa, and Australia, and the small chromosomes of the diploid species of America. Since the species that contributed the A genome to the allotetraploids has never been found in the New World, different viewpoints have been expressed on the time of origin of the allotetraploids. Some authors considered the origin to be very ancient (Skovsted, 1934a; Harland, 1935a, 1939; Stebbins, 1947; Valiček, 1978) while others considered it to be very recent (Davie, 1935; Hutchinson *et al.*, 1947; Hutchinson, 1959; Sherwin, 1970; Johnson, 1975; Heyerdahl, 1979).

According to Skovsted (1934a) and Valiček (1978) the origin of the allotetraploids occurred at the time when the areas of distribution of the two original A and D diploid parental forms overlapped prior to the origin of the rift. Origin of the allotetraploids in the area of overlap of the two progenitor genomes but after the rifting of the continents was also suggested. Since *G. arboreum* was cultivated across southeastern and eastern Asia and as far East as the Philippines and the D genome species were confined to Western coastal areas of the New World, Harland (1935a) proposed that the allotetraploids originated in the late Cretaceous or early Tertiary in the Polynesian Islands by migration of *G. arboreum* across a trans-Pacific land bridge where it hybridized with a D species. An Arctic route during the early Tertiary period was proposed by Stebbins (1947) for the arrival of A species in North America. Merrill (1954) initially proposed an Antarctic transfer of the Asiatic parent in Tertiary times, but he concluded that it is more plausible to accept an introduction and hybridization shortly after 1500 AD. Such hypotheses are no longer tenable (Stephens, 1947; Gerstel, 1953a; Hutchinson, 1959; Phillips, 1963; Fryxell, 1965a, 1979).

Mauer (1954) places origin of the hybrid in the Eocene, not involving an Old World diploid but between diploids of the New World—a speculative hypothesis not supported by any cytogenetical data. Davie (1935) expressed a similar view, but with the hybridization occurring in the recent past, either under conditions of Pre-Columbian aboriginal cultivation or in the wild.

A recent origin was also suggested by Hutchinson *et al.* (1947) and Hutchinson (1959). They assumed that a *G. arboreum* cotton was introduced into the New World by man over a Pacific route, where, under cultivation, it hybridized with an American wild diploid species, probably *G. raimondii.* A Pacific transfer of *G. arboreum* was discarded when Gerstel (1953a) reported that the A genome of *G. herbaceum,* rather than that of *G. arboreum,* is closer to the A subgenome of the allotetraploids. An Atlantic transfer was suggested by the existence of a wild form of *G. herbaceum* (var. *africanum*) in Southern Africa (Gerstel, 1953a; Phillips, 1963), the only truly wild type of the Asiatic cottons. The wild A genome form var. *africanum* is regarded as the ancestor of the cultivated A genome species (Hutchinson, 1954).

In view of Gerstel's evidence for an Atlantic meeting, Hutchinson (1962) contends that an origin of the allotetraploid in eastern South America indicates an ancient event in which the two wild diploids came in contact by natural spread. Since it is possible to interpret the present distribution of the wild species by the theory of continental drift, Hutchinson postulated that a wild type akin to *G. herbaceum* var. *africanum* existed east of the Andes hybridized with *G. raimondii.* The natural occurrence of an A genome species in the New World is yet to be documented.

A recent and biphyletic origin of the allotetraploids was proposed by Sherwin (1970), who presumed that no wild forms of the New World cottons exist and that a cultivated *G. herbaceum* was transported to northern South America either by man or by rafting in abandoned water craft or sealed in gourds. Because seed of several species of *Gossypium* including the allotetraploids have hard seed coats and saltwater tolerance (Stephens, 1958a,b, 1966; Fryxell, 1965a), dispersal by oceanic drift has been considered as a means of transport of the A genome to the New World. Stephens (1966), however, concluded from tests that seed of *G. herbaceum* var. *africanum* and its relative *G. anomalum* lack sufficient saltwater tolerance to survive a transatlantic crossing.

Johnson (1975) also believes that man transported *G. herbaceum* to the tropical regions of the New World, where it was cultivated and formed natural hybrids with more than one D genome species. Transport of the A genome species across the Atlantic during the cultural expansion period from about 3000–1200 BC by man from the Middle East, who produced the allotetraploid by hybridizing the introduced form with an American wild diploid was proposed by Heyerdahl (1979). It is noted that the theory of Hutchinson *et al.* (1947) is cited to reinforce most of those later arguments for a recent origin.

A very recent origin of the allotetraploids involving human transport of the A genome progenitor is not seriously considered a likely event by those familiar with the comparative cytogenetics and taxonomy of *Gossypium* species. Taxonomic diversity alone has been cited as clear evidence for an ancient origin (Mangelsdorf and Oliver, 1951). Historical evidence against a recent introduction of a *G. herbaceum* by man to the New World was cited by Phillips (1974). He points out that the earliest archeological specimens of cotton in the New World are dated at 4000–3000 BC which is 2000 years before known archeological cotton in Africa, 1000 years before agriculture in Africa, and 3000 years before man made long-distance voyages in the African area. Apparently there are differences in opinions regarding the time of oceanic voyages by man (see Heyerdahl, 1979).

Based on cytogenetic data, Phillips (1963) concluded that the allotetraploids are neither ancient nor recent in origin. Earlier studies had shown that the A and D subgenomes of the allotetraploids, primarily the former, had diverged very little during their evolutionary history from the progenitor genomes represented in the Asiatics and *G. raimondii*. It was argued that the high degree of structural similarity of the A and D subgenomes of the allotetraploids with the A and D genomes of the respective diploids is not compatible with an ancient origin. Neither did a recent origin appear tenable since the allotetraploids contain a vast amount of morphological and physiological variability, are genetically, and to some degree cytologically, differentiated from each other (disjunct species), exhibit a high amount of genetic diploidization (monofactoral segregation), and are rather widely distributed on islands of the Caribbean and the Pacific.

Thus, faced with these conflicting observations, Phillips (1963) concluded, as did Fryxell (1965a, 1979), who in addition stressed the importance of the several pairs of sibling species and the association of the allotetraploids with littoral habitats, that the allotetraploids arose, not in very ancient or very recent times, but most likely in the Pleistocene following an Atlantic transfer of the progenitor A genome species to the New World. The credibility of this theory is open to question, since considerable weight is given to the absence of cryptic chromosome rearrangements in the A and D genomes of the diploids and the allotetraploids. The theory assumes that cryptic structural changes invariably accumulate in a timewise process over geological time. Thus, their absence is therefore assumed to be indicative of a more recent origin of the tetraploids. An extensive review of the evidence for the hypothesis that cryptic structural differentiation has occurred in the chromosomes of *Gossypium* is given by Stephens (1950a). However, Stephens (1961) and Fryxell (1979) consider that the theory of cryptic structural differentiation as a major mechanism in the evolution of *Gossypium* is outdated and should be abandoned. The evidence for genome stability was pointed out in Section IV,C, and will be further emphasized in later sections.

XI. Mono- vs Polyphyletic Origin of the Allotetraploids

The five allotetraploid species *G. barbadense, G. hirsutum, G. tomentosum, G. mustelinum,* and *G. lanceolatum* have the same chromosome end arrangement (Gerstel and Sarvella, 1956; Endrizzi, 1966a; Hasenkampf and Menzel, 1980). Fryxell (1979) has reclassified *G. hirsutum* "race" palmeri as a new species, *G. lanceolatum,* based partly on the results of seed protein patterns (Johnson, 1975), but this ranking needs experimental verification. Recently, hybrids of *G. hirsutum* × *G. darwinii* were examined cytologically and they had regular pairing with 26 bivalents at metaphase I (J. E. Endrizzi, unpublished). Thus, it is highly probable that the A genomes of the six allotetraploid species had a common origin.

Cytological analyses of hybrids between the D genome species and the allotetraploids show that the D genome does not contain any fixed chromosomal structural changes like those of the A genome (Brown and Menzel, 1952). Chromosomal interchanges have been reported in some hybrid combinations, but these structural changes are believed to be sporadic and transitory, like any other mutation. They were discussed under the section on chromosome structural changes and genome stability (Section IV,C).

Based on the large body of cytological data involving the allotetraploids, most students of *Gossypium* evolution today postulate a monophyletic origin for the allotetraploids. However, for one reason or another, several individuals have proposed a polyphyletic origin (Davie, 1933, 1935; Mauer, 1938, 1954; Kammacher, 1959, 1960; Sherwin, 1970; Parks *et al.,* 1975; Johnson, 1975).

Because of the fact that several diploids existed in the New World,

Davie (1933, 1935) suggested that they probably intercrossed to produce different allotetraploids. The extreme genetic differences between the allotetraploid species, coupled with their geographical distribution, led Mauer (1938, 1954) to suggest, like Davie, that the allotetraploids had independent origins involving hybridization between different New World diploids. Such proposals, like some others, are not supported by cytogenetical data and are therefore only of historical interest.

Differences in chromosome pairing behavior in the F_1 hybrids of *G. hirsutum* and *G. barbadense* with *G. raimondii* formed the basis for Kammacher's (1959, 1960) conclusion that *G. hirsutum* is less "diploidized" and therefore of more recent origin than *G. barbadense*. The ancestor of *G. barbadense* was believed to have arisen in South America by hybridization between the A genome species and a *G. raimondii*-like genome species. Later, a second hybridization was assumed to have occurred in Central America between the A genome species and a D genome species closely related to *G. raimondii* to produce *G. hirsutum*. It is well known that pairing and chiasma formation in hybrids can show seasonal and environmental variation, as most recently demonstrated in cotton by Douglas and Brown (1971) and Menzel and Brown (1978b); therefore, it is rather doubtful that chromosome pairing data of that kind are sufficient to warrant speculation on the mono- vs polyphyletic origin of the allotetraploids.

According to Sherwin (1970), a *G. herbaceum*-like cotton was transported, either by rafting or by man, to northern South America, where it was cultivated and hybridized, possibly with *G. raimondii,* to produce *G. barbadense*. Humans then carried the diploid A genome species north to Mexico, where it hybridized with either *G. gossypioides* or *G. trilobum* to produce *G. hirsutum*. Overlooked is the fact that involvement of the D genome of *G. gossypioides* in the origin of the allotetraploids is excluded on cytological grounds alone (Table 12). Neither is the D genome of *G. trilobum* likely to be involved since it is closely related to *G. thurberi* (Hutchinson *et al.,* 1947; Fryxell, 1965c; Fryxell and Parks, 1967; Johnson and Thein, 1970; Parks *et al.,* 1975; Valiček, 1978).

Parks *et al.* (1975) found the flavonoid compounds of flowers to be strongly correlated with genome groups. The D genome species, however, could be separated into two groups on the basis of yellow vs cream flower color. The four allotetraploid species, *G. hirsutum, G. barbadense, G. tomentosum,* and *G. mustelinum,* were chromatographically very similar except for the absence of the flavonol rutinosides in the yellow-flowered species, *G. barbadense* and *G. tomentosum*. The quantities of a large number of the flavonoid pigments varied, and the patterns of *G. hirsutum* and *G. mustelinum* on one hand differed from those of *G. barbadense* and *G. tomentosum* on the other. In view of these differences and from comparison of their patterns with the patterns exhibited by the D species, it was speculated that *G. barbadense* and *G. tomentosum* could possibly have arisen from a combination of *G. klotzschianum* and an A species, while *G. hirsutum* originated from a combination of *G. raimondii* and an A species. However, they admit that there is no direct evidence for the involvement of *G. klotzschianum* since they did not succeed in synthesizing hybrids between an Asiatic species and *G. klotzschianum*. However, Hasenkampf and Menzel (1980) have pointed out that the speculation of Parks *et al.* (1975) for the involvement of *G. klotzschianum* is not supported by cytological data because the chromosomes of *G. tomentosum* rather than those of *G. mustelinum* are more similar to the chromosomes of *G. hirsutum*.

Johnson's (1975) polyphyletic hypothesis was based primarily on the electrophoretic banding patterns of seed proteins. The two A genome species, *G. herbaceum* and *G. arboreum,* had identical banding patterns, whereas the D genome species were divided into two groups. One, designated D_β, comprised *G. raimondii, G. lobatum, G. aridum, G. laxum,* and *G. gossypioides* and the other, D_ξ, consisted of *G. thurberi, G. trilobum, G. davidsonii, G. klotzschianum, G. armourianum,* and *G. harknessii* (Johnson and Thein, 1970).

The seed protein banding patterns of a number of accessions of the five allotetraploid species of *G. hirsutum, G. barbadense, G. tomentosum, G. darwinii,* and *G. mustelinum* (*G. caicoense* in Johnson) were identical except for some accessions of *G. hirsutum* "race" palmeri. Six of 44 palmeri accessions had a banding pattern of seed proteins identical to those of the other five taxa, which was similar to the banding pattern simulating a 1:1 mixture of the protein extract of *G. herbaceum* and any one of the D genome species in the D_β group. The remaining 38 palmeri accessions had identical banding patterns, which were distinctly different from those of the other group of allotetraploids. This pattern simulated a 1:1 mixture of protein extract of *G. herbaceum* and a species of the D_ξ genome group. This difference in the seed proteins in conjunction with some plant characters common to palmeri and some of the D species led Johnson to conclude that the "race" is a distinct species *G. palmeri* (*G. lanceolatum* in Fryxell, 1979), and that the allotetraploids arose from perhaps three independent hybridization events in a center of early American civilization.

According to Johnson (1975) *G. barbadense,* with the $AAD_\beta D_\beta$ genomes, originated in South America as a hybrid between *G. herbaceum* and *G. raimondii* of the D_β group. *G. palmeri* is a hybrid between *G. herbaceum* and the Mexican species *G. trilobum* of the D_ξ group. Later, *G. hirsutum* arose in Southern Mexico and Central America by hybridization of a *G. palmeri* prototype with the $AAD_\xi D_\xi$ genomes and one or more prototypes with the $AAD_\beta D_\beta$ genomes. From the Meso-American center of origin, *G. tomentosum* was carried by man or ocean currents to Hawaii.

Johnson attributed the diversity existing in *G. hirsutum* to introgression from *G. barbadense* as noted by Stephens (1967); and to account for structural similarity of the chromosomes of all allotetraploids, he assumed that the A genome parent carried the two reciprocal interchanges and that both interchanges became homozygous in the descendents of each hybridization event. It is emphasized here that any theory for a polyphyletic origin must account for all allotetraploid species having the same chromosome end arrangement, which in the ancestral A genome involved four breaks in the same position of four arms and identical exchanges of the four segments.

In a study of seed proteins by gel electrophoresis of species of *Gossypium,* not including "palmeri," Cherry *et al.* (1970) did not find the major distinctions observed by Johnson. They found some minor differences within genome groups, with the D species showing the greatest variability. Nevertheless, they concluded that each genome group exhibited its own banding patterns. Differences were also noted between *G. hirsutum, G. barbadense,* and *G. tomentosum,* but their banding patterns were more similar to a combination of *G. herbaceum* and *G. raimondii* than to any other A plus D combination.

For several reasons, it seems doubtful that differences in seed proteins, as well as flavonoid compounds, are sufficient in themselves to draw major conclusions on the origin of the allotetraploids and for the construction of phylogenetic relationships.

It is a foregone conclusion that genomes similar to those of *G. trilobum* and *G. gossypioides* were not involved in the origin of the allotetraploids.

Analyzing seed of four accessions of *G. thurberi* collected in four different mountain ranges in Southern Arizona, Cherry and Katterman (1971) reported high molecular polymorphism, which they attributed to allelic differences between the populations. Thus, it is most probable that differences in floral and seed constituents between the D genome species arose during their diversification. Likewise, similar differences could have arisen in the D subgenome in the diversification of the allotetraploids.

Protein analyses serve many useful functions in analyzing the genetics of species and populations, but it is questionable whether protein differences can serve as a reliable criterion for determining phylogenetic relationships of species within a genus. A classic example in which the protein data show very little difference between two widely separated species is that of man (*Homo sapiens*) and chimpanzee (*Pan troglodytes*) (King and Wilson, 1975). Very little difference was found in the unique DNA sequences and the proteins coded by them in these two mammalian species; the differences noted were comparable to that between sibling species. Yet it is quite obvious that man and chimpanzee are distinct species in separate genera, differing in their general anatomy and way of life. If such diverse forms as these do not show differences in their proteins, then how reliable are the protein differences between species within a single genus for determining whether allotetraploids arose from single or multiple ancestral hybrid types?

The possibility of a biphyletic origin was discussed by Phillips (1963, 1964). He points out that if *G. hirsutum* and *G. barbadense* have different D subgenomes due to a biphyletic origin, then the frequency of multivalents and the segregation for a given locus should be different in hexaploid hybrids involving these two species and the same American diploid species. However, both hexaploid hybrids have quite similar multivalent frequencies and segregation ratios (Table 12), indicating that both species stem from a single genomic allotetraploid.

XII. New Evidence for the Origin of the Allotetraploids

Solid data from taxonomy, genetics, and cellular cytogenetics appear to be exhausted in constructing a plausible hypothesis of the sequence of evolutionary events for the origin of the allotetraploids, but data on the molecular cytogenetics of *Gossypium* genomes are yet to be fully exploited. Recent research in *Gossypium* has provided information on the DNA content and on the kinetic components of chromosomal DNA and their relationships to the structural organization of chromatin that are crucial to any hypothesis on the origin and evolution of species of *Gossypium*.

A feature of evolutionary significance distinguishing the different genome groups of the diploid species is the size of their chromosomes, as discussed earlier. The most pronounced difference in size is between the small D chromosomes on one hand and the much larger chromosomes of the remaining six genomes on the other. The mean length of the somatic metaphase chromosomes of the A genome of the Asiatic cottons is between 2.4 and 2.7 μm, and the mean lengths of the A and D subgenomes of the allotetraploids are, respectively, 2.25 and 1.25 μm (Davie, 1933; Skovsted, 1934a; Edwards, 1977). This approximate two-fold difference in size between A and D genome chromosomes has been noted in meiotic as well as somatic studies by others (Mikhailova, 1938; Endrizzi and Phillips, 1960; Edwards *et al.*, 1974; Mursal and Endrizzi, 1976; Mursal, 1978). Moreover, this increase in size is distributed rather uniformly throughout the genome, i.e., there is a gradation in size from the longest to the shortest chromosomes, as can be seen in the somatic idiograms of *G. herbaceum, G. thurberi,* and *G. hirsutum* presented by Mikhailova (1938), which are reproduced in Fig. 6 and in the idiograms or karyograms of *G. arboreum, G. herbaceum, G. anomalum, G. sturtianum, G. stocksii, G. bickii,* and *G. hirsutum* (Arutjunova, 1936; Abraham, 1940b; Jacob, 1942; Wouters, 1948; Afzal, 1949; Bose, 1953–1955; Edwards, 1977, 1979; Abdullaev and Lazareva, 1975; Edwards and Mirza, 1979).

In (AD) haploids and in the A_2D_5 hybrids as many as seven and 12 bivalents, respectively, associated predominantly by a single chiasma can be found at metaphase I. The AD bivalents are asymmetrical in size and it is apparent in these bivalents that the homoeologs of corresponding rank size have paired and formed chiasmata (Endrizzi and Phillips, 1960; Mursal and Endrizzi, 1976; Mursal, 1978). Such associations can be seen also at pachytene and diakinesis (Mursal and Endrizzi, 1976).

The relationship between collective chromosome size and nuclear DNA content in *Gossypium* species has been established by two independent methods: cytospectrophotometric measurements of Feulgen-stained chromosomes (Edwards *et al.*, 1974; Edwards and Endrizzi, 1976; Kadir, 1976) and chemical measurements of total nuclear DNA (Katterman and Ergle, 1970). Since chromosomes are structurally uninemic, the differences in genome size are attributable to a longitudinal increase or decrease in the DNAs of individual chromosomes.

Since the classic studies of Britten and Kohne (1967, 1968), it is now well established from numerous reports that the eukaryotic chromosome is normally composed of four major sequence classes of DNA: (1)

VVVV vvVVVVV'
VVVVVVVVVVVV'vvvvvvVVVVVV
VVVVVVVVVVVV'

FIG. 6. Idiograms of *G. herbaceum* (top), *G. hirsutum* (middle), and *G. thurberi* (bottom) chromosomes. The chromosomes of the A genome are almost twice the size of the chromosomes of the D genome, but note the near uniformity in graduation of chromosomes between the two sets. (Reprinted from Mikhailova, 1938, by courtesy of the Editor, Journal of the Academy of Sciences, USSR.)

single copy, (2) moderately repeated (interspersed repetitive), (3) highly repeated (clustered repetitive) and (4) foldback or palindromic sequences. The fraction of the total DNA classed as single copy defines sequences presented only once in the haploid genome. This class is normally dispersed through the chromosomes and includes those sequences that code for polypeptides, enzymes, and other cellular products involved in metabolic and structural functions. The two classes characterized as repetitive are those DNA sequences that are repeated many times in relation to the single copy sequences. The moderately repetitive fraction is made up of sequences that range in average length from 300 base pairs, interspersed with single copy sequences averaging 1000 base pairs (*Xenopus* pattern), to 5600 base pairs, interspersed with single copy sequences averaging 10,000 base pairs (*Drosophila* pattern). The moderately repetitive sequences are generally repeated 10^2–10^4 times, found primarily contiguous with single copy DNA, and dispersed in a highly ordered manner in the chromosomes. Some moderately short repetitive sequences are transcribed along with the contiguous structural genes but are not translated for the most part. This repetitive class is believed to serve in coordinating gene activity during development (Davidson and Britten, 1979).

The highly repetitive fraction normally consists of short sequences of DNA repeated about 10^6 times, and it is clustered in long segments mainly in the constitutive heterochromatic regions of the chromosomes. Evidence indicates that this class of DNA is not transcribed. A small fraction of the total DNA consists of foldback sequences, which renature instantaneously and are grouped in a class referred to as zero binding. These short, inverted repeats are scattered throughout the genome and may be involved in gene regulation.

The extremely long DNA molecule and its associated proteins are packaged in a hierarchial series of coils, forming the chromatin structure normally seen in the stages of chromosome behavior during the cell cycle. The clustered or highly repetitive DNA may serve primarily for orderly packaging of the supercoiled chromatin. It is generally assumed that the chromatin structure is a modulator of gene activity. Therefore, since the repetitive DNAs are an integral part of the chromosome, they may have major effects in timing gene and chromatin responses to mechanisms regulating gene activity.

Recently, the DNAs of *G. arboreum, G. herbaceum, G. thurberi, G. raimondii,* and *G. hirsutum* have been analyzed (Walbot and Dure, 1976; Wilson *et al.,* 1976; Geever, 1980). In the studies of Walbot and Dure and of Geever, the fractional components of single copy and different classes of repetitive DNA were analyzed. Their results are briefly summarized in Table 13, and for the most part, only the data relative to the evolution of the A and D genomes and origin of the allotetraploid will be discussed.

Only the genome of *G. hirsutum* was analyzed by Walbot and Dure (Table 13). In addition to *G. hirsutum,* Geever also analyzed *G. herbaceum* var. *africanum* and *G. raimondii,* the two diploid species with genomes closest to the A and D subgenomes of the allotetraploids (Table 13).

Walbot and Dure estimated the haploid genomic size of *G. hirsutum* to be 0.795 pg, whereas Geever estimated it to be 0.950 pg. The discrepancy in values may be due to certain assumptions made in computing genome size from parameter measurements (R. F. Geever, personal communication). The analytical methods do not consider polyploidy in kinetic associations and normally give a value intermediate to that of the two putative diploids. The two studies also found different amounts of single copy and repetitive DNAs. Walbot and Dure's study may overestimate the former class and underestimate the latter.

A point of major interest in the results of Walbot and Dure is the nature of the interspersion pattern in *G. hirsutum.* They estimated that at least 80% of the genome consists of both a short and long interspersion of single copy and moderately repetitive DNA. Approximately 8% of the genome consists of single copy DNA interspersed with much longer sequences of repetitive DNA. This interspersion pattern is a longer period spacing than the *Xenopus* type. However, it is more similar to that of *Xenopus* than to the long-period interspersion pattern of the *Drosophila* type. Since at least 80% of the A and D subgenomes of *G. hirsutum* consists of interspersion of single copy and moderately repetitive DNA, it is likely that the genomes of other gen-

TABLE 13
Summary Data of Genome Organization (Haploid DNA Content)
of *Gossypium* Species[a]

Species	Total DNA (pg)	Component	DNA content (pg)	(%)
SC and rep. DNA				
G. raimondii[b]	0.680	Foldback	0.054	8
		Moderately rep.	0.204	30
		SC	0.422	62
G. herbaceum[b]	1.050	Foldback	0.053	5
var. *africanum*		Highly rep.	0.168	16
		Moderately rep.	0.441	42
		SC	0.389	37
G. hirsutum[b]	0.95	Foldback	0.038	4
		Highly rep.	0.238	25
		Moderately rep.	0.257	27
		SC	0.418	44
G. hirsutum[c]	0.795	Highly rep.	0.064	8
		Moderately	0.215	27
		SC	0.480	60
Interspersion pattern				
G. hirsutum[c]	0.795	Palindrome and highly rep.		8
		1250 NT moderately rep. element + 1800 NT SC element:		
		rep.		25
		SC		36
		1250 NT moderately rep. element + 4000 NT SC element:		
		rep.		5
		SC		16
		SC < 6000 NT from nearest moderately rep. element		8[d]

[a] NT, nucleotides; SC, single copy; rep., repetitive.
[b] From Geever (1980).
[c] From Walbot and Dure (1976).
[d] SC.

omic groups will likewise have high levels of interspersion of these two classes of DNA.

Geever (1980) reported the genome size of *G. herbaceum* to be 1.05 pg and that of *G. raimondii* to be 0.68 pg, the former being about 1.5 times greater than the latter. In comparing the amounts of single copy and repetitive DNA between the two diploid genomes, both genomes have basically the same amount of single copy DNA, 0.42 and 0.39 pg, but differ greatly in their quantities of repetitive DNA. *G. herbaceum* has 0.44 pg and *G. raimondii* has 0.20 pg of moderately repetitive DNA. In addition, the *G. herbaceum* genome has a significant fraction of highly repetitive DNA, which probably exists also in the genome of *G. raimondii* but at a low level that is kinetically undetectable by the analytical methods used.

These data show that the two genomes have essentially the same quantity of single copy DNA, but that the genome of *G. herbaceum* has twice the amount of moderately repetitive DNA and a greater amount of highly repetitive DNA than the genome of *G. raimondii*. Similarly, Wilson *et al.* (1976) recorded greater quantities of repetitive DNA in the A genome of *G. arboreum* than in the genome of *G. thurberi*.

Thus, it is obvious that the only quantitative difference between the chromosomal DNAs of the A and D diploid genomes is in their repetitive sequences, and that there has been a fairly large shift in this class of DNA in the differentiation of these two genomes. Furthermore, it can be deduced that this substantial change in repetitive DNAs occurred rather uniformly throughout the chromosome complement of the two species (Fig. 6) and in all other genomic groups as well, since the chromosomes of each show a general graduation in size from the longest to the shortest (Arutjunova, 1936; Abraham, 1940b; Jacob, 1942; Wouters, 1948; Afzal, 1949; Bose, 1953–1955; Edwards, 1977, 1979; Edwards and Mirza, 1979). It is highly unlikely that such a uniform increase or decrease in size affecting all 13 chromosomes of each of the seven genome groups could come about by the gradual accumulation of random cryptic structural changes involving inversion, translocations, and deletions or additions occurring over a long evolutionary time scale. It seems more likely that such changes were brought about by saltatory fluctuations in the repetitive DNAs.

Most of the descriptive evidence on the evolution of quantitative and qualitative changes in the genome size of species in eukaryotes shows that saltatory fluctuations have occurred in the repetitive DNAs (Dover and Flavell, 1982; Nagl *et al.*, 1983). Prime examples are demonstrated in the diploid species of *Lathyrus* and the Gramineae where the total nuclear DNA varies from 12 to 22 pg in the former and from 3.6 to 8.8 pg in the latter. These differences in DNA content are due almost wholly to an increase in the repetitive fractions, the single copy fraction showing only a small change among the species (Narayan and Rees, 1976, 1977; Narayan, 1982; Flavell, 1982). The genomes of *G. herbaceum* and *G. raimondii* parallel the observations in *Lathyrus* and the Gramineae as well as of species in other genera. All other genomes of *Gossypium* show a uniform change in chromosome size; thus it is reasonable to assume that all genomes of *Gossypium* have similar quantities of single copy DNA but vary in their quantities and qualities of repetitive DNAs depending on genomic size.

Fryxell (1971) has shown that the species showing the least evolutionary advancement occur in the D genome group, indicating that the D genomes are equivalent to the ancestral progenitor genome of the genus in their DNA content and organization. It follows then that the six groups of genomes with larger chromosomes evolved from a D genome-like progenitor or a genome derived from the D-like genome by saltatory amplification of the repeated sequences.

In a series of publications, Lima de Faria (summarized 1980) showed that the eukaryotic chromosome has an ordered organization in which the chromosome regions have a molecular interdependence as opposed to a mechanical and random relationship between DNA segments. Only those rearrangements that do not alter appreciably the established relationship of the molecular organization within the chromosome field have the opportunity to become fixed during evolution.

Amplification of repetitive DNA would differentiate cytologically the chromosomes between the genome groups and alter the regulatory and communicative functions of specific DNA sequences as well as synapsis or chromosome affinity. The specific DNA sequences consisting of the structural genes and related transcriptional units would still be maintained in the same linear order and in their optimal DNA territory in the genomes that differ in size. However, the activity of identical genes of two genomes such as A and D, for example, would likely differ in space and time, regulated in part by differential rates of condensation and decondensation of chromatin, which is dictated by the different quantities and types of repetitive DNA associated with each functional unit. Or more simply, because chromosome structure has a functional role in the regulation of genes, the reaction norms of the genetic systems of the genomes of different groups would differ.

A macroevolutionary change at the genome level in *Gossypium* as discussed above readily explains the differential chromosome affinity between genomes, thereby eliminating the necessity for accounting for the assumed inconsistencies in genome differentiation in the evolutionary history of the A and D genomes of the diploid and tetraploid species based on the theory of cryptic structural hybridity. In addition to the large-scale genomic changes in repetitive DNA's noted here in *Gossypium,* studies of frequencies in chromosome pairing in wheat and related genera indicate that divergence in their genomes is not structural but due to differentiation in the sequence of nucleotides (Dvořák, 1981; Dvořák and McGuire, 1981). Such differences may also account for intragenomic differentiation in *Gossypium.*

In the past attempts in constructing a hypothesis for the origin of the allotetraploids, it has been assumed that the following sequence of evolutionary events must be considered if one accepts the modern synthesis' concept of speciation (Stephens, 1947): (1) long geographical isolation of the American and Asiatic groups, during which their genomes became cytologically differentiated; (2) geographical reunion of the two, followed by interspecific hybridization and polyploidy; and (3) geographical reisolation of the two parental species and genetic differentiation of the allotetraploids. The first two are "double-event" rather than "single-event" processes (Stephens, 1947). However, according to our hypotheses, the origin of the allotetraploids could possibly be nearly like a "single-event" process.

Assuming that the rare event of saltatory amplification of repetitive DNA occurred throughout the chromosomes of a D genome-like progenitor and became fixed (homozygous) forming an A genome, such a plant form would have an immediate built-in isolating mechanism. In the newly formed species the "chromosome field" would be maintained even though the two genomes would differ in their gross structural organization and in gene and chromosome regulation. The only adjustment of any significance that may have been necessary in the newly evolved form would be the development of a more compatible nuclear–cytoplasmic interacting system.

If such an event did occur, the two species would satisfy the three primary requirements for the origin of an allotetraploid: propinquity, cross-compatibility, and chromosome differentiation (Hutchinson *et al.,* 1945). Thus, within a relatively short geological time period a progenitor and the hybrid allotetraploid species could become established as independent, true-breeding taxa. The only factors limiting the time of hybridization and polyploidy would be genetic stabilization of the A genome species, population size, amount of cross-pollination, and extent of pollen competition in mixed pollinations.

No fossil records of *Gossypium* are known (Fryxell, 1979), but it is presumed that the genus is very old and that the distribution of the two major genomic chromosome size groups, large chromosomes in the Old World and small chromosomes in the New World, was due to continental drift that occurred during the Cretaceous period. Thus, according to the present hypothesis, the allotetraploids are ancient in origin, occurring perhaps during Cretaceous times in the westward land amassment.

In conclusion, the above proposal for the origin of the amphidiploid species can be summarized as follows: (1) the origin of an A genome species occurred by macroevolution from a progenitor D genome; (2) hybridization and polyploidy occurred within a short geological time period following the formation of the A genome parent; (3) the origin of the amphidiploid probably occurred during the Cretaceous period; (4) diploidization of the gene regulatory system of the hybrids led to the formation of at least five distinct amphidiploid species; and finally (5) the generation of genetic diversity that exists primarily in *G. barbadense* and *G. hirsutum*. The last two events are discussed in the next section.

Available records show that the major initial evolutionary diversification of the diocotyledonous angiosperms probably occurred during the Lower Cretaceous. From mid-Cretaceous period onward, they underwent a rapid rate of divergence and expansion (Takhtajan, 1969; Beck, 1976). By the beginning of the Cretaceous period, the North Atlantic Ocean had begun to form, separating the North American land mass from the conjoined land mass comprising South America and Africa. If the distribution of the species of *Gossypium* is due to continental drift, the genus must therefore have arisen early in angiosperm evolution and diversification (Hawkes and Smith, 1965), which is not untenable, since the Malvaceae is among the most primitive families of dicots and is assumed to have emerged rather early in angiosperm evolution (Sporne, 1954, 1976). Also, Stebbins (1947) contends that the New World allotetraploids of *Gossypium* are among the oldest allopolyploids.

XIII. Diploidization of the Allotetraploids

Cytological and genetic data showed that the A genomes of *G. herbaceum, G. arboreum,* and the natural allotetraploids have remained highly homologous, whereas the D genomes of *G. raimondii* and the natural allotetraploids have undergone some divergence since the origin of the allotetraploids (Phillips, 1963) (Table 12). Phillips suggests that this may be due to the direction of the cross, i.e., the presence of the D genome in the cytoplasm of the A species of the ancestral allotetraploid, and "during the 'shakedown' of the raw amphidiploid the D genome might have been genetically and chromosomally more unstable than the A genome, leading to more rapid genetical and cytological diploidization of the D genome." It is interesting to note that cytoplasmic male sterility is highly stable in cotton lines with a *G. hirsutum* nuclear genome in a D species cytoplasm, whereas those in which the *G. hirsutum* nuclear genome is in the cytoplasm of an Asiatic species (A genome) are unstable and exhibit various levels of pollen fertility depending on climatic conditions (Meyer, 1973, 1975). Since the genomes have remained quite stable during their evolutionary histories, the cytoplasmic male-sterile results indicate that the A genome progenitor was indeed the female parent in formation of the allotetraploids. The same cross direction was proposed by Hutchinson *et al.* (1947) but for different reasons (see also Meyer, 1971).

During the "shakedown" in the genetic stabilization of the raw amphidiploid, it is assumed that diploidization or loss of duplicate gene expression has occurred at many loci (Harland, 1932, 1934, 1937b, 1939; Silow, 1946; Hutchinson, 1946b; Stephens, 1951a,b; Rhyne, 1955, 1962b; Giles, 1962), which is supported by the fact that a majority of the mutants of *G. hirsutum* and *G. barbadense* show monofactorial inheritance (Table 3). Diploidization of different loci of a duplicate set in two allotetraploid species was first demonstrated by Stroman and Mahoney (1925) for genes controlling chloroplast development, which were studied more extensively by Harland (1932, 1934, 1937b). For example, some varieties of *G. barbadense* have the unimeric genotype of $Chl_1Chl_1 \ chl_2 \ chl_2$ whereas some varieties of *G. hirsutum* have the other unimeric genotype, $chl_1chl_1 \ Chl_2Chl_2$. It is well known that genetic breakdown occurs in the F_2 and later generations of interspecific hybrids (Harland, 1933, 1936b, 1941; Hutchinson *et al.*, 1947; Stephens, 1949b). The interspecific F_1 hybrids are vigorous and highly fertile; however, the segregants show considerable depression in vigor consisting of morphologically unbalanced, depauperate, and unproductive types. The types that became established from inbreeding are indistinguishable from the parent species. Such behavior does not occur in intraspecific crosses. It is evident in the genetic breakdown of the interspecific hybrids, that an important genetic element in the A and D subgenomes that also became diploidized was the dimeric set of regulatory genes. If mutations leading to diploidization are random, it is conceivable that several different combinations of unimeric regulatory systems could evolve from the dimeric combination of the original allotetraploid. Thus, the F_2 breakdown of interspecific hybrids indicates that each of the allotetraploids has evolved a different diploidized system of gene regulation (Endrizzi, 1966a). Variation within an allotetraploid species could be due primarily to diploidization and allelic differences of the structural genes.

There are at least four mechanisms by which diploidization of a dimeric set of genes may occur: (1) suppression or inactivation, (2) mutation to a recessive amorphic or hypomorphic allelomorph (Stephens, 1951a), (3) an intermediate stage of divergence in function (Silow, 1946; Stephens, 1951a), and (4) dosage and interaction (Dalton, 1965). The last two involve a system in which the normal allele of one of the duplicate loci cannot mask the mutant allele at the other duplicate locus, as shown in genetic studies with the cluster (cl_1) and short brance (cl_2) loci. It should be emphasized that diploidization can occur at either one of at least two levels of gene expression: within the structural gene itself and within the DNA regions regulating the activity of the gene and its products.

XIV. Suppression of Homoeologous Pairing in the Allotetraploids

Avivi and Feldman (1980) reviewed the evidence for the orderly arrangement of chromosomes in the nuclei of plants. Ashley (1979) reported that the three pairs of chromosomes of *Ornithogalum virens* maintain a specific sequential arrangement in the nucleus. At one side of the nucleus, specific centromeric ends are associated, and at the other side, specific telemeric ends are associated. It was demonstrated by Bennett (1982) that the individual chromosomes of the haploid genome in mitotic cells of barley and rye are highly ordered according to the size of their arms. In the allohexaploid wheat, Avivi *et al.* (1982a,b) determined that the three genomes tend to occupy different positions in the nucleus and that the homologous chromosomes of a genome are closer to one another than the homoeologs. Thus, it is apparent that during the cell cycle, the chromosomes, including the homoeologs, maintain a definite spatial arrangement with respect to one another that is controlled by the centromeres, telomeres, and the size of the arms. The cellular events in somatic and meiotic cells regulating these relationships are genetically controlled.

Avivi and Feldman (1980) maintain that plants with only homologous sets of chromosomes as in diploids and autoploids have a convergent spindle, which brings all centromeres to a single polar region. In allohexaploid wheat, a gene (*Ph*) induces a divergent spindle, which results in the movement of homologous sets to separate polar regions, thereby preventing multivalent formation. It was contended that all polyploid plants exhibiting regular bivalent pairing have a gene(s) similar to wheat for a divergent spindle.

It has been argued that the allotetraploid *Gossypium* species do not require a gene analogous to that in wheat for regulating bivalent pairing (Endrizzi, 1962; Gerstel, 1963; Mursal and Endrizzi, 1976).

In *Gossypium* the diploid A × D hybrids have a mean of about six bivalents (13 possible) (Table 6) whereas the raw amphidiploid 2(A × D) hybrid has a mean of about 25 bivalents (26 possible) at metaphase I (Endrizzi, 1962). A similar behavior in chromosome pairing occurs in the diploid and spontaneous amphidiploid F_1 hybrids of *G. davidsonii* (D genome) × *G. anomalum* (B genome) (Brown, 1951). Thus, upon doubling the two sets of differentiated chromosomes, pairing is immediately limited almost entirely to homoeologs. Pairing and segregation data in hexaploid hybrids have shown that the A and D genomes of the diploids and those of the allotetraploids have remained highly homologous during their evolutionary histories. Thus, it is reasonable to assume that the diploidlike pairing we see in the raw 2(AD) hybrid existed when the ancestral allotetraploid first formed.

Unlike haploids in wheat, haploids of *G. hirsutum* show a high frequency of homoeologous pairing at pachytene. The cotton haploids have a mean frequency of 10 bivalents (13 possible) at pachytene, but only a mean frequency of 0.80 bivalents remain at metaphase I. If a *Ph* type of gene existed in *G. hirsutum,* pairing of homoeologs during prophase should have been prevented or at least occurred at a very low frequency, unless the gene is hemizygous-inefficient, which is not the case in wheat. The immediate expression of almost complete homologous pairing in the raw amphidiploid hybrid and the lack of prevention of synapsis of homoeologous chromosomes during prophase in haploids suggest that a gene functioning like that in wheat for regulating bivalent pairing did not evolve in the allotetraploid species. Pairing of chromosomes in the two cotton plant types above suggest that the major evolutionary changes that occurred in the ancestral allotetraploids were the genetic enhancement of those processes preventing the proper alignment of homoeologous chromosomes and/or subsequent events necessary for chiasma formation.

Since presynaptic condensation of the chromosomes excludes gene-for-gene matching, synapsis must reside in specific regions of the chromosomes. There is good evidence suggesting that each chromosome has synaptic sites that have been identified and labeled as zygotene DNA or "Z-DNA" regions (Stern and Hotta, 1973). These regions are constant features of each chromosome, localized in specific regions, and independent of sites of genetic exchange. One of the primary functions of the Z-DNA regions is in facilitating strand matching, or synapsis. Synapsis of chromosomes occurs at zygotene, when they are already highly condensed, and it is estimated that less than 1% of the total DNA of a chromosome is located in the peripheral regions of the chromosomes undergoing synapsis (Stern and Hotta, 1973). For proper alignment and synapsis to occur in these regions, allowing for proper formation of a synaptonemal complex, these synaptic sites must occupy a fixed spatial relationship with each other in the two homologs, which is obtained in synapsis of two normal homologs. However, in the case of synapsis between two homoeologs unequal in size, such as A and D chromosomes, it is unlikely that the synaptic sites would occupy direct opposing positions between the homoeologs. Only rarely would the two sites in A and D homoeologs come in contact, thereby increasing the probability of strand exchange in the recombinational sites. This relation applies to any intergenomic combinations that differ in chromosome size (Table 6). Therefore, it is not unreasonable to assume that the amplification of repetitive DNA not only shifted the synaptic sites between the two homoeologs, but may have also shifted the genomic chromosomal arrangement in the interphase nucleus of the amphidiploid. Evidence lending support to this concept exists in the classic study by Bennett (1982) of the disposition of chromosomes at mitotic metaphase in barley and rye.

Bennett (1982) proposes that there are rules governing the spatial ordering of chromosomes in the nucleus. In diploids, the basic unit is the haploid genome, within which the individual chromosomes are arranged in a highly ordered sequence according to the size of their arms. Major structural changes that disrupt the orderly spatial sequence of the chromosomes and their arms will be at a selective disadvantage. Thus, for any one species or genome, there will be a strong tendency to conserve the shapes of the individual chromosomes and, therefore, the haploid karyotype in evolution. In the event of an increase or decrease in the C value of the genome, selection will occur for a balanced increase or decrease in the DNA content of each chromosome and chromosome arm so that the altered genome, though differing in C value, will retain the basic structural pattern of the initial karyotype. The genomes of *Gossypium* manifest this kind of DNA change in their chromosomes.

It is suggested that the amplification of repetitive DNA in one genomic set, much of which is located in the centromeric region, also enabled each chromosome set in the amphidiploid to maintain or form essentially separate and independent positions in the mitotic spindle apparatus and concomitantly to initiate a divergent spindle, which would align the centromeres of each set at separate polar regions. If there is a highly specific affinity between a centromere and the telomeres and given polar sites, then it is conceivable that a change in the highly repetitive DNA (heterochromatic regions that are localized primarily around the centromeres and perhaps the telomeres) could change the affinity between the centromere and the telomeres and their initial polar site. Conjoint, and perhaps more important, is the major role played by the arm lengths in determining the most balanced spatial arrangement of chromosomes. Since the chromosomes of the A subgenome are approximately twice the size of the D subgenome chromosomes, and since the increase in size is essentially uniform throughout the A subgenome chromosomes, the most balanced arm length associations would naturally occur within each of the genomic groups. These biological mechanisms could establish different assemblages of the A and D sets of chromosomes in the mitotic apparatus, thus initiating a divergent spindle and suppression of homoeologous pairing.

Detailed cytological studies comparable to those performed on wheat, barley, and rye on chromosome arrangements at interphase and metaphase are needed in *Gossypium* diploid species, the natural allotetraploid and its haploids, and the A × D and 2 (A × D) species hybrids. It may be possible to test for the presence of a *Ph*-like gene in cotton with use of the translocations, since they may provide a means for obtaining homozygous deficiencies for segments of the chromosomes.

XV. Concluding Remarks

Cotton is cultivated essentially worldwide in the tropical and subtropical regions and contributes significantly to the economy of many countries. For that reason, a considerable amount of labor and funds are normally directed to many kinds of applied research projects for the economic improvement of the crop. Only a small number of research projects at a few institutions have been devoted specifically to the accumulation of data on basic genetics and cytology of the genus. Nevertheless, it is apparent in the present review that considerable progress has been made over the past several decades in developing a body of knowledge on the taxonomy, cytology, genetics, and cytogenetics of the species of *Gossypium*.

It is an accepted fact that an increase in basic information increases the opportunity for making worthwhile contributions for improving the cultivated cottons. Some of the contributions of basic genetics to commercial production were pointed out in the review. Wider application can be anticipated for the future as genetic information increases.

It would be desirable to reinitiate studies on the qualitative genetics of the A genome species so as to develop as extensively as possible the 13 linkage groups of the two species. This would not only provide information of immense value for determining the similarities of the diploid A genomes and the A subgenome of the allotetraploids, but it would also provide a more defined genetic source for eventual transfer to the cultivated cotton. It remains to be determined, but we believe as discussed in this review that uniformity in the linear sequence of genes will be found between the A genomes of the diploids and the allotetraploids.

A vast majority of mutants of the cultivated allotetraploids display monofactorial inheritance, establishing that the varieties have a unimeric genotype. A number of studies have shown that plant forms exist that have different unimeric genotypes, i.e., $D_1D_1d_2d_2$ and $d_1d_1D_2D_2$ for example. A research program for screening for such genotypes, primarily within the race stocks (Anonymous, 1974), would probably double the number of segregating loci for genetic studies.

With the advent of the new technology of genetic engineering and its potential for improving the commercial cottons by inter- and intragenomic transfer of desirable genetic segments, the basic genetic analyses should have even greater application in the future. The successful application of genetic engineering is greatly enhanced by the availability of fundamental knowledge of the genetic organization of the chromosomes gained through the classical genetic and cytogenetic approaches. Thus, to utilize the full potential of the new technology, it is of utmost importance that the classical approaches to the genetic analysis of the chromosomes of cotton be augmented.

The extent of success of genetic engineering largely depends also on the knowledge of the molecular structure of the chromosomes. Therefore, more detailed studies on the complexity, diversity, and amplification of the DNA sequences within and between species would be widely applicable to a number of problems. Such data will provide a better understanding of the mechanisms that initiated divergence of the genomes of the genus. Knowledge of the molecular structure of the chromosomes, even though it is related to the degree of chromosome synapsis and exchange, will be far more definitive in determining the feasibility of intergenomic transfer of genetic segments.

It is yet to be determined whether transposable elements exist in *Gossypium* genomes. If found, they would be useful in mediating DNA rearrangements such as the duplication, mutation, and diversity of genetic loci. They would also provide a mechanism for the intergenomic transfer of genes.

Unlike many other genera of plants, the genus *Gossypium* is very favorable for obtaining basic information on a wide range of problems on the origin and evolution of species.

The genus is considered to be rather ancient in origin and is composed of 35 diploid and six allotetraploid species. The diploid species are divided into seven genome groups in which there is at least a twofold difference in C-value of the smallest and largest genomes. Considerable information already exists on the taxonomic and genomic affinities of the diploid and allotetraploid species. Over 100 loci have been identified and new ones will be identified in the future. Homoeologous linkage groups are being constructed for the A and D subgenomes of *G. hirsutum* and *G. barbadense*. Monosomes, telosomes, and translocation stocks have been developed that, *in toto,* identify the 26 chromosomes of the allotetraploids. The former two types are being used to locate genes on specific chromosomes and the telosomes permit the positioning of genes in relation to the centromere in an arm of the chromosome. Studies have been conducted on the gross molecular organization of the genomes of two diploids and one allotetraploid. This kind of research provides an entirely different, and most promising, approach to gaining in-depth information on genome organization that can answer a number of questions on the tetraploids' origin and evolution, as well as have practical application.

With the continual accumulation of data from the classical genetic and cytogenetic studies and with the recent initiation of research on molecular cytogenetics, the future looks very promising for the accumulation of information that will be useful to both those who are interested in the genetics, cytology, and evolution of the genus and those involved in practical plant breeding.

ACKNOWLEDGMENTS

We are grateful to Drs. M. Y. Menzel, D. T. Ray, L. S. Stith, and F. D. Wilson for reading the manuscript and for helpful suggestions. We thank Grant and Debbie Ramsay for preparation of the figures. Part of this work was supported by Regional Research Project S-77. This is publication No. 476 of the Arizona Agricultural Experiment Station.

REFERENCES

Abdullaev, A. A., and Lazareva, O. N. (1975). Chromosome morphology of two species of cotton. *Dokl. Akad. Nauk Uzb. USSR* **6,** 48–50.

Abraham, P. (1940a). Cytological studies in *Gossypium*. I. Chromosome behavior in the interspecific hybrid *G. arboreum* × *G. stocksii*. *Indian J. Agric. Sci.* **10,** 285–298.

Abraham, P. (1940b). Morphology of the somatic chromosomes of three Asiatic cottons. *Indian J. Agric. Sci.* **10,** 299–302.

Afzal, M. (1949). "Growth and Development of the Cotton Plant and its Improvement in the Punjab." Govt. Printer West Punjab, Lahore.

Afzal, M., and Hutchinson, J. B. (1933). The inheritance of "lintless" in Asiatic cottons. *Indian J. Agric. Sci.* **3,** 1124–1132.

Afzal, M., and Singh, S. (1939). The genetics of a petaloid mutant in cotton. *Indian J. Agric. Sci.* **9,** 787–790.

Allison, D. C., and Fisher, W. D. (1964). A dominant gene for male-sterility in Upland cotton. *Crop Sci.* **4,** 548–549.

Andries, J. A., Jones, J. E., Sloane, L. W., and Marshall, J. G. (1969). Effects of Okra leaf shape on boll rot, yield, and other important characters of Upland cotton, *Gossypium hirsutum* L. *Crop Sci* **9,** 705–710.

Andries, J. A., Jones, J. E., Sloane, L. S., and Marshall, J. G. (1970). Effects of Super Okra leaf shape on boll rot, yield, and other characters of Upland cotton, *Gossypium hirsutum* L. *Crop Sci* **10,** 403–407.

Anonymous (1957). Report of the international committee on genetic symbols and nomenclature. *Union Int. Sci. Biol., Ser. B* No. 30, pp. 1–6.

Anonymous (1974). The regional collection of *Gossypium* germplasm. *U.S., Agric. Res. Serv., Hyattsville, Md. [Rep] ARS-H* **ARS-H-2,** 1–105.

Anonymous (1981). Preservation and utilization of germplasm in cotton—1968–1980. *South. Coop. Res. Bull.* **256.**

Arutjunova, L. G. (1936). An investigation of chromosome morphology in the genus *Gossypium*. *C. R. (Dokl.) Acad. Sci. URSS.* **3,** 37–40.

Ashley, T. (1979). Specific end-to-end attachment of chromosomes in *Ornithogolum virens. J. Cell Sci.* **38,** 357–367.

Avivi, L., and Feldman, M. (1980). Arrangement of chromosomes in the interphase nucleus of plants. *Hum. Genet.* **55,** 281–295.

Avivi, L., Feldman, M., and Brown, M. (1982a). An ordered arrangement of chromosomes in the somatic nucleus of common wheat, *Triticum aestivum* L. I. Spatial relationships between chromosomes of the same genome. *Chromosoma* **86,** 1–16.

Avivi, L., Feldman, M., and Brown, M. (1982b). An ordered arrangement of chromosomes in the somatic nucleus of common wheat, *Triticum aestivum* L.II. Spatial relationships between chromosomes of different genomes. *Chromosoma* **86,** 17–26.

Bahavandoss, M., and Veluswamy, P. (1969). An interspecific hybrid between *Gossypium arboreum* and *G. raimondii*. *Indian J. Genet. Plant Breed.* **29**, 262–267.

Bahavandoss, M., Jayaraman, N., Chinnadurai, K., and Narayanan, S. S. (1973). A study of the F₁ hybrid between *Gossypium arboreum* L. and *G. davidsonii*. Kellogg. *Madras Agric. J.* **60**, 1581–1586.

Balasubrahmanyan, R. (1947). The inheritance of two chlorophyll deficients in Asiatic cottons. *Conf. Cotton Grow. Probl. 3rd, India, I.C.C.C., 1946*, pp. 97–99.

Balasubrahmanyan, R. (1950). Inheritance of "meristic variant" in cottons. *Indian J. Genet. Plant Breed.* **10**, 62–66.

Balasubrahmanyan, R. (1951). A mutant in Asiatic cotton. *Curr. Sci.* **20**, 73.

Balasubrahmanyan, R., and Santhanam, V. (1950a). The inheritance of "dwarf" mutants in *G. arboreum* race *indicum*. *Indian J. Genet. Plant Breed.* **10**, 56–61.

Balasubrahmanyan, R., and Santhanam, V. (1950b). Inheritance of sparse lint mutant in Cocanadas cotton. *Curr. Sci.* **19**, 60–61.

Balasubrahmanyan, R., and Santhanam, V. (1951a). Inheritance of "pistillate" in cotton. *Curr. Sci.* **20**, 17.

Balasubrahmanyan, R., and Santhanam, V. (1951b). Inheritance of "crinkled leaf"—a new abnormal mutant in Asiatic cotton. *Curr. Sci.* **20**, 46.

Balasubrahmanyan, R., and Santhanam, V. (1952). Inheritance of short lint mutant in Cocanadas cotton. *Curr. Sci.* **21**, 16–17.

Balasubrahmanyan, R., Santhanam, V., and Mayandi Pillai, S. (1949). Inheritance of three new characters in "Cocanadas" cotton. *Conf. Cotton Grow. Probl., 4th, India, I.C.C.C., 1949*, pp. 1–11; also *Indian Cott. Grow. Rev.* **4**, 154–166 (1950).

Balasubrahmanyan, R., Mudaliar, V. R., and Santhanam, V. (1950). Inheritance of lint colour in Cocanadas cotton. *Indian J. Genet. Plant Breed.* **10**, 67–71.

Balls, W. L. (1906). Studies of Egyptian cotton. *Yearb. Khediv. Agric. Soc., 1906* pp. 29–89.

Banerji, I. (1929). The chromosome numbers of Indian cottons. *Ann. Bot. (London)* **43**, 603–607.

Baranov, P. (1930). Work of the cytoanatomical laboratory of N.I.Kh.I. during the growing period of 1930. *Bull. Sci. Res. Inst. Cotton Cult., Tashkent* **5**, 7–17 (in Russian), *Plant Breed. Abstr.* **2**, 197–198 (1932).

Barrow, J. R., and Davis, D. D. (1974). Gl₂ˢ—a new allele for pigment glands in cotton. *Crop Sci.* **14**, 325–326.

Barrow, J. R., and Dunford, M. P. (1974). Somatic crossing over as a cause of chromosome multivalents in cotton. *J. Hered.* **65**, 3–7.

Beal, J. M. (1928). A study of the heterotypic prophases in the microsporogenesis of cotton. *Cellule* **38**, 247–268.

Beasley, J. O. (1940a). The production of polyploids in *Gossypium*. *J. Hered.* **31**, 39–48.

Beasley, J. O. (1940b). The origin of American tetraploid *Gossypium* species. *Am. Nat.* **74**, 285–286.

Beasley, J. O. (1942). Meiotic chromosome behavior in species, species hybrids, haploids and induced polyploids of *Gossypium*. *Genetics* **27**, 25–54.

Beasley, J. O., and Brown, M. S. (1942). Asynaptic *Gossypium* plants and their polyploids. *J. Agric. Res. (Washington, D.C.)* **65**, 421–427.

Beasley, J. O., and Brown, M. S. (1943). The production of plants having an extra pair of chromosomes from species hybrids of cotton. *Rec. Genet. Soc. Am.* **12**, 43.

Beck, C. B., ed. (1976). "Origin and Early Evolution of Angiosperms." Columbia Univ. Press, New York.

Bennett, M. D. (1982). Nucleotypic basis of the spatial ordering of chromosomes in eukaryotes and the implications of the order for genomic evolution and phenotypic variation. *Syst. Assoc. Spec. Vol.* **20**, 239–261.

Bose, S. (1953–1955). VIII. Cytogenetical investigations in cotton, with special reference to chromosome morphology of five types belonging to *Gossypium arboreum* L. *Trans. Bose Res. Inst.* **19**, 67–71.

Bowman, D. T., and Weaver, J. B., Jr. (1979). Analyses of a dominant male-sterile character in Upland cotton. II. Genetic studies. *Crop Sci.* **19**, 628–630.

Britten, R. J., and Kohne, D. E. (1967). Nucleotide sequence repetition in DNA. *Year Book—Carnegie Inst. Washington* **65**, 78–106.

Britten, R. J., and Kohne, D. E. (1968). Repeated sequences in DNA. *Science* **161**, 529–540.

Brown, H. B., and Cotton, J. R. (1937). "Round leaf" cotton. Notes on the appearance and behavior of a peculiar new strain. *J. Hered.* **28**, 45–48.

Brown, M. S. (1950). Cotton from Bikini. *J. Hered.* **41**, 115–121.

Brown, M. S. (1951). The spontaneous occurrence of amphiploidy in species hybrids of *Gossypium*. *Evolution* **5**, 25–41.

Brown, M. S. (1958). The division of univalent chromosomes in *Gossypium*. *Am. J. Bot.* **45**, 24–32.

Brown, M. S. (1966). Attributes of intra- and interspecific aneuploidy in *Gossypium*. *In* "Chromosome Manipulation and Plant Genetics" (R. Riley and K. R. Lewis, eds.), pp. 98–112. Oliver & Boyd, Edinburgh.

Brown, M. S. (1967). The identification of A and D genome chromosomes in translocations by means of species crosses. *Tex. J. Sci.* **19**, 423.

Brown, M. S. (1980). Identification of the chromosomes of *Gossypium hirsutum* L. by means of translocations. *J. Hered.* **71**, 266–274.

Brown, M. S., and Endrizzi, J. E. (1964). The origin, fertility and transmission of monosomics in *Gossypium*. *Am. J. Bot.* **51**, 108–115.

Brown, M. S., and Menzel, M. Y. (1950). New trispecies hybrids in cotton. *J. Hered.* **41**, 291–295.

Brown, M. S., and Menzel, M. Y. (1952). The cytology and crossing behavior of *Gossypium gossypioides*. *Bull. Torrey Bot. Club* **79**, 110–125.

Brown, M. S., Menzel, M. Y., Hasenkampf, C. A., and Naqi, S. (1981). Chromosome configurations and orientations in 58 heterozygous translocations in *Gossypium hirsutum*. *J. Hered.* **72**, 161–168.

Carr, G. D. (1980). Experimental evidence for saltational chromosome evolution in *Calycadenia pauciflora* Gray (Asteraceae). *Heredity* **45**, 107–112.

Carver, W. A. (1929). The inheritance of certain seed, leaf, and flower characters in *Gossypium hirsutum* and their genetic interrelations. *J. Am. Soc. Agron.* **21**, 467–480.

Cherry, J. P., and Katterman, F. R. H. (1971). Nonspecific esterase isozyme polymorphism in natural populations of *Gossypium thurberi*. *Phytochemistry* **10**, 141–145.

Cherry, J. P., Katterman, F. R. H., and Endrizzi, J. E. (1970). Comparative studies of seed proteins of species of *Gossypium* by gel electrophoresis. *Evolution* **24**, 431–447.

Dalton, L. G. (1965). An analysis of the duplicate fruiting branch loci, cluster and short branch, in *Gossypium*. Ph.D. Dissertation, North Carolina State University, Raleigh.

daSilva, F. P., Endrizzi, J. E., and Stith, L. S. (1981). Genetic study of restoration of pollen fertility of cytoplasmic male-sterile cotton. *Rev. Bras. Genet.* **4,** 411–426.

Davidson, E. H., and Britten, R. J. (1979). Regulation of gene expression: Possible role of repetitive sequences. *Science* 204, 1052–1059.

Davie, J. H. (1933). Cytological studies in the Malvaceae and certain related families. *J. Genet.* **28,** 33–67.

Davie, J. H. (1935). Chromosome studies in the Malvaceae and certain related families. II. *Genetica* **17,** 487–498.

Denham, H. J. (1924a). The cytology of the cotton plant. I. Microspore formation in Sea Island cotton. *Ann. Bot. (London)* **38,** 407–432.

Denham, H. J. (1924b). The cytology of the cotton plant. II. Chromosome numbers of Old and New World cottons. *Ann. Bot. (London)* **38,** 433–438.

Dilday, R. H., and Waddle, B. A. (1974). Inheritance and linkage relationship of strap leaf mutants in cotton. *Crop Sci.* **14,** 387–390.

Dilday, R. H., Kohel, R. J., and Richmond, T. R. (1975). Genetic analysis of leaf differentiation mutants in Upland cotton. *Crop Sci.* **15,** 393–397.

Douglas, C. R. (1968). Abortive pollen: A phenotypic marker of monosomics in Upland cotton, *Gossypium hirsutum. Can. J. Genet. Cytol.* **10,** 913–915.

Douglas, C. R. (1972). Relationship of seed weight to cytotype of monosomic progeny in cotton. *Crop Sci.* **12,** 530–531.

Douglas, C. R., and Brown, M. S. (1971). A study of triploid and $3X-1$ aneuploid plants in the genus *Gossypium. Am. J. Bot.* **58,** 65–71.

Dover, G. A., and Flavell, R. B., eds. (1982). "Genome Evolution," Syst. Assoc. Spec. Vol. No. 20. Academic Press, New York.

Duncan, D. N., and Pate, J. B. (1967). Inheritance and uses of golden crown virescent in cotton. *J. Hered.* **58,** 237–239.

Dvořák, J. (1981). Chromosome differentiation in polyploid species of *Elytrigia,* with special reference to the evolution of diploid-like chromosome pairing in polyploid species. *Can. J. Genet. Cytol.* **23,** 287–303.

Dvořák, J., and McGuire, P. E. (1981). Nonstructural chromosome differentiation among wheat cultivars, with special reference to differentiation of chromosomes in related species. *Genetics* 97, 391–414.

Edwards, G. A. (1977). The karyotype of *Gossypium herbaceum* L. *Caryologia* **30,** 369–374.

Edwards, G. A. (1979). Genomes of the Australian wild species of cotton. I. *Gossypium sturtianum,* the standard for the C genome. *Can. J. Genet. Cytol.* **21,** 363–366.

Edwards, G. A., and Endrizzi, J. E. (1976). Cell size, nuclear size and DNA content relationships in *Gossypium. Can. J. Genet. Cytol.* **17,** 181–186.

Edwards, G. A., and Mirza, M. A. (1979). Genomes of the Australian wild species of cotton. II. The designation of a new G genome for *Gossypium bickii. Can. J. Genet. Cytol.* **21,** 367–372.

Edwards, G. A., Endrizzi, J. E., and Stein, R. (1974). Genome DNA content and chromosome organization in *Gossypium. Chromosoma* **47,** 309–326.

Edwards, G. A., Brown, M. S., Niles, G. A., and Naqi, S. A. (1980). Monosomics in cotton. *Crop Sci.* **20,** 527–528.

El-Nockrashy, A. S., Simmons, J. G., and Frampton, V. L. (1969). A chemical survey of seeds of the genus *Gossypium. Phytochemistry* **8,** 1949–1958.

Endrizzi, J. E. (1962). The diploid-like cytological behavior of tetraploid cotton. *Evolution* **16,** 325–329.

Endrizzi, J. E. (1963). Genetic analysis of six primary monosomes and one tertiary monosome in *Gossypium hirsutum*. *Genetics* **48**, 1625–1633.

Endrizzi, J. E. (1966a). Additional information on chromosomal structural changes and differentiation in *Gossypium*. *J. Ariz. Acad. Sci.* **4**, 28–34.

Endrizzi, J. E. (1966b). Use of haploids in *Gossypium barbadense* L. as a source of aneuploids. *Curr. Sci.* **35**, 34–35.

Endrizzi, J. E. (1974). Alternate-1 and alternate-2 disjunctions in heterozygous reciprocal translocations. *Genetics* **77**, 55–60.

Endrizzi, J. E. (1975). Monosomic analysis of 23 mutant loci in cotton. *J. Hered.* **66**, 163–165.

Endrizzi, J. E., and Bray, R. (1980). Cytogenetics of disomics, monotelo- and monoisodisomics and ml_1 st_1 mutants of chromosome 4 of cotton. *Genetics* **94**, 979–988.

Endrizzi, J. E., and Brown, M. S. (1962). Identification of a ring of four and two chains of three chromosomes from the *Gossypium arboreum–hirsutum* ring of six. *Can. J. Genet. Cytol.* **4**, 458–468.

Endrizzi, J. E., and Brown, M. S. (1964). Identification of monosomes for six chromosomes in *Gossypium hirsutum*. *Am. J. Bot.* **51**, 117–120.

Endrizzi, J. E., and Kohel, R. J. (1966). Use of telosomes in mapping three chromosomes in cotton. *Genetics* **54**, 535–550.

Endrizzi, J. E., and Phillips, L. L. (1960). A hybrid between *Gossypium arboreum* L. and *G. raimondii* Ulbr. *Can. J. Genet. Cytol.* **2**, 311–319.

Endrizzi, J. E., and Ramsay, G. (1979). Monosomes and telosomes for 18 of the 26 chromosomes of *Gossypium hirsutum*. *Can. J. Genet. Cytol.* **21**, 531–536.

Endrizzi, J. E., and Ramsay, G. (1980). Identification of ten chromosome deficiencies in cotton. *J. Hered.* **71** 45–48.

Endrizzi, J. E., and Ramsay, G. (1983a). The inheritance of the H_1, H_2, and Sm_2 genes in cotton. *Crop Sci.* **23**, 449–452.

Endrizzi, J. E., and Ramsay, G. (1983b). The association of the Fg Ia linkage group within the short arm of chromosome 3 in cotton. *J. Hered.* **74**, 388–390.

Endrizzi, J. E., and Stein, R. (1975). Association of two marker loci with chromosome 1 in cotton. *J. Hered.* **66**, 75–78.

Endrizzi, J. E., and Stith, L. S. (1970). Association of two marker genes with chromosome 1 of cotton. *Agron. Abstr.* p. 9.

Endrizzi, J. E., and Taylor, T. (1968). Cytogenetic studies of N Lc_1 yg_2 R_2 marker genes and chromosome deficiencies in cotton. *Genet. Res.* **12**, 295–304.

Endrizzi, J. E., McMichael, S. C., and Brown, M. S. (1963a). Chromosomal constitution of "Stag" plants in *Gossypium hirsutum* "Acala 4-42." *Crop Sci.* **3**, 1–3.

Endrizzi, J. E., Richmond, T. R., Kohel, R. J., and Brown, M. S. (1963b). Monosomes—a tool for developing better cottons. *Tex. Agric. Prog.* **9**, 9–11.

Endrizzi, J. E., Katterman, F. R. H., Fisher, W. D., and Stith, L. S. (1974). Extrachromosomal inheritance of a variegated chlorophyll mutant in cotton. *Egypt. J. Genet. Cytol* **3**, 277–284.

Endrizzi, J. E., Ray, D. T., and Gathman, A. C. (1983). Centromere orientation of quadrivalents of heterozygous translocations and an autoploid of *Gossypium hirsutum* L. *Genetics* **105**, 723–731.

Ergle, D. R., Katterman, F. R. H., and Richmond, T. R. (1964). Aspects of nucleic acid composition of *Gossypium*. *Plant Physiol.* **39**, 145–150.

Flavell, R. (1982). Sequence amplification, deletion and rearrangement: Major sources of variation during species divergence. *Syst. Assoc. Spec. Vol.* **20**, 301–323.

Fryxell, P. A. (1965a). Stages in the evolution of *Gossypium. Adv. Front. Plant Sci.* **10**, 31–56.

Fryxell, P. A. (1965b). A revision of the Australian species of *Gossypium* with observations on the occurrence of *Thespesia* in Australia (Malvaceae). *Ann. Mo. Bot. Gard.* **56**, 179–250.

Fryxell, P. A. (1965c). A further description of *Gossypium trilobum. Madrono* **18**, 113–118.

Fryxell, P. A. (1971). Phenetic analysis and the phylogeny of the diploid species of *Gossypium* L. (Malvaceae). *Evolution* **25**, 554–562.

Fryxell, P. A. (1979). "The Natural History of the Cotton Tribe." Texas A&M Univ. Press, College Station.

Fryxell, P. A., and Parks, C. R. (1967). *Gossypium trilobum:* An addendum. *Madrono* **19**, 117–123.

Galen, D. F., and Endrizzi, J. E. (1968). Induction of monosomes and mutations in cotton by gamma irradiation of pollen. *J. Hered.* **59**, 343–346.

Gates, R. R. (1938). The origin of cultivated cotton. *Emp. Cotton Grow. Rev.* **15**, 1–6.

Geever, R. F. (1980). The evolution of single-copy nucleotide sequences in the genomes of *Gossypium hirsutum* L. Ph.D. Dissertation, University of Arizona, Tucson.

Gerstel, D. U. (1953a). Chromosome translocations in interspecific hybrids of the genus *Gossypium. Evolution* **7**, 234–244.

Gerstel, D. U. (1953b). Genetic segregation of allopolyploids in the genus *Gossypium. Genetics* **38**, 664–665.

Gerstel, D. U. (1954). A new lethal combination in interspecific cotton hybrids. *Genetics* **39**, 628–639.

Gerstel, D. U. (1956). Segregation in new allopolyploids of *Gossypium*. I. The R_1 locus in certain New World wild American hexaploids. *Genetics* **41**, 31–44.

Gerstel, D. U. (1963). Evolutionary problems in some polyploid crop plants. *Hereditas, Suppl.* **2**, 481–504.

Gerstel, D. U., and Phillips, L. L. (1957). Segregation of new allopolyploids of *Gossypium*. II. Tetraploid combinations. *Genetics* **42**, 783–797.

Gerstel, D. U., and Phillips, L. L. (1958). Segregation of synthetic amphidiploids in *Gossypium* and *Nicotiana. Cold Spring Harbor Symp. Quant. Biol.* **23**, 225–237.

Gerstel, D. U., and Sarvella, P. (1956). Additional observations on chromosome translocations in cotton hybrids. *Evolution* **10**, 408–414.

Giles, J. A. (1962). The comparative genetics of *Gossypium hirsutum* L. and the synthetic amplidiploids, *Gossypium arboreum* L. × *Gossypium thurberi* Tod. *Genetics* **47**, 45–59.

Govande, G. K. (1944a). A new gene for lintlessness in Asiatic cottons. *Curr. Sci.* **13**, 15–16.

Govande, G. K. (1944b). Linkage relationship of the Li_d gene for lintlessness in Asiatic cottons. *Curr. Sci.* **13**, 321.

Govande, G. K. (1946). A new mutant in Asiatic cottons. *Curr. Sci.* **15**, 170.

Green, J. M. (1955). Frego bract, a genetic marker in Upland cotton. *J. Hered.* **46**, 232.

Green, J. M., and Brinkerhoff, L. A. (1956). Inheritance of three genes for bacterial blight resistance in Upland cotton. *Agron. J.* **48**, 481–485.

Griffee, F., and Ligon, L. L. (1929). Occurrence of "lintless" cotton plants and the inheritance of the character "lintless." *J. Am. Soc. Agron.* **21**, 711–717.

Harland, S. C. (1917). On the genetics of crinkled dwarf rogues in Sea Island cotton. *West Indian Bull.* **16**, 82–84.

Harland, S. C. (1920). Studies of inheritance in cotton. I. The inheritance of corolla colour. *West Indian Bull.* **18**, 13–19.

Harland, S. C. (1928). Cotton notes. *Trop. Agric.* **5**, 116–117.

Harland, S. C. (1929a). The genetics of cotton. I. The inheritance of petal spot in New World cottons. *J. Genet.* **20**, 365–385.

Harland, S. C. (1929b). The genetics of cotton. II. The inheritance of pollen colour in New World cottons. *J. Genet.* **20**, 387–399.

Harland, S. C. (1929c). The work of the Genetics Department of the Cotton Research Station, Trinidad. *Emp. Cotton Grow. Rev.* **6**, 304–314.

Harland, S. C. (1932). The genetics of cotton. VI. The inheritance of chlorophyll deficiency in New World cottons. *J. Genet.* **25**, 271–281.

Harland, S. C. (1933). The genetical conception of the species. *Mem. Acad. Sci. USSR* (4), 181–186.

Harland, S. C. (1934). The genetics of cotton. XI. Further experiments on the inheritance of chlorophyll deficiency in New World cottons. *J. Genet.* **29**, 181–195.

Harland, S. C. (1935a). The genetics of cotton. XII. Homologous genes for anthocyanin pigmentation in New and Old World cottons. *J. Genet.* **30**, 465–476.

Harland, S. C. (1935b). The genetics of cotton. XIV. The inheritance of brown lint in New World cottons. *J. Genet.* **31**, 27–37.

Harland, S. C. (1936a). Duplicate genes for corolla colour in *Gossypium barbadense* L. and *G. darwinii* Watt. *Z. Indukt. Abstamm. Vererbungsl.* **71**, 417–419.

Harland, S. C. (1936b). The genetical conception of the species. *Biol. Rev. Cambridge Philos. Soc.* **11**, 83–112.

Harland, S. C. (1937a). Homologous loci of wild and cultivated American cottons. *Nature (London)* **140**, 467–468.

Harland, S. C. (1937b). Chlorophyll deficiency and modifying factors in New World cottons. *Z. Vererbungsl.* **73**, 49–54.

Harland, S. C. (1939). "The Genetics of Cotton." Jonathan Cape, London.

Harland, S. C. (1940). New allopolyploids in cotton by the use of colchicine. *Trop. Agric.* **17**, 53–55.

Harland, S. C. (1941). Genetical studies in the genus *Gossypium* and their relationship to evolutionary and taxonomic problems. *Proc. Int. Congr. Genet., 7th, 1939* pp. 138–143.

Harland, S. C. (1944). The selection experiment with Peruvian Tanguis cotton. *Soc. Nac. Agrar., Inst. Cotton Genet. Bull.* No. 1 (*Plant Breeding Abstracts*, **15**, 661).

Harland, S. C., and Atteck, O. M. (1941). The genetics of cotton. XVIII. Transference of genes from diploid North American wild cottons (*Gossypium thurberi* Tod., *G. armourianum* Kearney, *G. aridum* comb. nov. Skovsted) to tetraploid New World cottons (*G. barbadense* L. and *G. hirsutum* L.). *J. Genet.* **42**, 1–19.

Hasenkampf, C. A., and Menzel, M. Y. (1980). Incipient genome differentiation in *Gossypium*. II. Comparison of 12 chromosomes of *G. hirsutum*, *G. mustelinum* and *G. tomentosum* using heterozygous translocations. *Genetics* **95**, 971–983.

Hau, B. (1981a). Additive lines of the species *G. hirsutum* L. I. Use of interspecific hybridization and the additive lines method in cotton breeding. *Cotton Fibres Trop., Engl. Ed.* **36**, 247–258.

Hau, B. (1981b). Additive lines incorporated into the species *G. hirsutum* L. II. Phenotypic description of certain monosomic additive lines from *G. anomalum* and *G. stocksii. Cotton Fibres Trop., Engl. Ed.* **36**, 285–296.

Hau, B. (1982). Additive lines in the species *G. hirsutum* L. III. Changes in a collection of

G. hirsutum additives derived from *G. anomalum* and *G. stocksii* after several generations of self-fertilization. *Cotton Fibres Trop., Engl. Ed.* **37**, 163–172.

Hau, B., Koto, E., and Schwendiman, J. (1980a). Déterminisme génétique de deux mutants du cotonnier capsule pileuse et fleur cleistogame. *Coton Fibres Trop. (Fr. Ed.)* **35**, 355–357.

Hau, B., Koto, E., and Schwendiman, J. (1980b). Examen du groupe de liaison III du cotonnier *Gossypium hirsutum:* Déscription d'un nouveau phénotype cluster et localisation des gènes. *Coton Fibres Trop. (Fr. Ed.)* **35**, 359–367.

Hau, B., Koto, E., and Schwendiman, J. (1981). Déscription d'une mutation induisant une feuille filiforme chez le cotonnier *G. hirsutum* L. *Coton Fibres Trop. (Fr. Ed.)* **36**, 205–208.

Hawkes, J. G., and Smith, P. (1965). Continental drift and the age of angiosperm genera. *Nature (London)* **207**, 48–50.

Heyerdahl, T. (1979). "Early Man and the Oceans." Doubleday, Garden City, New York.

Holder, D. S., Jenkins, J. N., and Maxwell, F. G. (1968). Duplicate linkage of glandless and nectariless genes in Upland cotton, *Gossypium hirsutum* L *Crop Sci.* **8**, 577–580.

Hutchinson, J. B. (1931). The genetics of cotton. Part IV. The inheritance of corolla colour and petal size in Asiatic cottons. *J. Genet.* **24**, 325–353.

Hutchinson, J. B. (1932a). The genetics of cotton. VII. Crumpled: A new dominant in Asiatic cottons produced by complementary factors. *J. Genet.* **25**, 281–291.

Hutchinson, J. B. (1932b). The genetics of cotton. VIII. The inheritance of anthocyanin pigmentation in Asiatic cottons. *J. Genet.* **26**, 317–339.

Hutchinson, J. B. (1934). The genetics of cotton. X. The inheritance of leaf shape in Asiatic *Gossypiums*. *J. Genet.* **28**, 437–513.

Hutchinson, J. B. (1935). The genetics of cotton. XV. The inheritance of fuzz and lintlessness and associated characters in Asiatic cottons. *J. Genet.* **31**, 451–470.

Hutchinson, J. B. (1946a). The crinkled dwarf allelomorph series in New World cottons. *J. Genet.* **47**, 178–207.

Hutchinson, J. B. (1946b). On the occurrence and significance of deleterious genes in cotton. *J. Genet.* **47**, 272–289.

Hutchinson, J. B. (1954). New evidence on the origin of the Old World cottons. *Heredity* **8**, 225–241.

Hutchinson, J. B. (1959). "The Application of Genetics to Cotton Improvement." Cambridge Univ. Press, London and New York.

Hutchinson, J. B. (1962). The history and relationships of the world's cotton. *Endeavour* **21**, 5–15.

Hutchinson, J. B., and Gadkari, P. D. (1935). Note on inheritance of sterility in cotton. *Indian J. Agric. Sci.* **5**, 554–558.

Hutchinson, J. B., and Gadkari, P. D. (1937). The genetics of lintlessness in Asiatic cottons. *J. Genet.* **35**, 161–175.

Hutchinson, J. B., and Ghose, R. L. M. (1937a). A note on two new genes affecting anthocyanin pigmentation in Asiatic cottons. *Indian J. Agric. Sci.* **7**, 873–876.

Hutchinson, J. B., and Ghose, R. L. M. (1937b). Petalody in cotton. *Curr. Sci.* **6**, 99–100.

Hutchinson, J. B., and Nath, B. (1938). A note on the occurrence of chlorophyll deficiency in *Gossypium arboreum*. *Indian J. Agric. Sci.* **8**, 425–427.

Hutchinson, J. B., and Silow, R. A. (1939). Gene symbols for use in cotton genetics. *J. Hered.* **30**, 461–464.

Hutchinson, J. B., Stephens, S. G. and Dodd, K. S. (1945). The seed hairs of *Gossypium*. *Ann. Bot. (London)* [N.S.] **9**, 361–367.

Hutchinson, J. B., Silow, R. A., and Stephens, S. G. (1947). "The Evolution of Gossypium." Oxford Univ. Press, London and New York.

Jacob, K. T. (1942). Studies in cotton. IV. Morphology of the somatic chromosomes of eight types of Asiatic cottons. *Trans. Bose Res. Inst.* **15**, 17–27.

Jayaraman, N., Narayanan, S. S., Chinnadurai, K., and Bahavandoss, M. (1973). The role of *Gossypium aridum* in the origin of the amphidiploids of cotton. *Madras Agric. J.* **60**, 1587–1591.

Jenkins, J. N., Maxwell, F. G., and Lafever, H. N. (1966). The comparative preference of insects for glanded and glandless cottons. *J. Econ. Entomol.* **59**, 352–356.

Johnson, B. L. (1949). Complementary factors for dark-red plant color in Upland cotton. *J. Agric. Res. (Washington, D.C.)* **78**, 535–543.

Johnson, B. L. (1975). *Gossypium palmeri* and a polyphyletic origin of the New World cottons. *Bull. Torrey Bot. Club* **102**, 340–349.

Johnson, B. L., and Thein, M. M. (1970). Assessment of evolutionary affinities in *Gossypium* by protein electrophoresis. *Am. J. Bot.* **57**, 1081–1092.

Jones, J. E., and Andries, J. A. (1969). The effect of frego bract on the incidence of cotton boll rot. *Crop Sci.* **9**, 426–428.

Justus, N. J., and Leinweber, C. L. (1960). A heritable partial male-sterile character in cotton. *J. Hered.* **51**, 191–192.

Justus, N. J., and Meyer, J. R. (1963). A female-sterile mutant in a double haploid of Upland cotton. *J. Hered.* **54**, 65–66.

Justus, N. J., Meyer, J. R., and Roux, J. B. (1963). A partially male-sterile character in Upland cotton. *Crop Sci.* **3**, 428–429.

Kadir, Z. B. Z. (1976). DNA evolution in the genus *Gossypium*. *Chromosoma* **56**, 85–95.

Kammacher, P. (1958). La production artificielle d'aberrations chromosomiques chez *Gossypium hirsutum* par les rayons X. *Coton Fibres Trop. (Fr. Ed.)* **13**, 1–22.

Kammacher, P. (1959). Rélations cytologiques entre l'espèce sauvage *Gossypium raimondii* Ulb. et les espèces cultivées de cotonniers *G. hirsutum* L. et *G. barbadense* L. *Coton Fibres Trop. (Fr. Ed.)* **14**, 1–4.

Kammacher, P. (1960). Observations cytologiques sur deux hybrides F_1 entre éspèces cultivées tetraploids de cotonniers et l'espèce diploid sauvage *Gossypium raimondii* Ulb. *Rev. Cytol. Biol. Veg.* **22**, 1–32.

Kammacher, P. (1968). New investigations of linkage group I of *Gossypium hirsutum*. *Coton Fibres Trop., Engl. Ed.* **23**, 179–183.

Kammacher, P., and Poisson, C. (1964). Sur les possibilités de transferer du material génétique du cotonnier sauvage *Gossypium anomalum* Wow. et Peyr. A l'espèce cultivée *G. hirsutum* L. *Coton Fibres Trop. (Fr. Ed.)* **19**, 243–264.

Kammacher, P., and Schwendiman, J. (1969). Addition au génome de l'espéce de cotonnier, *Gossypium hirsutum* de deux chromosomes de l'espèce sauvage *G. stocksii*. *Can. J. Genet. Cytol.* **11**, 169–183.

Kammacher, P., Poisson, C., and Schwendiman, J. (1967). Study of chromosomal localization of gene ms_3 of cotton pollinic sterility. *Coton Fibres Trop., Engl. Ed.* **22**, 417–420.

Katterman, F. R. H., and Endrizzi, J. E. (1973). Studies on the 70S ribosomal content of a plastid mutant in *Gossypium hirsutum* L. *Plant Physiol.* **51**, 1138–1139.

Katterman, F. R. H., and Ergle, D. R. (1970). A study of quantitative variations of nucleic acids in *Gossypium*. *Phytochemistry* **9**, 2007–2010.

Kearny, T. H. (1930). Cotton plants, tame and wild and genetics of cotton. *J. Hered.* **21**, 325–336, 375–415.

Kearney, T. H., and Harrison, G. J. (1927). Inheritance of smooth seeds in cotton. *J. Agric. Res. (Washington, D.C.)* **35**, 193–217.

Kelkar, S. G., Choudhari, S. P., and Hiremath, N. B. (1947). Inheritance of *Fusarium* resistance in Indian cottons. *Conf. Cotton Grow. Probl. India, I.C.C.C., 3rd 1946,* pp. 125–162.

Kestler, D. P., Katterman, F. R. H., and Endrizzi, J. E. (1977). Buoyant density determinations on chloroplast DNA in a variegated cytoplasmic mutant of *Gossypium hirsutum* L. *Biochem. Biophys. Res. Commun.* **76**, 720–727.

Khadilkar, T. R. (1946). A dwarf mutant in *Neglectum verum* cotton. *Curr. Sci.* **15**, 278–279.

Killough, D. T., and Horlacher, W. R. (1933). The inheritance of virescent yellow and red plant characters in cotton. *Genetics* **18**, 329–334.

King, M. C., and Wilson, A. C. (1975). Evolution at two levels in human and chimpanzee. *Science* **188** 107–116.

Knight, R. L. (1944). The genetics of blackarm resistance. IV. *G. punctatum* (Sch. & Thon.) crosses. *J. Genet.* **46**, 1–27.

Knight, R. L. (1945). Theory and application of the backcross technique in cotton breeding. *J. Genet.* **47**, 76–86.

Knight, R. L. (1947). The genetics of blackarm resistance. V. Dwarf-bunched and its relationship to B_1. *J. Genet.* **48**, 43–50.

Knight, R. L. (1948a). The genetics of blackarm resistance. VI. Transference of resistance from *G. arboreum* to *G. barbadense*. *J. Genet.* **48**, 359–369.

Knight, R. L. (1948b). The genetics of blackarm resistance. VII. *Gossypium arboreum* L. *J. Genet.* **49**, 109–116.

Knight, R. L. (1950). The genetics of blackarm resistance. VIII. *G. barbadense. J. Genet.* **50**, 67–76.

Knight, R. L. (1952a). The genetics of withering or deciduous bracteoles in cotton. *J. Genet.* **50**, 392–395.

Knight, R. L. (1952b). The genetics of jassid resistance in cotton. I. The genes H_1 and H_2. *J. Genet.* **51**, 46–66.

Knight, R. L. (1953a). The genetics of blackarm resistance. IX. The gene B_{6m} from *G. arboreum. J. Genet.* **51**, 270–275.

Knight, R. L. (1953b). The genetics of blackarm resistance. X. The gene B_7 from Stoneville 20. *J. Genet.* **51**, 515–519.

Knight, R. L. (1954a). "Abstract Bibliography of Cotton Breeding and Genetics—1900–1950." Commonw. Agric. Bur., Farnham Royal, Cambridge, England.

Knight, R. L. (1954b). The genetics of jassid resistance. IV. Transference of hairiness from *Gossypium herbaceum* to *G. barbadense. J. Genet.* **52**, 199–207.

Knight, R. L. (1954c). The genetics of blackarm resistance. XI. *Gossypium anomalum. J. Genet.* **52**, 466–472.

Knight, R. L. (1956). The genetical approach to disease resistance in plants. *Emp. Cotton Grow. Rev.* **33**, 191–196.

Knight, R. L. (1957). Blackarm disease of cotton and its control. *Plant Prot. Conf., Proc. Int. Conf., 2nd, 1956* pp. 53–59.

Knight, R. L. (1963). The genetics of blackarm resistance. XII. Transference of resistance from *Gossypium herbaceum* to *G. barbadense. J. Genet.* **58**, 328–346.

Knight, R. L., and Clouston, T. W. (1939). The genetics of blackarm resistance. I. Factors B_1 and B_2. *J. Genet.* **38**, 133–159.

Knight, R. L., and Sadd, J. (1953). The genetics of jassid resistance in cotton. II. Pubescent T611. *J. Genet.* **51**, 582–585.

Kohel, R. J. (1964). Inheritance of abnormal palisade mutant in American Upland cotton. *Gossypium hirsutum* L. *Crop Sci.* **4**, 112–113.

Kohel, R. J. (1965). Inheritance of accessory involucre mutant in American Upland cotton. *Gossypium hirsutum* L. *Crop Sci.* **5**, 119–120.

Kohel, R. J. (1967). Variegated mutants in cotton. *Crop Sci.* **7**, 490–492.

Kohel, R. J. (1972). Linkage tests in Upland cotton, *Gossypium hirsutum* L. II. *Crop Sci.* **12**, 66–69.

Kohel, R. J. (1973a). Genetic analysis of the open bud mutant in cotton. *J. Hered.* **64**, 237–238.

Kohel, R. J. (1973b). Genetic nomenclature in cotton. *J. Hered.* **64**, 291–295.

Kohel, R. J. (1973c). Analysis of irradiation induced virescent mutants and the identification of a new virescent mutant (v_5v_5,v_6v_6) in *Gossypium hirsutum* L. *Crop Sci.* **13**, 86–88.

Kohel, R. J. (1973d). Genetic analysis of the Crumpled mutant in cotton, *Gossypium hirsutum* L. *Crop Sci.* **13**, 384–386.

Kohel, R. J. (1973e). Genetic analysis of the depauperate mutant in cotton, *Gossypium hirsutum* L. *Crop Sci.* **13**, 427–428.

Kohel, R. J. (1978a). Monosomic analysis of cotton mutants. *J. Hered.* **69**, 275–276.

Kohel, R. J. (1978b). Linkage tests in Upland cotton. III. *Crop Sci.* **18**, 844–847.

Kohel, R. J. (1979). Gene arrangement in the duplicate linkage groups V and IX: Nectariless, glandless, and withering bract in cotton (*Gossypium hirsutum* L.). *Crop Sci.* **19**, 831–833.

Kohel, R. J. (1982). Crinkle-yellow, a new mutant in *Gossypium hirsutum* L. *J. Hered.* **73**, 382–383.

Kohel, R. J. (1983a). Genetic analysis of virescent mutants and the identification of virescents v_{12}, v_{13}, v_{14}, v_{15} and $v_{16}v_{17}$ in Upland cotton. *Crop Sci.* **23**, 289–291.

Kohel, R. J. (1983b). Genetic analysis of the yellow-veins mutant in cotton. *Crop Sci.* **23**, 291–293.

Kohel, R. J., and Benedict, C. R. (1971). Description and CO_2 metabolism of aberrant and normal chloroplasts in variegated cotton. *Gossypium hirsutum* L. *Crop Sci.* **11**, 486–488.

Kohel, R. J., and Douglas, C. R. (1974). Monosomic analysis of mutants of cotton. *Gossypium hirsutum* L. *Can. J. Genet. Cytol.* **16**, 299–231.

Kohel, R. J., and Lewis, C. F. (1962a). Inheritance of ragged mutant in American Upland cotton. *G. hirsutum* L. *Crop Sci.* **2**, 61–62.

Kohel, R. J., and Lewis, C. F. (1962b). Inheritance of veins-fused mutant in American Upland cotton. *G. hirsutum* L. *Crop Sci.* **2**, 174–175.

Kohel, R. J., and Richmond, T. R. (1971). Isolines in cotton: Effects of nine dominant genes. *Crop Sci.* **11**, 287–289.

Kohel, R. J., Lewis, C. F., and Richmond, T. R. (1965). Linkage tests in Upland cotton. *Gossypium hirsutum* L. *Crop Sci.* **5**, 582–585.

Kohel, R. J., Lewis, C. F., and Richmond, T. R. (1967). Isogenic lines in American Upland cotton, *Gossypium hirsutum* L.: Preliminary evaluation of lint measurements. *Crop Sci.* **7**, 67–70.

Kohel, R. J., Richmond, T. R., and Lewis, C. F. (1970). Texas marker. 1. Description of a genetic standard for *Gossypium hirsutum* L. *Crop Sci.* **10**, 670–671.

Kohel, R. J., Quisenberry, J. E., and Benedict, C. R. (1974). Fiber elongation and dry weight changes in mutant lines of cotton. *Crop Sci.* **14**, 471–474.

Kohel, R. J., Endrizzi, J. E., and White, T. G. (1977a). An evaluation of *Gossypium barbadense* L. chromosomes 6 and 17 in the *G. hirsutum* L. genome. *Crop Sci.* **17**, 404–406.

Kohel, R. J., Lewis, C. F., and Christiansen, M. N. (1977b). The identification of a new mutant and linkage group in cotton. *J. Hered.* **68**, 65–66.

Kokuev, V. I. (1935). Inheritance of certain agronomic and morphological characters in cotton. *Sredaz Nauchno-Issled. Kh. I. Tashkent* **80** (*Plant Breeding Abstracts* **6**, 580).

Kottur, G. L. (1927). A mutant in cotton. *Nature (London)* **119**, 747.

Lagiere, R. (1959). "La Bactériose du Cotonnier, *Xanthomonas malvacearum* (E. F. Smith) Dowson, dans le monde et en République Centrafricaine." Inst. Rech. Cott. Rex. Exot., Paris.

Leake, H. M. (1911). Studies on Indian cotton. *J. Genet.* **1**, 202–272.

Lee, J. A. (1962). Genetical studies concerning the distribution of pigment glands in the cotyledons and leaves of Upland cotton. *Genetics* **47**, 131–142.

Lee, J. A. (1965). The genomic allocation of the principal foliar-gland loci in *Gossypium hirsutum* and *Gossypium barbadense*. *Evolution* **19**, 182–188.

Lee, J. A. (1966). Genetics of the smooth stem character in *Gossypium hirsutum* L. *Crop Sci.* **6**, 497–498.

Lee, J. A. (1968). Genetical studies concerning the distribution of trichomes on the leaves of *Gossypium hirsutum* L. *Genetics* **60**, 567–575.

Lee, J. A. (1972). An example of increased recombination in *Gossypium*. *Crop Sci.* **12**, 114–116.

Lee, J. A. (1976). Transfer of a smooth leaf allele from *Gossypium barbadense* L. to *Gossypium hirsutum* L. *Crop Sci.* **16**, 601–602.

Lee, J. A. (1978). Allele determining rugate fruit surface in cotton. *Crop Sci.* **18**, 251–254.

Lee, J. A. (1981a). Genetics of D_3 complementary lethality in *Gossypium hirsutum* and *G. barbadense*. *J. Hered.* **72**, 299–300.

Lee, J. A. (1981b). A new linkage relationship in cotton. *Crop Sci.* **21**, 346–347.

Lee, J. A. (1982). Linkage relationships between Le and Gl alleles in cotton. *Crop Sci.* **22**, 1211–1213.

Lefort, P. L. (1970). Complemental studies on the localization of the ms_3 gene of partial pollinic sterility —its linkage with yg_1. *Coton Fibres Trop., Engl. Ed.* **25**, 311–314.

Lewis, C. F. (1954). The inheritance of cup leaf in cotton. *J. Hered.* **45**, 127–128.

Lewis, C. F. (1958). Genetic studies of a leaf mosaic mutant causing somatic instability in cotton. *J. Hered.* **49**, 267–271.

Lewis, C. F. (1960). The inheritance of mottled leaf in cotton. *J. Hered.* **51**, 209–212.

Lewis, C. F., and Richmond, T. R. (1960). The genetics of flowering response in cotton. *Genetics* **45**, 79–85.

Lewis, F. H. (1966). Speciation in flowering plants. *Science* **152**, 167–172.

Lima-de-Faria, A. (1980). Classification of genes, rearrangements and chromosomes according to the chromosome field. *Hereditas* **93**, 1–46.

Longley, A. E. (1933). Chromosomes of *Gossypium* and related genera. *J. Agric. Res. (Washington, D.C.)* **46**, 217–227.

McLendon, C. A. (1912). Mendelian inheritance in cotton hybrids. *Ga. Agric., Exp. Stn., Bull.* **89**, 141–228.

McMichael, S. C. (1942). Occurrence of the dwarf-red character in Upland cotton. *J. Agric. Res. (Washington, D.C.)* **64**, 477–481.

McMichael, S. C. (1954). Glandless boll in Upland cotton and its use in the study of natural crossing. *Agron. J.* **46**, 527–528.

McMichael, S. C. (1960). Combined effects of the glandless genes gl$_2$ and gl$_3$ on pigment glands in the cotton plant. *Agron. J.* **46**, 385–386.

McMichael, S. C. (1965). Inheritance of aberrant stigmas in the flowers of Upland cotton. *J. Hered.* **56**, 21–33.

McMichael, S. C. (1970). Yuma glandless, an allelomorph of glandless-one in cotton. *Gossypium hirsutum* L. *Crop Sci.* **10**, 202–203.

Malek-Hedayat, S. (1981). The relationship between seed weight and 13 monosomic and two monotelodisomic chromosomes in *Gossypium hirsutum*. M.S. Thesis, University of Arizona, Tucson.

Mangelsdorf, P. C., and Oliver, D. L. (1951). Whence came maize to Asia? *Bot. Mus. Leafl., Harv. Univ.* **14**, 263–291.

Mauer, F. M. (1938). On the origin of cultivated species of cotton. A highly fertile triple hybrid (*Gossypium barbadense* × *G. thurberi*) × *G. arboreum. Bull. Acad. Sci. USSR, Ser. Biol.* pp. 695–709; *Plant Breed. Abstr.* **9**, 318–319. (1939).

Mauer, F. M. (1954). "Origin and Systematics of Cotton." Izv. Akad. Nauk Uzb. SSR, Tashkent (in Russian) (from English summary provided by S. S. Kanash, Tashkent Cotton Station, 1956).

Menzel, M. Y. (1955). A cytological method for genome analysis in *Gossypium. Genetics* **40**, 214–223.

Menzel, M. Y. (1982). Why no monosomes have been identified for eleven cotton chromosomes: An educated guess. *Proc. Beltwide Cotton Prod. Res. Conf., 1982* pp. 138–139.

Menzel, M. Y., and Brown, M. S. (1952). Viable deficiency-duplications from a translocation in *Gossypium hirsutum. Genetics* **37**, 678–692.

Menzel, M. Y., and Brown, M. S. (1954a). The significance of multivalent formation in three-species *Gossypium* hybrids. *Genetics* **39**, 546–557.

Menzel, M. Y., and Brown, M. S. (1954b). The tolerance of *Gossypium hirsutum* for deficiencies and duplications. *Am. Nat.* **88**, 407–418.

Menzel, M. Y., and Brown, M. S. (1978a). Genetic length and breakpoints in twelve chromosomes of *Gossypium hirsutum* involved in ten reciprocal translocations. *Genetics* **88**, 541–558.

Menzel, M. Y., and Brown, M. S. (1978b). Reciprocal chromosome translocations in *Gossypium hirsutum*. Arm location of breakpoints and recovery of duplications and deficiencies. *J. Hered.* **69**, 383–390.

Menzel, M. Y., Brown, M. S., and Naqi, S. (1978). Incipient genome differentiation in *Gossypium.* I. Chromosomes 14, 15, 16, 19 and 20 assessed in *G. hirsutum, G. raimondii,* and *G. lobatum* by means of seven A–D translocations. *Genetics* **90**, 133–149.

Menzel, M. Y., Hasenkampf, C. A., and Stewart, J. McD. (1982). Incipient genome differentiation in *Gossypium.* III. Comparison of chromosomes of *G. hirsutum* and Asiatic diploids using heterozygous translocations. *Genetics* **100**, 89–103.

Meredith, W. R., Jr., Ranny, C. E., Laster, M. L., and Bridge, R. R. (1973). Agronomic potential of nectariless cotton (*Gossypium hirsutum* L.). *J. Environ. Qual.* **2**, 141–144.

Merrill, E. D. (1954). The botany of Cook's voyages. *Chron. Bot.* **14**, 161–384.

Meyer, J. R. (1957). Origin and inheritance of D_2 smoothness in Upland cotton. *J. Hered.* **48**, 249–250.

Meyer, J. R., and Meyer, V. G. (1961). Origin and inheritance of nectariless cotton.*Crop Sci.* **1**, 167–169.

Meyer, V. G. (1971). Cytoplasmic effects on anther numbers in interspecific hybrids of cotton. *J. Hered.* **62**, 77–78.

Meyer, V. G. (1973). A study of reciprocal hybrids between Upland cotton (*G. hirsutum* L.) and experimental lines with cytoplasms from seven other species. *Crop Sci.* **13**, 439–444.

Meyer, V. G. (1975). Male sterility from *Gossypium harknessii*. *J. Hered.* **66**, 23–27.

Mikhailova, K. A. (1938). Chromosome morphology of cotton. *C. R. (Dokl.) Acad. Sci. URSS* **19**, 181–184.

Murray, J. C. (1965). A new locus for glanded stem in tetraploid cotton. *J. Hered.* **56**, 42–44.

Murray, J. C., and Brinkerhoff, L. A. (1966). Inheritance of pale-green color mutant in cotton. *Crop Sci.* **6**, 375–376.

Mursal, I. El Jack (1978). Chromosome association in haploids and species hybrids of *Gossypium*. *J. Hered.* **69**, 413–416.

Mursal, I. El Jack, and Endrizzi, J. E. (1976). A reexamination of the diploid-like meiotic behavior of polyploid cotton. *Theor. Appl. Genet.* **47**, 171–181.

Nagl, W., Jeanjour, M., Kling, H., Kühner, S., Michels, I., Müller, T., and Stein, B. (1983). Genome and chromatin organization in higher plants. *Biol. Zentralbl.* **102**, 129–148.

Nakatomi, S. (1931). Hybridization between Old World and New World cotton species and the chromosome behavior of the pollen mother-cells in the F_1 hybrid. *Jpn. J. Bot.* **5**, 371–384.

Narayan, R. K. J. (1982). Discontinuous DNA variation in the evolution of plant species: The genus *Lathyrus*. *Evolution* **36**, 877–891.

Narayan, R. K. J., and Rees, H. (1976). Nuclear DNA variation in *Lathyrus*. *Chromosoma* **54**, 141–154.

Narayan, R. K. J., and Rees, H. (1977). Nuclear DNA divergence among *Lathyrus* species. *Chromosoma* **63**, 101–107.

Nath, B. (1942). Genetics of petal colour in Asiatic cottons. *Indian J. Genet. Plant Breed.* **2**, 43–49.

Nikolajeva, A. (1923). A hybrid between Asiatic and American cotton plant *Gossypium herbaceum* L. and *Gossypium hirsutum* L. *Bull. Appl. Bot. Plant Breed.* **13**, 117–134 (cited from Skovsted, 1933).

Parks, C. R., Ezell, W. L., Williams, D. E., and Dreyer, D. L. (1975). VII. The application of flavonoid distribution to taxonomic problems in the genus Gossypium. *Bull. Torrey Bot. Club* **102**, 350–361.

Patel, G. B., Munchi, Z. A., and Patel, G. T. (1947). Genetics of some mutations in *herbaceum* cottons of Gujerat (Bombay Province). *Conf. Cotton Grow. Probl. India, I.C.C.C., 3rd, 1946*, pp. 87–97.

Percival, A. E., and Kohel, R. J. (1974). Genetic analysis of virescent mutants in cotton, *Gossypium hirsutum* L. *Crop Sci.* **14**, 439–440.

Percival, A. E., and Kohel, R. J. (1976). New virescent cotton mutant linked with the marker gene Yellow petals. *Crop Sci.* **16**, 503–505.

Percival, A. E., Kohel, R. J., and Dilday, R. H. (1976). A dominant round leaf mutant in cotton. *Crop Sci.* **16**, 794–796.

Phillips, L. L. (1960). The cytogenetics of speciation in Asiatic cotton. *Genetics* **46**, 77–83.

Phillips, L. L. (1962). Segregation in new allopolyploids of *Gossypium*. IV. Segregation in New World × Asiatic and New World × wild American hexaploids. *Am. J. Bot.* **49**, 51–57.

Phillips, L. L. (1963). The cytogenetics of *Gossypium* and the origin of New World cottons. *Evolution* **17**, 460–469.

Phillips, L. L. (1964). Segregation in new allopolyploids of *Gossypium*. V. Multivalent formation in New World × Asiatic and New World × wild American hexaploids. *Am. J. Bot.* **51**, 324–329.

Phillips, L. L. (1966). The cytology and phylogenetics of the diploid species of *Gossypium*. *Am. J. Bot.* **53**, 328–335.

Phillips, L. L. (1974). Cotton (*Gossypium*). In "Handbook of Genetics" (R. C. King, ed.), Vol. 2, pp. 111–133. Plenum, New York.

Phillips, L. L. (1976). Interspecific incompatibility in *Gossypium*. III. The genetics of tumorigenesis in hybrids of *G. gossypioides*. *Can. J. Genet. Cytol.* **18**, 365–369.

Phillips, L. L., and Gerstel, D. U. (1959). Segregation in new allopolyploids of *Gossypium*. III. Leaf shape segregation in hexaploid hybrids of New World cottons. *J. Hered.* **50**, 103–108.

Phillips, L. L., and Strickland, M. A. (1966). The cytology of a hybrid between *Gossypium hirsutum* and *G. longicalyx*. *Can. J. Genet. Cytol.* **8**, 91–95.

Poisson, C. (1967). Sur les possibilités de transfert de matériel génétique du cotonnier sauvage *Gossypium anomalum* Wav. et Peyr. A l'espèce cultivée *G. hirsutum* L. II. Création de lignées d'addition à 27 paires de chromosomes. *Coton Fibres Trop. (Fr. Ed.)* **22**, 401–415.

Poisson, C. (1968). Preliminary report on a monosome of *Gossypium hirsutum* corresponding to linkage group I. *Coton Fibres Trop., Engl. Ed.* **23**, 183–186.

Poisson, C. (1970). Contribution à l'étude de l'hybridation interspécifique dans le genre *Gossypium:* Transfert de matériel génétique de l'espèce sauvage diploide *Gossypium anomalum* à l'espèce cultivèe tetraploide *G. hirsutum*. *Coton Fibres Trop. (Fr. Ed.)* **25**, 449–488.

Poisson, C. (1971). Etude des lignées de substitution triées de populations on de lignées d'addition. *Coton Fibres Trop. (Fr. Ed.)* **26**, 162–197.

Poisson, C., Schwendiman, J., and Kammacher, P. (1969). Mise en évidence d'une homologie chromosomique entre *Gossypium anomalum* Wav. et Peyr. et *G. stocksii* Mast. *Coton Fibres Trop. (Fr. Ed.)* **24**, 469–470.

Prokhanov, Y. I. (1947). The conspectus of a new system of cottons (*Gossypium* L.). *Bot. Zh. SSSR* **32**, 61–78 (cited from Fryxell, 1965a).

Quisenberry, J. E., and Kohel, R. J. (1970). Genetics of the virescent-4 mutant in cotton. *J. Hered.* **61**, 212–214.

Ramanathan, V., and Balasubrahmanyan, R. (1933a). Inheritance of pollen colour in Asiatic cottons. *Indian J. Agric. Sci.* **3**, 1116–1123.

Ramanathan, V., and Balasubrahmanyan, R. (1933b). Lint colour in Asiatic cottons. *Curr. Sci.* **2**, 128.

Ramanathan, V., and Balasubrahmanyan, R. (1938). Some effects of X-rays on Uppam and Karunganni cottons. *Conf. Sci. Res. Workers Cotton, India, I.C.C.C., 1st, 1937,* pp. 393–400.

Ramanathan, V., and Sankaran, R. (1934). Petalody in cotton. *Indian J. Agric. Sci.* **4,** 938–942.

Ramey, H. H. (1966). Historical review of cotton variety development. *Proc. Cotton Improve. Conf., 18th, 1966,* pp. 310–326.

Ramiah, K. (1945). Anthocyanin genetics of cotton and rice. *Indian J. Genet. Plant Breed.* **5,** 1–14.

Ramiah, K., and Nath, B. (1943). Genetics of single lobed leaf mutant in cotton. *Indian J. Genet. Plant Breed.* **3,** 89–98.

Ramiah, K., and Nath, B. (1944). Genetics of two new anthocyanin patterns in Asiatic cottons. *Indian J. Genet. Plant Breed.* **4,** 23–24.

Ramiah, K., and Nath, B. (1947). Studies on the cotton boll with special reference to *G. arboreum. Conf. Cotton Grow. Probl., India, I.C.C.C., 3rd, 1946,* pp. 100–106.

Ramiah, K., and Paranjpe, V. J. (1944). The occurrence and inheritance of a new type of hairiness in Asiatic cottons. *Curr. Sci.* **13,** 158–160.

Ray, D. T. (1982). A tester-set of translocations in *Gossypium hirsutum* L. *J. Hered.* **73,** 429–433.

Rhyne, C. L. (1955). The inheritance of yellow-green, a possible mutation in cotton. *Genetics* **40,** 235–245.

Rhyne, C. L. (1957). Duplicated linkage groups in cotton. *J. Hered.* **48,** 59–62.

Rhyne, C. L. (1958). Linkage studies in *Gossypium.* I. Altered recombination in allotetraploid *G. hirsutum* L. following linkage group transference from related diploid species. *Genetics* **43,** 822–834.

Rhyne, C. L. (1960). Linkage studies in *Gossypium.* II. Altered recombination values in a linkage group of allotetraploid *G. hirsutum* L. as a result of transferred diploid species genes. *Genetics* **45,** 673–681.

Rhyne, C. L. (1962a). Inheritance of the glandless leaf phenotype of Upland cotton. *J. Hered.* **53,** 115–123.

Rhyne, C. L. (1962b). Diploidization in *Gossypium hirsutum* as indicated by glandlessstem and boll inheritance. *Am. Nat.* **96,** 265–276.

Rhyne, C. L. (1965a). Duplicate linkage blocks in glandless-leaf cotton. *J. Hered.* **56,** 247–252.

Rhyne, C. L. (1965b). The anamalous behavior of the ghost spot of *Gossypium anomalum* in amphidiploid *Gossypium hirsutum. Genetics* **51,** 689–698.

Rhyne, C. L. (1971). Indehiscent anther in cotton. *Cotton Grow. Rev.* **48,** 194–199.

Rhyne, C. L., and Rhyne, P. (1972). Linkage of indehiscent anthers and lack of leaf nectaries in *Gossypium hirsutum* L. *Cotton Grow. Rev.* **49,** 57–60.

Richmond, T. R. (1951). Procedures and methods of cotton breeding with special reference to American cultivated species. *Adv. Genet.* **4,** 213–245.

Richmond, T. R., and Kohel, R. J. (1961). Analysis of a completely male-sterile character in American Upland cotton. *Crop Sci.* **1,** 397–401.

Sarvella, P. (1958). Multivalent formation and genetic segregation in some allopolyploid *Gossypium* hybrids. *Genetics* **43,** 601–619.

Sarvella, P., and Stojanovic, B. J. (1968). Amino acid analysis of the species in the genus *Gossypium. Can. J. Genet. Cytol.* **10,** 362–368.

Saunders, J. H. (1961a). "The Wild Species of *Gossypium*—and their Evolutionary History." Oxford Univ. Press, London and New York.

Saunders, J. H. (1961b). The mechanism of hairiness in *Gossypium.* I. *Gossypium hirsutum. Heredity* **16,** 331–348.

Saunders, J. H. (1963). The mechanism of hairiness in *Gossypium* II. *Gossypium bar-badense*—the inheritance of stem hair. *Emp. Cotton Grow. Rev.* **40**, 104–106.

Saunders, J. H. (1965a). The mechanism of hairiness in *Gossypium*. III. *Gossypium barbadense*—the inheritance of upper leaf laminar hair. *Emp. Cotton Grow. Rev.* **42**, 15–25.

Saunders, J. H. (1965b). The mechanism of hairiness in *Gossypium*. IV. The inheritance of plant hair length. *Emp. Cotton Grow. Rev.* **42**, 26–32.

Saunders, J. H. (1965c). Genetics of hairiness transferred from *Gossypium raimondii* to *G. hirsutum*. *Euphytica* **14**, 276–282.

Saunders, J. H. (1969). Changes in recombination and genetic disturbance in transferring the H_2Lc_2 linkage group from *Gossypium tomentosum* to *G. barbadense* and *G. hirsutum*. *Genet. Res.* **13**, 1–15.

Schwendiman, J. A. (1974). Mise en évidence de trois nouvelles homoéologies chromosomiques entre *Gossypium anomalum* et *G. stocksii*. *Can. J. Genet. Cytol.* **16**, 871–881.

Schwendiman, J. A. (1975). Les modifications induites par la substitution complète de la paire de chromosome A_6 de *Gossypium hirsutum* par l' homologue de *G. barbadense*. *Coton Fibres Trop. (Fr. Ed.)* **30**, 283–291.

Schwendiman, J. A. (1977). Improvement of *Gossypium hirsutum* cotton by interspecific hybridization: Utilization of the species *G. barbadense* and *G. stocksii*. *Coton Fibres Trop., Engl. Ed.* **32**, 239–255.

Schwendiman, J. A., Koto, E., and Hau, B. (1980). Reflections on the evolution of chromosome pairing in cotton allohexaploids (*G. hirsutum* × *G. stocksii* and *G. hirsutum* × *G. longicalyx*) and on the taxonomic position of *G. longicalyx*. *Coton Fibres Trop., Engl. Ed.* **35**, 269–275.

Seyam, S. M., and Brown, M. S. (1973). Types of disjunction in interchange heterozygotes of *Gossypium hirsutum* L. *Egypt. J. Genet. Cytol.* **2**, 322–330.

Sharif, A., and Islam, A. S. (1970). Cytogenetical studies of some backcross derivatives of *Gossypium hirsutum* × *G. anomalum*. *Can. J. Genet. Cytol.* **12**, 454–460.

Sheetz, R. H., and Weaver, J. B., Jr. (1980). Inheritance of a fertility enhancer factor from Pima cotton when transferred into Upland cotton with *Gossypium harknessii* Brandegee cytoplasm. *Crop Sci.* **20**, 272–275.

Sherwin, K. H. (1970). Winds across the Atlantic—possible African origins for some Pre-Columbia New World cultigens. *Res. Rec. Univ. Mus. S. Ill. Univ. Meso-Am. Stud.* **6**, 1–33.

Shoemaker, D. N. (1908). A study of leaf characters in cotton hybrids. *Rep. Am. Breed. Assoc.* **5**, 116–119.

Sikka, S. M., and Joshi, A. B. (1960). Genetics. *In* "Cotton in India," Vol. I, pp. 403–466. Indian Central Cotton Comm., Bombay.

Silow, R. A. (1939). The genetics of leaf shape in diploid cottons and the theory of gene interaction. *J. Genet.* **38**, 229–276.

Silow, R. A. (1941). The comparative genetics of *Gossypium anomalum* and the cultivated Asiatic cottons. *J. Genet.* **42**, 259–358.

Silow, R. A. (1944). The inheritance of lint colour in Asiatic cottons. *J. Genet.* **46**, 78–115.

Silow, R. A. (1945). Further data on the inheritance of lint colour in the Old World cultivated cottons. *J. Hered.* **36**, 62–64.

Silow, R. A. (1946). Evidence on chromosome homology and gene homology in the amphidiploid New World cottons. *J. Genet.* **47**, 213–221.

Silow, R. A., and Yu, C. P. (1942). Anthocyanin pattern in Asiatic cottons. *J. Genet.* **43**, 249–284.

Simpson, D. M. (1947). Fuzzy leaf in cotton and its association with short lint. *J. Hered.* **38**, 153–156.

Skovsted, A. (1933). Cytological studies in cotton. I. The mitosis and the meiosis in diploid and triploid Asiatic cotton. *Ann. Bot. (London)* **47**, 227–258.

Skovsted, A. (1934a). Cytological studies in cotton. II. Two interspecific hybrids between Asiatic and New World cottons. *J. Genet.* **28**, 407–424.

Skovsted, A. (1934b). Cytogenetics in relation to plant breeding in cotton. *Emp. Cotton Grow. Corp., Rep. Conf., 2nd, 1934,* pp. 46–49.

Skovsted, A. (1935a). Cytological studies in cotton. III. A hybrid between *Gossypium davidsonii* Kell. and *G. sturtii* F. Muell. *J. Genet.* **30**, 397–405.

Skovsted, A. (1935b). Some new interspecific hybrids in the genus *Gossypium* L. *J. Genet.* **30**, 447–463.

Skovsted, A. (1935c). Chromosome numbers in the Malvaceae I. *J. Genet.* **31**, 263–296.

Skovsted, A. (1937). Cytological studies in cotton. IV. Chromosome conjugation in interspecific hybrids. *J. Genet.* **34**, 97–134.

Smith, E. G. (1942). Inheritance of smooth and pitted bolls in Pima cotton. *J. Agric. Res. (Washington, D.C.)* **64**, 101–103.

Sporne, K. R. (1954). Statistics and the evolution of dicotyledons. *Evolution* **8**, 55–64.

Sporne, K. R. (1976). Character correlations among the angiosperms and the importance of fossil evidence in assessing their significance. *In* "Origin and Early Evolution of the Angiosperms" (C. B. Beck, ed.), pp. 312–329. Columbia Univ. Press, New York.

Stebbins, G. L., Jr. (1947). Evidence on rates of evolution from the distribution of existing and fossil plant species. *Ecol. Monogr.* **17**, 149–158.

Stephens, S. G. (1942). Colchicine-produced polyploids in *Gossypium.* I. An autotetraploid Asiatic cotton and certain of its hybrids with diploid species. *J. Genet.* **44**, 227–295.

Stephens, S. G. (1944a). Phenogenetic evidence for amphidiploid origin of New World cottons. *Nature (London)* **153**, 53–54.

Stephens, S. G. (1944b). The genetic organization of leaf-shape development in the genus *Gossypium. J. Genet.* **46**, 28–51.

Stephens, S. G. (1945). A genetic survey of leaf shape in New World cottons—a problem in critical identification of alleles. *J. Genet.* **46**, 313–330.

Stephens, S. G. (1947). Cytogenetics of *Gossypium* and the problem of the origin of New World cottons. *Adv. Genet.* **1**, 431–442.

Stephens, S. G. (1949a). Genome analysis in amphidiploids. A new method of allocating genes to their correct subgenomes. *J. Hered.* **40**, 102–104.

Stephens, S. G. (1949b). The cytogenetics of speciation in *Gossypium.* I. Selective elimination of the donor parent genotype in interspecific backcrosses. *Genetics* **34**, 627–637.

Stephens, S. G. (1950a). The internal mechanism of speciation in *Gossypium. Bot. Rev.* **16**, 115–149.

Stephens, S. G. (1950b). The genetics of "corky." II. Further studies on its genetic basis in relation to the general problems of interspecific isolating mechanisms. *J. Genet.* **50**, 9–20.

Stephens, S. G. (1951a). Possible significance of duplication in evolution. *Adv. Genet.* **4**, 247–265.

Stephens, S. G. (1951b). "Homologous" genetic loci in *Gossypium*. *Cold Spring Harbor Symp. Quant. Biol.* **16**, 131–134.

Stephens, S. G. (1954a). Interspecific homologies between gene loci in *Gossypium*. I. Pollen color. *Genetics* **39**, 701–711.

Stephens, S. G. (1954b). Interspecific homologies between gene loci in *Gossypium*. II. Corolla color. *Genetics* **39**, 712–723.

Stephens, S. G. (1955). Linkage in Upland cotton. *Genetics* **40**, 903–917.

Stephens, S. G. (1958a). Salt water tolerance of seed of *Gossypium* species as a possible factor in seed dispersal. *Am. Nat.* **90**, 83–92.

Stephens, S. G. (1958b). Factors affecting seed dispersal in *Gossypium* and their possible evolutionary significance. *N. C., Agric. Exp. Stn., Tech. Bull.* **131**, 1–32.

Stephens, S. G. (1961). Recombination between supposedly homologous chromosomes of *Gossypium barbadense* L. and *G. hirsutum* L. *Genetics* **46**, 1483–1500.

Stephens, S. G. (1966). The potential for long range oceanic dispersal of cotton seeds. *Am. Nat.* **100**, 199–210.

Stephens, S. G. (1967). Evolution under domestication of New World cottons (*Gossypium* spp.). *Cienc. Cult. (Sao Paulo)* **19**, 118–134.

Stephens, S. G. (1974a). Geographic and taxonomic distribution of anthocyanin genes in New World cottons. *J. Genet.* **61**, 128–141.

Stephens, S. G. (1974b). The use of two polymorphic systems, nectary fringe hairs and corky alleles, as indicators of phylogenetic relationships in New World cottons. *Biotropica* **6**, 194–201.

Stern, H., and Hotta, Y. (1973). Biochemical controls of meiosis. *Annu. Rev. Genet.* **7**, 37–66.

Stewart, J. McD. (1979). Use of ovule cultures to obtain interspecific hybrids of *Gossypium*. *In* "Plant Tissue Culture" (J. T. Barber, ed.) pp. 44–56. Published by: Southern Section Amer. Soc. Plant Physiol.

Stewart, J. McD., and Hsu, C. L. (1978). Hybridization of diploid and tetraploid cottons through in-ovule embryo culture. *J. Hered.* **69**, 404–408.

Stroman, G. N., and Mahoney, C. H. (1925). Heritable chlorophyll deficiencies in seedling cotton. *Tex., Agric. Exp. Stn., [Bull.]* **333**, 5–19.

Takhtajan, A. (1969). "Flowering Plants: Origin and Dispersal." Smithson. Inst. Press, Washington, D.C.

Tayel, M. A. F., Bird, L. S., and Smith, J. D. (1973). Linkage between blight resistance and yellow pollen color genes. *J. Hered.* **64**, 208–212.

Thadani, K. I. (1923). Linkage relations in the cotton plant. *Agric. J. India* **18**, 572–579.

Tiranti, I. N. (1966). Descripcion citmorfologica de los hibridos F$_1$ de *Gossypium robinsonii* × *G. klotzschianum* y por su var. *davidsonii*. *Inst. Nac. Tecnol. Agropec. Bol.* **39**, 4–20.

Tiranti, I. N. (1967). "Algodon crespo," a new mutation in *Gossypium hirsutum* L. *Bot. Genet. Inst. Fitotec, Castelar* **3**, 29–33.

Tiranti, I. N., and Endrizzi, J. E. (1963). Identification of three rings of four from a ring of six chromosomes in cotton. *Can. J. Genet. Cytol.* **5**, 374–379.

Turcotte, E. L., and Feaster, C. V. (1963). Inheritance of a cream petal mutant in Pima cotton, *Gossypium barbadense* L. *Crop Sci.* **3**, 563–564.

Turcotte, E. L., and Feaster, C. V. (1964). Inheritance of a mutant with a rudimentary stigma and style in Pima cotton, *Gossypium barbadense* L. *Crop Sci.* **4**, 377–378.

Turcotte, E. L., and Feaster, C. V. (1965). The inheritance of rugate leaf in Pima cotton. *J. Hered.* **56**, 234–236.

Turcotte, E. L., and Feaster, C. V. (1966). A second locus for pollen color in Pima cotton, *Gossypium barbadense* L. *Crop Sci.* **6**, 117–119.

Turcotte, E. L., and Feaster, C. V. (1967). Semigamy in Pima cotton. *J. Hered.* **58**, 55–57.

Turcotte, E. L., and Feaster, C. V. (1969). Semigametic production of haploids in Pima cotton. *Crop Sci.* **9**, 653–655.

Turcotte, E. L., and Feaster, C. V. (1973). The interaction of two genes for yellow foliage in cotton. *J. Hered.* **64**, 231–232.

Turcotte, E. L., and Feaster, C. V. (1975a). Effects of R_1, a gene for red plant color, on American Pima cotton. *Crop Sci.* **15**, 875–876.

Turcotte, E. L., and Feaster, C. V. (1975b). Inheritance of semigamy in American Pima cotton, *Gossypium barbadense* L. *Agron. Abstr.* **75**, 65.

Turcotte, E. L., and Feaster, C. V. (1978). Inheritance of three genes for plant color in American Pima cotton. *Crop Sci.* **18**, 149–150.

Turcotte, E. L., and Feaster, C. V. (1979). Linkage tests in American Pima cotton. *Crop Sci.* **19**, 119–120.

Turcotte, E. L., and Feaster, C. V. (1980). Inheritance of a leaf mutant in American Pima cotton. *J. Hered.* **71**, 134–135.

Turcotte, E. L., and Feaster, C. V. (1983). Inheritance of Golden-veins mutant in American Pima cotton. *J. Hered.* **74**, 213–214.

Valiček, P. (1978). Wild and cultivated cottons. *Coton Fibres Trop., Engl. Ed.* **33**, 363–387.

Valiček, P. (1981). Some notes on the morphology characteristics of *Gossypium ellenbeckii* (Gürke) Mauer. *Coton Fibres Trop., Engl. Ed.* **36**, 157–162.

Venkoba Rao, M., and Ramachandran, C. K. (1943). The mode of inheritance of "dwarf bushy" type in *G. herbaceum*. *Madras Agric. J.* **31**, 28–29.

Vieira da Silva, J. B., and Poisson, C. (1969). Solubilization d'enzymes hydrolytiques chez *Gossypium hirsutum, G. anomalum* et des derivés de l'hybridization entre ces deux espèces. *Can. J. Genet. Cytol.* **11**, 582–586.

Vijayaraghavan, G., Iyengar, N. K., and Venkoba Rao, M. (1936). A heritable case of female sterility in *herbaceum* cotton. *Madras Agric. J.* **24**, 365–368.

Walbot, V., and Dure, L. S., III (1976). Developmental biochemistry of cotton seed embryogenesis and germination. VII. Characterization of the cotton genomes. *J. Mol. Biol.* **101**, 503–536.

Wannamaker, W. K. (1957). The effects of plant hairiness of cotton strains on boll weevil attack. *J. Econ. Entomol.* **50**, 418–423.

Ware, J. O. (1929). Inheritance of lint percentage in cotton. *J. Am. Soc. Agron.* **21**, 876–894.

Ware, J. O., Benedict, L. I., and Rolfe, W. H. (1947). A recessive naked seed character in Upland cotton. *J. Hered.* **38**, 313–319.

Weaver, J. B. (1968). Analysis of a double recessive completely male-sterile cotton. *Crop Sci.* **8**, 597–600.

Weaver, J. B., and Ashley, T. (1971). Analysis of a dominant gene for male sterility in Upland cotton. *Gossypium hirsutum*. *Crop Sci.* **11**, 596–598.

Weaver, J. B., and Weaver, J. B., Jr. (1977). Inheritance of pollen fertility restoration in cytoplasmic male-sterile cotton. *Crop Sci.* **17**, 497–499.

Weaver, J. B., Jr., and Weaver, J. B. (1979). Cracked root mutant in cotton. Inheritance and linkage with fertility restoration. *Crop Sci.* **19**, 307–309.

Webber, J. M. (1934a). Chromosome number and meiotic behavior in *Gossypium*. *J. Agric. Res. (Washington, D.C.)* **49**, 223–237.

Webber, J. M. (1934b). Cytogenetic notes on cotton and cotton relatives. *Science* **80,** 268–269.

Webber, J. M. (1935). Interspecific hybridization in *Gossypium* and the behavior of F₁ plants. *J. Agric. Res. (Washington, D.C.)* **51,** 1047–1070.

Webber, J. M. (1939). Relationships in the genus *Gossypium* as indicated by cytological data. *J. Agric. Res. (Washington, D.C.* **58,** 237–261.

White, T. G., and Endrizzi, J. E. ((1965). Tests for the association of marker loci with chromosomes in *Gossypium hirsutum* L. by the use of aneuploids. *Genetics* **51,** 605–612.

White, T. G., Richmond, T. R., and Lewis, C. F. (1967). Use of cotton monosomes in developing interspecific substitution lines. *Crops Res. USDA Rep. ARS* 34-91.

Wilson, F. D., and Fryxell, P. A. (1970). Meiotic chromosomes of *Cienfuegosia* species and hybrids and *Hampea* species (Malvaceae). *Bull. Torrey Bot. Club* **97,** 367–376.

Wilson, F. D., and Kohel, R. J. (1970). Linkage of green-lint and okra leaf genes in a reciprocal translocation stock of Upland cotton. *Can. J. Genet. Cytol.* **12,** 100–104.

Wilson, J. T., Katterman, F. R. H., and Endrizzi, J. E. (1976). Analysis of repetitive DNA in three species of *Gossypium*. *Biochem. Genet.* **14,** 1071–1075.

Wilson, R. L., and Wilson, F. D. (1975). Effects of pilose, pubescent, and smooth cottons on cotton leaf perforator. *Crop. Sci.* **15,** 807–809.

Wouters, W. (1948). Contribution à l'étude taxonomique et caryologique due genre *Gossypium* et application à l'amélioration due cotonnier au Congo Belge. *Publ. Inst. Natl. Agron. Congo Belge, Ser. Sci.* No. 34, pp. 1–383.

Youngman, W., and Pande, S. C. (1927). Occurrence of branched hairs in cotton and upon *Gossypium stocksii*. *Nature (London)* **119,** 845.

Yu, C. P. (1939a). The inheritance and linkage relations of yellow seedling, a lethal gene in Asiatic cotton. *J. Genet.* **39,** 61–68.

Yu, C. P. (1939b). The inheritance and linkage relations of curly leaf and virescent bud, two mutants in Asiatic cotton. *J. Genet.* **39,** 69–77.

Yu, C. P. (1941). The genetical behavior of three virescent mutants in Asiatic cotton. *J. Am. Soc. Agron.* **33,** 756–768.

Zhurbin, A. J. (1930). Chromosamen bein bastard *G. herbaceum* × *G. hirsutum*. *Aus. N.J.H.J.* No. 5 (cited from Skovsted, 1934a).

INDEX

A

Allotetraploids
Gossypium spp.
  diploidization, 347–349
  homologous pairing suppression,
    349–352
  origin
    genomic, 327–331
    mono- vs polyphyletic, 335–339
    new evidence, 339–347
    time and place of, 332–335
Aneuploids
  frequency in *Gossypium hirsutum*
    seeds, monosomic lines, 315–316
  production by translocations in *Gos-*
    sypium spp., 305–306
Antigens
  surface of human hepatitis B virus,
    expression in yeast, 137–138
Aspergillus nidulans
  transformation
    applications, 153
    genomic library, 154
    methods, 152–153
Autosomes
  during *Drosophila hydei* spermatocyte
    stage, 208–209

B

Bacteriophages
  molecular cloning, 84
Biotechnology
  fungal transformation and, 125

C

Caffeine
  effects on radiation injuries in *Schizo-*
    saccharomyces pombe, 27–33, 35,
    38–40, 50–51
Cell cycle
  of *Schizosaccharomyces pombe*
    growth
      DNA synthesis, 6–7
      mitochondrial functions, 7–8
      protein synthesis, 6–7
      RNA synthesis, 6–7
    meiotic, 9–18
      mating types, 11–13
      morphology, 9–11
      mutants of, 10–14
      physiology, 9–11
      recombinations, 14–18
    phases G_1 and G_2, 4
    radiation resistance during, 19–22
    regulation system, 5–6
    vegetative mitotic, 4–6
Cell fusion
  in fungal gene transfer, 79–80
Cell synchronization
  in *Schizosaccharomyces pombe*
    induction methods, 19
    selection techniques, 18–19
Centromeres
  fungal, isolation and characterization,
    112–114
Chloroplasts
  genome polymorphism, 244–245

mechanism, 155–160
with recombination, 98–101
without recombination, 98–99
in higher organisms, 77–79
of *Mucor racemosus*, 154–155
of *Neurospora crassa*
with cloned DNA, 145–147
with total DNA preparations, 140–143
transformant genetics, 143–144
of *Podospora anserina*
with recombinant plasmids, 150–154
Schizosaccharomyces pombe
gene isolation for, 53–54
plasmid vectors for, 53–54
system for, 154
of yeast
by plasmid carrying yeast and *E. coli*
genes, 147, 152
Translocations
Gossypium hirsutum
homozygous lines, 300–301
Gossypium spp., 296–307
effect on
chromosome structure, 306–307
genome stability, 306–307
origin and identification, 296–303
use for analysis of
chromosome deficiency, 305–306
chromosome orientation, 306
gene location, 303–305
incipient genome differentiation, 305
Transcription
in *Drosophila hydei* primary spermatocytes, 192, 194–199
Transposable elements, *see* Transposons
Transposons
fungal
nucleotide sequences, 122–124
population dynamics, 246–250
in *Drosophila melanogaster*, 247–249
in *Escherichia coli*, 248, 250
Trisomes
Gossypium spp., 326

U

Ultraviolet (uv)
Schizosaccharomyces pombe sensitive
mutant injuries, 22–26
dark repair pathways, 26–35, *see
also* Dark repair
recovery after, 35–40, *see also* DNA

V

Vectors
in molecular cloning, 80–86
bacteriophages, 84
cosmids, 84–85
plasmids, 81–84
versatile for fungi, construction, 138–139

Y

Yeast
chromosomes
mapping by new methods, 120–121
DNA
insertion sequences, 122–124
gene expression
of *E. coli*, 127–132
higher eukaryotic, 134–136
human interferon, 137
of *Neurospora crassa*, 132–134
nif from *Klebsiella pneumoniae*, 130, 132
gene molecular cloning, 90–93
genome complexity, 76
plasmid combined with those from *E.
coli*
characteristics, 148–150
replication in yeast and bacteria, 138–139
transformants
donor DNA effect on, 101–103
methods for detection, 88–89, 94–95
transformation
biotechnology and, 125
frequency, 96
minichromosomes during, 101–102, 113–114
by plasmid carrying yeast and *E. coli*
genes, 147, 152
recombination and, 124–125